Multiplicity in Unity

Interspecific Interactions

A Series Edited by John N. Thompson

Multiplicity in Unity

Plant Subindividual Variation and Interactions with Animals

CARLOS M. HERRERA

THE UNIVERSITY OF CHICAGO PRESS   CHICAGO AND LONDON

CARLOS M. HERRERA is professor of research and an evolutionary ecologist at Estación Biológica de Doñana, part of the Consejo Superior de Investigaciones Científicas (CSIC) in Seville.

The University of Chicago Press, Chicago 60637
The University of Chicago Press, Ltd., London
© 2009 by The University of Chicago
All rights reserved. Published 2009
Printed in the United States of America
18 17 16 15 14 13 12 11 10 09   1 2 3 4 5

ISBN-13: 978-0-226-32793-8
ISBN-13: 978-0-226-32794-5
ISBN-10: 0-226-32793-0
ISBN-10: 0-226-32794-9

Library of Congress Cataloging-in-Publication Data

Herrera, Carlos M.
  Multiplicity in unity : plant subindividual variation and interactions with animals / Carlos M. Herrera.
     p. cm.
  Includes bibliographical references and index.
  ISBN-13: 978-0-226-32793-8 (cloth : alk. paper)
  ISBN-13: 978-0-226-32794-5 (pbk. : alk. paper)
  ISBN-10: 0-226-32793-0 (cloth : alk. paper)
  ISBN-10: 0-226-32794-9 (pbk. : alk. paper) 1. Plants—Variation. 2. Animal-plant relationships. 3. Plant genetics. I. Title.
  QK983.H47 2009
  581.3'5—dc22

                                                                    2009010049

⊗ The paper used in this publication meets the minimum requirements of the American National Standard for Information Sciences—Permanence of Paper for Printed Library Materials, ANSI Z39.48–1992.

Contents

Preface

Desvarío laborioso y empobrecedor el de componer vastos libros; el de explayar en quinientas páginas una idea cuya perfecta exposición oral cabe en pocos minutos.
—Jorge Luis Borges, foreword to *El jardín de senderos que se bifurcan*

Values for biological phenomena are often condensed into means. Theoretically, organisms dealing with those values can "expect" the mean value and adapt for it . . . In reality, organisms encounter values one by one, so if variance is high the mean may be irrelevant.
—P. Feinsinger, "Variable Nectar Secretion in a *Heliconia* Species Pollinated by Hermit Hummingbirds"

My interest in the ideas discussed in this book can be traced back to the already remote past when I first became involved in the study of interactions between frugivorous birds and fleshy-fruited Mediterranean plants. Trained as an ornithologist, I was accustomed to using just one measurement to characterize the bill-length phenotype of an individual bird. It struck me as a nuisance that individual fruits produced by the same plant often varied so widely in many important respects, and when I came to characterize the fruit size of a single bush, I had to collect and measure a well-planned subsample of the many hundreds or even thousands of fruits simultaneously available. The same practical trouble struck me again as I shifted to study interactions between pollinators and insect-pollinated plants, this time because I had been taught to consider flowers as the epitome of invariability and constancy. To my dismay, however, flowers also vary, and I had to make repeated measurements on a plant whenever I wanted to properly characterize its floral phenotype. These practical difficulties taught me that, in contrast to the majority of animals, plants generally possess a distinct within-individual component of phenotypic variance that should ideally be taken into consideration.

Later on, as I became more interested in the problem, I also realized that such within-plant variance could be surprisingly large for certain flower, fruit, and seed traits. How, then, to characterize the flower or fruit phenotype of such inconstant individuals? Like others, and for want of a better alternative, I just kept ignoring within-plant variance—sweeping it under the rug, or to be more precise, under the mean, as this is the statistic routinely used to represent (not to describe, to be sure) the phenotypic traits of the reiterated organs produced by a plant. My latest experience regarding within-plant variation, and the one that finally sparked me to write this book, was my discovery that in some species the number of pollen grains on the stigma, and of pollen tubes in the style (two important parameters related to maternal pollination success and the likelihood of microgametophyte competition) are far more variable among the different flowers borne by the same plant, or even the same inflorescence, than among conspecific individuals in the same population. Could there be, after all, some interesting biology hidden behind the familiar nuisance of within-plant variance, routinely brushed under the rug of the mean?

This book addresses this deceptively simple question, paying particular attention to the specific context of plant-animal interactions, as this is the field in ecology with which I am most familiar. Some sparse remarks on the potential significance of within-individual variance in features of reiterated plant structures may be found in the ecological literature of the last few decades, often made from an evolutionary perspective and considering the possible adaptive value of variance. This book, however, purposely follows a different path. Rather than adopting a hypothesis-driven stance and asking from the outset whether subindividual variation in organ traits resulting from the modular architecture of plants is adaptive or not, I start from first principles and leave adaptive and evolutionary considerations for the closing chapter. As will be shown, within-plant variation in organ traits is a universal phenomenon caused by a complex web of mechanisms and with an astounding variety of ramifying consequences, many of which have not been properly acknowledged. In the introductory chapter I briefly define the issue and set the stage. The following chapters examine what features vary among reiterated organs of the same plant (chapter 2), what the magnitude of such variation is in the different types of organs (chapter 3), and how it is temporally and spatially organized (chapter 4). The complex mechanisms, both genetic and ontogenetic, that originate such variation are considered next (chapters 5 and 6). The three sine qua nons for within-plant variation possessing some evolutionary relevance—namely that (1) within-plant variance in organ traits is an individual attribute possessing a genetic basis, (2) animals may respond to such variation, and (3) individual differences in extent and characteristics of variation may translate into differences in plant fitness—are considered in turn in chapters 7 to 9. Finally, chapter 10 provides a synthesis of the possible evolutionary implications of within-plant variation in organ traits.

This book was started in 2003 largely as a spare-time project, and writing has proceeded intermittently since then as time allowed. The order of chapters approximately follows the temporal writing sequence. Although I have attempted to keep an eye on the literature relevant to already-finished chapters, and some colleagues have generously sent articles or drawn my attention to useful references as they have come out, it is possible that some recent investigations relevant to the earlier chapters may have been overlooked. My sincere apologies to those authors whose recent work has been not given adequate credit.

Some of the topics dealt with here have not been specifically considered in earlier experimental or field studies; hence I had difficulty finding relevant published information that could be useful to support or disprove some of the expectations I and others had. There are also very few published raw data that can be used to obtain estimates of variability for different organs and traits, excepting some raw data sets that appeared in print before the habit of compressing data into summary statistics arose. I have had to use my own unpublished raw data or reanalyze published data sets, and to ask colleagues for raw data or analyses not given in their original publications. For their invariably quick and generous responses to my requests, I am deeply indebted to Julio Alcántara, Conchita Alonso, María J. Bañuelos, Raul Bonal, Marisa Buide, Cala Castellanos, Jorge Castro, Cecilia Díaz-Castelazo, Andrew Doust, José M. Fedriani, José L. Garrido, José M. Gómez, Javier Guitián, Pablo Guitián, Benjamin Harlow, Javier Herrera, Asier R. Larrinaga, Antonio Manzaneda, Mónica Medrano, Luis Navarro, José R. Obeso, Víctor Parra-Tabla, Beatriz Pías, Miguel Salvande, Alfonso M. Sánchez-Lafuente, and Pedro A. Tíscar. Conchita Alonso often assisted in locating literature references. Michael Dohm, Alex Kacelnik, and Diana Tomback contributed useful correspondence and discussion, and Fernando Hiraldo provided constant encouragement and removed distracting stumbling blocks as far as he could. José L. Garrido, Javier Guitián, and Pablo Guitián provided accommodation, companionship, and a congenial atmosphere during two stays at the

University of Santiago de Compostela in the early days of writing. Conchita Alonso, Pablo Guitián, and José R. Obeso read parts of the book and provided useful comments and discussion, and Raquel Alejandre carefully cross-checked literature citations. Special mention goes to Paul Wilson, who carefully read the whole book with his usual sharp eye and broad insight, suggested many improvements, spotted a number of weak points, and helped to improve the language. Susan Mazer provided a thoughtful review of the manuscript and suggested many improvements. The book also owes a very special debt to the friend who facilitated access to the amazing online resources of the library of a large American university, most of which were unavailable at the rather modest library of my home institution. Family and friends were always sympathetic toward the varied side effects of book writing. I am most grateful to all of them for their forbearance, generosity, and continued support. Christie Henry, from the University of Chicago Press, always conveyed the message that I should rate quality over quickness, which was a real relief during some impasses. Joann Hoy's able copyediting of the manuscript improved the organization of some sections and greatly helped polish my English. Last but not least, I am most grateful to John Thompson for inviting me to write the book, providing useful suggestions, and making me feel comfortable despite missed deadlines.

CHAPTER ONE

Introduction

[A] plant produces a considerable number of structures of one kind. This simple feature can explain a major difference in the variation patterns exhibited by plants and animals.
—D. G. Lloyd, "Variation Strategies of Plants in Heterogeneous Environments"

Modularity: A Defining Feature of Land Plants

Plants and animals are "most remote in the scale of nature" (Darwin 1859, 73). A myriad of features set them apart. Some of the differences are quite apparent, such as the sharp contrast between the sessile habit and photosynthetic ability of the majority of plants, and the mobile and heterotrophic lifestyle characteristic of animals. Most major differences between the two groups, however, are considerably less obvious, which is not to imply that they cannot be equally crucial from an evolutionary or ecological perspective, as repeatedly emphasized by numerous authors who have over the years dissected, enumerated, or discussed the long list of differences between plants and animals, and highlighted the manifold consequences of being a plant (Stebbins 1950; Bradshaw 1972; Levin 1978; Jerling 1985; Klekowski 1988a; Meyerowitz 2002; Borges 2005). Possibly some of the most profound differences between plants and animals stem from their contrasting developmental modes (Jerling 1985; Klekowski 1988a; Walbot 1996), and there is now considerable evidence from whole-genome sequencing and experiments in developmental genetics indicating that plants and animals have independently evolved development (Meyerowitz 2002). In animals, reproductive and somatic cell lineages diverge early in embryogenesis. In contrast, plant reproductive or vegetative structures may indistinctly arise from the same meristems, implying that they share a common cell lineage and that individual cells retain their whole potential for differentiation until late in development. In other words, plants lack a differentiated germline, and August Weismann's doctrine of separation of soma and germ clearly does not hold for them (Klekowski 1988a; Poethig 1989). Apart from its genetic implications (e.g., the possible transmission of somatic-cell mutations to gametes; Klekowski 1988a; Walbot 1996), the indefinite developmental totipotency of meristematic plant cells is a distinctive character making possible another crucial trait, namely the modular construction of plant bodies by continual organogenesis and the reiterated production of homologous structures. The elemental structural subunit that is repeatedly produced over a plant's lifetime, the metamer, consists of an internode and a node bearing one or several leaves plus associated axillary meristems. The reiteration of this basic subunit results in the production of a higher-level module, such as a stem or branch (Hallé et al. 1978).

Reiteration of homologous, functionally equivalent structures is a truly quintessential, ancestral feature of the body plan of higher plants. The significance of the repetition of elements in the construction of plants was already envisaged by classical Greek philosopher Theophrastus more than two thousand years ago, who noted that such repetition was "of the essence of a plant" (cited in White 1984). This intuitive perception has been confirmed in modern times by paleontological and phylogenetic evidence showing that the main features ultimately responsible for plant modularity were already present at a very early stage in the evolution of land plants, and are a property shared by the whole lineage. The key developmental innovations responsible for modularity and functional repetition included well-defined sporophytic apical meristems that allow the production of vegetative or reproductive organs, and the capacity for the proliferation of shoot meristems that allow branching of the sporophyte body and a concomitant multiplication of meristems (Graham et al. 2000). These two decisive changes in the body plan, which led to a more complex architectural framework on which all subsequent morphological diversification was based, took place between the divergence of the modern bryophyte and vascular plant lineages in the late Ordovician, about 450 mya (Kenrick and Crane 1997; Graham et al. 2000). By the Middle Devonian (380 mya), tall tree fern–like plants with clearly discernible "modern," reiterative modular architecture similar to that of tree ferns, cycads, and palms were already abundant in terrestrial ecosystems (Stein et al. 2007). In addition to allowing for increased body size and, consequently, improved ability to compete for light, sporophyte branching (in contrast to gametophyte branching, as in bryophytes) provided the capacity to pursue growth if some apical meristems were lost or damaged by herbivores. This possibility, acting in combination with meristem dormancy, another early key innovation (Kenrick and Crane 1997), most likely represented a decisive evolutionary breakthrough that contributed to the rapid spread and diversification of land plants.

Multiplicity and Phenotypic Variation

The ecological and evolutionary implications of plant modularity have frequently been highlighted following White's pioneering treatment (1979, 1984) of plant individuals as metapopulations of repeated modules. With few exceptions, these investigations have approached the study of plant modularity by adopting either demographically or physiologically oriented perspectives. Following Watkinson and White's demographically oriented treatment of the consequences of modular construction in plants (1985), subsequent studies of modularity have often focused on the consequences of the differential growth and survival of modular subunits, particularly in species characterized by extensive clonal proliferation (Tuomi and Vuorisalo 1989; Otto and Orive 1995; Otto and Hastings 1998; Pineda-Krch and Poore 2004). These aspects lie beyond the scope of this book, and will not receive particular consideration here. Physiologically inspired studies of modularity have examined its influence on spatial patterns of within-plant distribution of water, photosynthates, and other substances such as secondary compounds and hormones (Watson 1986; Marshall 1996; Price et al. 1996). These studies are reviewed in chapter 6, because compartmentalization of plant bodies into a series of relatively independent "integrated physiological units" (Watson 1986), a phenomenon known as "sectoriality," is one of the main mechanisms that can generate within-plant variation in the phenotypic features of reiterated organs (Orians and Jones 2001).

Another consequence of plant modularity, and the one this book is directly concerned with, is the appearance of a distinctive source of phenotypic variance, namely the within-plant or subindividual component. An inescapable consequence of the multiplicity of modules is variation in the characteristics of the copies of the same organ (leaves, flowers, fruits, seeds) produced on different modules of the same plant. Such variation may be large or small, but it will always exist insofar as variance is an emergent property that inevitably arises whenever a collection of equivalent copies of a given structure are produced. Perfect copy identity is not achieved even in tightly controlled, fully automated industrial processes that are purposely designed to produce identical copies of a given item, as evidenced by the need of tight postproduction quality controls. As previously suggested by Suomela and Ayres (1994), and shown at length in the following chapters, within-individual variance in organ traits is an emergent property of plant individuals: the outcome of modular construction through reiteration of elemental subunits (metamers) and the associated repetition of homologous structures performing the same function (leaves, flowers, fruits, seeds). The existence of a subindividual component of phenotypic variance sets plants apart from the vast majority of animals (Lloyd 1984). This does not mean, however, that modular repetition of structures by the same genotype is, strictly speaking, a feature unique to plants. It is shared by some animals, such as corals, hymenopteran societies, and aphid parthenogenetic progenies. In addition, a subindividual component of variance is associated with repeated behaviors and structures that vary within individual animals over space or time (e.g., alarm calls, hairs, feathers, blood cells, eggs, clutch size). Some of the implications, predictions, and methodological suggestions developed in this book specifically for plants could therefore be easily extrapolated to animals, as suggested in the epilogue.

Within-Plant Variance: Beyond Statistical Nuisance

The study of variation is central to the biological sciences, and, as stressed long ago by William Bateson (1894, vi), "to collect and codify the facts of variation is . . . the first duty of the naturalist." In natural systems, biologically significant variations occur at each of several hierarchically nested spatial scales including, from top to bottom, biomes, ecosystems, communities, species, metapopulations, populations, and individual organisms. Interest in patterns and processes of spatial and temporal variations occurring at each of these organizational levels has ultimately provided the impetus for nearly all ecological and evolutionary work done so far. At the bottom end of that continuum lies variation among conspecific individuals from the same population, involving differences in survival, fecundity, phenotypic traits, and genetic makeup. These differences provide the basic raw material for natural selection and microevolutionary change; hence the individual (genotypic) level has been customarily deemed the lowest one in the above hierarchy at which variations may possess some ecological and/or evolutionary significance. This is perhaps one of the reasons why variations occurring at the subindividual (i.e., sub-genotypic) level have been traditionally granted so little biological significance by botanists, ecologists, and plant biologists in general, as detailed in chapters 2 and 3. Two conspicuous exceptions to this prevailing attitude of neglect toward within-plant variation in phenotypic traits, however, deserve mention here.

The subindividual component of phenotypic variance in organ traits of land plants was acknowledged and carefully quantified more than a century ago by an influential group of biometricians and plant morphologists who had statistician Karl Pearson as their visible leader (Pearson 1901; Fry 1902; Gain 1904; Harris 1916). These researchers aimed to use very detailed quantitative data on within-plant variation in organ traits as an indirect way of obtaining insights on the nature of the heredity of these traits, following Pearson's theory of homotyposis (1901). Shortly after the theory they were aiming to support fell into disrespect and oblivion, interest in the type of empirical data that was being collected in its support also faded away, as described in chapter 2. The lack of interest in the phenomenon of within-plant variation in organ traits that prevailed for nearly all the 20th century thus was not because it was demonstrably uninteresting, but because the theory with which it had become associated was dismissed.

Agronomists have examined subindividual variation in phenotypic plant traits in considerable detail for decades. Not unexpectedly, their motivations have been predominantly economical rather than biological. On one hand, variability in the characteristics (e.g., size, shape, nutritional quality) of crop products such as fruits or grains may have an important economic impact on postharvesting processing, and uniform crops are invariably preferred over variable ones. Chocolate manufacturers, for example, prefer cocoa beans of uniform size for processing (Glendinning 1963), and variance in oat kernel size has nontrivial costs to the oat-milling industry. Processing oats for human food generally involves size separation of kernels into different streams to optimize dehulling efficiency, so the greater the variance in seed size the larger the number of seed-size classes required and the higher the total milling costs (Doehlert et al. 2004). Since an important fraction of total variance in the features of crop products is due to within-plant variation (particularly so, although not exclusively, when pure genetic lines are used), considerable effort and resources have been invested by agronomists to evaluate the magnitude of this costly source of variation and to understand its origin, in order to attempt to reduce its impact on total crop variance. On the other hand, agronomists have been long aware that the existence of an extensive within-plant component of phenotypic variance in crop plants demands carefully designed sampling plans if accurate estimates of per-plot averages for product yield or quality are to be obtained (Sites and Reitz 1950; Williams 1962; Kondra and Downey 1970; Fick and Zimmerman 1973; Audergon et al. 1993; Miles et al. 1996), an aspect with obvious economic implications. For example, accurate assessment of overall fruit yield in kiwifruit or apple orchards requires designing relatively complex, stratified sampling schemes that can account satisfactorily for the large within-plant component of variance in individual fruit size (Habib et al. 1991; de Silva and Ball 1997; de Silva et al. 2000). In accordance with the interest of agronomists on the subject, a voluminous literature has accumulated in the last few decades describing the magnitude, spatial patterns, mechanisms responsible, and correlates of within-plant variation in organ traits for a variety of commercially valuable species, particularly grain crops and fruit trees (de Silva et al. 2000; Bramble et al. 2002; Liu et al. 2005). I take advantage throughout the book of this extensive literature to compensate for the nearly complete absence of equivalent data from wild plants. My usage is restricted, however, to aspects for which a reasonable similarity between wild and cultivated species can be assumed. This holds, for example, for the spatial organization of subindividual variation (chapter 4), the organismal mechanisms accounting for subindividual variation (chapter 6), and the genetic basis of variability-related traits (chapter 7).

From time to time, ecologists have also acknowledged the high levels of within-plant variation exhibited by some organ traits in wild plants, and suggested that the phenomenon may have ecological implications deserving closer examination (Suomela and Ayres 1994; Markham 2002). There has been, however, surprisingly little work done to quantify and compare the relative proportions of variation occurring at the within- and among-individual levels (e.g., Herrera and Soriguer 1983; Michaels et al. 1988; Obeso and Herrera 1994; Williams and Conner 2001). With few exceptions, the within-plant level of phenotypic variation has been systematically ignored in most investigations of wild plants, and its possible ecological and evolutionary consequences remain largely unexplored to date (but see, e.g., Whitham 1981; Feinsinger 1983; Field 1983; Suomela

1996; Gripenberg and Roslin 2005). Even in the few instances where within-plant variation was studied in some detail, its possible ecological or evolutionary interest has often been dismissed summarily (e.g., Midgley et al. 1991). Like agronomists, ecologists have also often treated the phenomenon merely as a statistical nuisance demanding more careful sampling designs (Zimmerman and Pyke 1986; Freeman and Wilken 1987). In recent years, however, there has been increasing recognition that, in addition to a statistical nuisance, within-plant variance in organ traits may be an ecologically relevant phenomenon to both the plants themselves and the animals that interact with them (e.g., herbivores, pollinators; Orians and Jones 2001; Biernaskie et al. 2002; Orians et al. 2002; Biernaskie and Cartar 2004; Shelton 2000, 2004; Herrera, Pérez, and Alonso 2006; Canto et al. 2007).

The prevailing neglect by plant evolutionary ecologists of the within-plant level of phenotypic variation in organ traits may be attributed to the following three, generally unstated (and untested) assumptions, among others: (1) subindividual variation in most organ traits is generally small in comparison to variation among individuals, and largely or entirely reflects random developmental noise; (2) phenotypic differences between homologous organs produced by the same plant are environmental in origin and lack a genetic basis; and, accordingly, (3) the subindividual level of phenotypic variation is "invisible" to natural selection, which operates on plant mean phenotypes alone. This prevailing attitude has both a long tradition and a distinguished pedigree, as revealed by the following quotations. Haldane (1932, 19) stated that "differences within a clone are not inherited. They are the best example of what is called fluctuating variability, due to differences of environment, not transmissible by inheritance, and therefore irrelevant for the problem of evolution." (Twenty-five years later, Haldane had apparently changed his view on this issue and conferred some evolutionary significance to within-plant variability; see chapter 7 for details.) Stebbins (1950, 74) provides another example in the same vein: "Environmental modification is a source of variability which must be kept in mind by the evolutionist because it affects every individual we see in nature. But as a direct factor in evolutionary divergence it is not significant."

In the chapters that follow I develop the thesis that the multiplicity of homologous structures arising from plant modularity gives rise to a subindividual level of phenotypic differences among organs of the same plant involving a constellation of phenotypic traits, differences whose quantitative importance is often similar or even greater than that of phenotypic differences among individual means. The existence of phenotypic variation at the subindividual scale can have diverse ecological implications for the interaction between plants and the animals that use reiterated organs as food, including herbivores, flower visitors, frugivores, and seed predators. Animal consumers will respond to subindividual variability in organ traits, and, depending on its magnitude and the nature of the factors causing it, they may eventually become selective agents of patterns and levels of subindividual phenotypic variation through a variety of mechanisms. In this way, animals interacting with plants may ultimately condition, constrain, or modify plant ontogenetic patterns, developmental stability, and the extent to which feasible phenotypic variants are expressed by individuals.

CHAPTER TWO

Which Traits Vary within Plants?

Many different features vary across reiterated structures of the same plant.

That plants are variable organisms, and that such variability might be one of their most distinctive features in comparison to animals, has been emphasized many times in the botanical literature since the classical accounts of the Greek philosopher Theophrastus. More than two thousand years ago, in book I of his *Enquiry into Plants*, this pioneer botanist wrote about heterophylly. The leaves "of the abele ivy and of the plant called kroton are unlike one another and of different forms. The young leaves in these are round, the old ones angular, and eventually all the leaves assume that form. On the other hand, in the ivy, when it is young, the leaves are somewhat angular, but when it is older, they become rounder" (Theophrastus 1916, 69).

Much more recently, around the turn of the 19th century, plant variability captured the interest of biometricians, and remarkably detailed investigations of patterns and extent of variability of different plant organs were produced. Outstanding among these contributions was Karl Pearson's 100-page memoir devoted to quantitatively documenting the concept of "homotyposis," or the similarity of undifferentiated like parts ("homotypes") of the same organism. There he provided extensive data on the relative levels of within- and between-plant variability in meristic and continuous characters of leaves, flowers, and fruits from many species, and developed methods to measure variability (Pearson 1901). Subsequently in the earliest issues of *Biometrika*, which Pearson edited, there appeared many detailed quantitative analyses of intraspecific variability in leaf, fruit, flower, and inflorescence features for a variety of plants (e.g., Weldon 1901; Yule 1902; Tower 1902; Gain 1904; Robbins 1908; Harris 1909b, 1910, 1911, 1916; Simpson 1914). Around the same period, publications adopting similar approaches appeared, often in North American scientific journals such as *American Naturalist* and *American Journal of Botany* (Lucas 1898; Goodspeed and Clausen 1915). Variation in the number of pistils and stamens in the flowers of *Ficaria ranunculoides* (Weldon 1901), number of sepals in *Anemone nemorosa* (Yule 1902), number of flowers in the inflorescence of *Adoxa moschatellina* (Whitehead 1902), and flower size in *Nicotiana* (Goodspeed and Clausen 1915) were some of the topics meticulously addressed by these early studies. These pioneering biometrical investigations brought two significant novelties to the study of plant variability. On one hand, contemporaneous state-of-the-art statistical techniques such as correlation and regression were profusely used for the first time in investigations of plant variation. On the other hand, they were exclusively concerned with normal, non-monstrous plant variability. Perhaps this does not sound particularly focused nowadays, but early in the 20th century it represented a turning point. Up to that time, investigations of plant variability had been impregnated with strongly typological notions, and had focused predominantly on naming and cataloguing "the principal deviations from the usual construction of plants," as reads the subtitle of Masters' extensive treatise on "vegetable teratology" (1869). Published reports of floral anomalies or monstrosities, exemplifying the classic, typological approach to the study of plant variation, kept appearing from time to time in the botanical literature for most of the first half of the 20th century (for reports on floral anomalies see, e.g., McCrea 1924; Arber 1931; Halket 1932; Kausik 1938; Bond 1941; Saunders 1941). In retrospect, perhaps one of the most relevant contributions of those early biometricians to plant biology was to reveal that variation in morphological, continuous metric characters around the "modal" or "normal type" for the species is a normal, unexceptional condition widely found in natural populations in the wild, *both* within and among individual plants.

Pearson was a strong opponent of Mendelian genetics (Norton 1975), and the central tenets and evolutionary ideas underlying his work and that of his followers were proven downright erroneous in the following decades (Fisher 1918; Haldane 1957). Pearson's publications and personal influence had been decisive in arousing the interest of biometricians in gathering the quantitative data needed to test his ideas on the origin and maintenance of variation in natural populations. As Pearson's ideas fell into disrepute, there was an understandable sharp decline in the number of quantitative studies of natural variability of plant organs from 1930 to 1950 (e.g., Lowndes 1931; Baten 1935; Tansley 1948). In addition, as stressed by Pearl (1936, 662) in an obituary published shortly after Pearson's death, "in the early heyday of Mendel's rediscovered work, [Pearson] unwisely questioned the accuracy and validity of the experimental results that the Mendelian experimenters were getting, and of which he had a somewhat less than adequate first-hand knowledge. In consequence his influence with the biologists was for a time weakened. With similar lack of soberly poised judgment the Mendelists endeavoured to throw to the wolves the whole body of observed biometric facts about heredity." The neglect of intraspecific variability by plant biologists for most of the first half of the 20th century is thus hardly surprising.

During the second half of the 20th century, studies explicitly addressing within-plant variability in morphological or functional features of plant organs almost exclusively focused on the particular case whereby plants produce alternative, clearly differentiated forms of the same structure (e.g., species with cleistogamous and chasmogamous flowers, Campbell et al. 1983; heterophyllous plants with two or more distinct leaf types on the same individual, Wells and Pigliucci 2000). This sort of discrete within-plant polymorphism affecting leaves, flowers, or fruits has generally been the only category of within-plant variation acknowledged in reviews of plant phenotypic plasticity (e.g., Bradshaw 1965; Sultan 1987). In contrast, quantitative investigations specifically aimed at describing and quantifying patterns of within-plant variation in meristic or continuous traits of structures that do not exhibit any obvious, discrete polymorphism have remained remarkably scarce (but see Dronamraju 1961; Roy 1963; Huether 1968, 1969; Ellstrand 1983; Ellstrand and Mitchell 1988; Williams and Conner 2001). This does not mean that the fact that "plants display remarkable morphological variability, both among individuals of the same species and among organs of the same plant" (Ellstrand 1983, 119) has gone unacknowledged in the literature, but rather that variability itself has rarely occupied the center stage in recent investigations. The occurrence of continuous, nonpolymorphic within-plant variation in an enormous array of features of leaves, flowers, fruits, and seeds is implicit in many botanical and ecological investigations. Plant biologists are well aware of continuous within-plant variation in characters of quite diverse kinds, but when explicitly referring to this variation, they tend to consider it more as a statistical nuisance whose influence on sampling designs must be properly accounted for, rather than as a genuinely interesting biological phenomenon (e.g. Wood 1972; de Silva and Ball 1997; Velasco et al. 1998; Rowe and Cadisch 2002). This slant toward disregard may be traced back at least to Baten (1936), who in one of the first investigations specifically devoted to within-plant variation in floral traits wrote in the closing sentence of the paper's summary that "one should be very careful when taking a random sample of flowers, for flowers at different positions on certain plants are different. Distributions pertaining to them should not be mixed."

In this chapter, I review published and, to a lesser degree, unpublished information to illustrate the variety of morphological and functional features of plant reiterated structures that exhibit significant amounts of variation within individuals. My objective is to show that, far from being a biological curiosity or an infrequent phenomenon involving just a few organ traits, within-plant variation affects virtually every organ trait that has ever been studied at a sufficiently detailed spatial resolution. To highlight the variety of traits involved, I try to present a comprehensive account of subindividually variable features, rather than a detailed treatment of each of the features considered. I consider instances of both continuous and discrete (polymorphic) within-plant variation, although with a slant toward continuous variation, because this kind of variation has traditionally received less attention. The main purpose is to provide just a catalogue of features that exhibit subindividual variation, rather than quantification of such variation, an aspect that chapter 3 deals with. Here and in subsequent chapters, I focus on plant organs and structures with determinate growth (e.g., leaves, fruits, flowers, seeds), leaving aside those ramifying features of plants that have an indeterminate growth (e.g., stem diameter, internode length). At this point, it is important to emphasize that many of the subindividually variable features mentioned in this chapter are later shown (chapters 8 and 9) to be potentially influential in different kinds of plant interactions with animals.

Leaf Traits

Perhaps because of the ease of preservation of certain features, leaves have traditionally received more attention from researchers than any other reiterated structure from the viewpoint of within-individual variability. An important part of Pearson's pioneering memoir on plant variability (1901) dealt with variation in leaf features. Studies have most often focused on variation in external morphology and structural features, particularly leaf form and size, and much less frequently on less obvious functional or physiological traits like nutrient content, composition of secondary metabolites, water-use efficiency, and photosynthetic characteristics. Taken together, all these investigations have identified a broad array of leaf traits that vary among leaves of the same individual plant.

Leaf Form

There are probably few plants whose leaves are all uniform in shape (Sparks and Postlethwait 1967), and variation in leaf form has been thoroughly investigated by biometricians, plant anatomists, and plant ecologists for more than a century (e.g., Fry 1902; Harris 1909a; Johnson 1926). Extensive documentation accumulated over this long period demonstrates that, quite often, individual plants bear leaves of different forms. This phenomenon, which is generally designated as heterophylly (etymologically meaning "varied leaves") has been reported from at least 56 genera from 42 plant families (table 2.1). Within-individual variation in leaf form may be originated by a variety of mechanisms (reviews in Ashby 1948; Wells and Pigliucci 2000), and diverse modalities of heterophylly have been recognized. The contrasting leaf types can be produced along longitudinal axes (positional heterophylly; Font Quer 1979), irregularly scattered over the plant (vague heterophylly; Font Quer 1979), produced at different times of the growing season (seasonal heterophylly; Winn 1999a), or associated with different ontogenetic stages of the plant (developmental or heteroblastic heterophylly; Ashby 1948). Heterophylly can also be plastic or nonplastic (Wells and Pigliucci 2000). Regardless of origin or modality, the widespread occurrence of heterophylly denotes that coexistence on the same individual plant of leaves differing in form is far from exceptional in nature.

Heterophylly is widespread (although not universal; Sculthorpe 1967) among aquatic plants, having been reported from most predominantly or exclusively aquatic plant families, and from aquatic taxa belonging to predominantly nonaquatic families (table 2.1). Individuals of heterophyllous aquatic plants generally bear two or more morphologically distinct leaf variants, generally corresponding to aerial, floating, and submersed leaves (fig. 2.1). The exaggerated heterophylly exhibited by some

TABLE 2.1 **Examples of sensu lato heterophylly (i.e., including positional, seasonal, heteroblastic, plastic, and nonplastic modalities) in aquatic and terrestrial plants.**

Family	Predominant habitat type		Reference
	Aquatic	Terrestrial	
Acanthaceae	Hygrophila		Sculthorpe 1967
Aceraceae		Acer	Powell et al. 1982; Steingraeber 1982
Alismataceae*	Echinodorus, Luronium, Sagittaria		Sculthorpe 1967
Apiaceae	Apium, Eryngium		Sculthorpe 1967; Webb 1984
Aponogetonaceae*	Aponogeton		Sculthorpe 1967
Araceae		Monstera, Philodendron, Syngonium	Ray 1987, 1990
Araliaceae		Hedera, Pseudopanax	Robbins 1960; Clearwater and Gould 1993
Begoniaceae		Begonia	McLellan 1993
Betulaceae		Betula	Clausen and Kozlowski 1965
Cabombaceae*	Cabomba		Sculthorpe 1967
Callitrichaceae*	Callitriche		Deschamp and Cooke 1985
Campanulaceae		Cyanea	Givnish et al. 1994
		Phyteuma	Wheeler and Hutchings 2002
Cecropiaceae		Pourouma	Kincaid et al. 1998
Ceratophyllaceae*	Ceratophyllum		Sculthorpe 1967
Cercidiphyllaceae		Cercidiphyllum	Titman and Wetmore 1955
Convolvulaceae		Ipomoea	Njoku 1956
Cucurbitaceae		Cucurbita	Jones 1993
Dipsacaceae		Knautia	C. M. Herrera unpubl.
Fabaceae		Cyamopsis, Ulex	Millener 1961; Sparks and Postlethwait 1967
Haloragaceae*	Myriophyllum, Proserpinaca		Schmidt and Millington 1968; Sculthorpe 1967; Kane and Albert 1982
Hamamelidaceae		Liquidambar	Smith 1967
Hippuridaceae*	Hippuris		Bodkin et al. 1980
Hydrocharitaceae*	Ottelia		Sculthorpe 1967
Juncaceae	Juncus		Bradshaw 1965
Lamiaceae		Dicerandra, Prunella	Winn 1996b; C. M. Herrera unpubl.
Lauraceae		Sassafras	Ghent 1973
Loranthaceae		Amyema	Kuijt 1980
Moraceae		Morus	Fry 1902
Myrtaceae	Melaleuca	Eucalyptus	Johnson 1926; Lockhart 1996
Nymphaeaceae*	Nuphar		Titus and Sullivan 2001
Oleaceae		Nyctanthes	Roy 1963
Onagraceae	Ludwigia		Kuwabara et al. 2001
Plantaginaceae	Littorella		Robe and Griffiths 1998
Polygonaceae	Polygonum	Muehlenbeckia	Mitchell 1971; Bruck and Kaplan 1980

Pontederiaceae*	*Eichhornia,*		Richards and Lee 1986;
	Heteranthera		Horn 1988
Potamogetonaceae*	*Potamogeton*		Pearsall and Hanby 1925
Ranunculaceae	*Ranunculus*	*Delphinium*	Ashby 1948; Bostrack and
			Millington 1962; Cook and
			Johnson 1968
Salicaceae		*Populus*	Critchfield 1960; Curtis and
			Lersten 1978
Scrophulariaceae	*Limnophila*		Ram and Rao 1982
Tropaeolaceae		*Tropaeolum*	Whaley and Whaley 1942
Violaceae		*Viola*	Winn 1999a
Vitaceae		*Parthenocissus*	Critchfield 1970

Note: This compilation is incomplete and only intended to illustrate the widespread occurrence of heterophylly among angiosperms. Families exclusively or predominantly associated with aquatic habitats are denoted by asterisks.

aquatic plants, along with the fact that the functionality, ecological significance, and physiological mechanisms of such within-plant morphological disparity are relatively straightforward and well resolved in this group (reviewed in Sculthorpe 1967; Wells and Pigliucci 2000), probably explain why aquatic plants are almost invariably used to exemplify the heterophyllous condition (e.g., Schlichting and Pigliucci 1998; Pigliucci 2001). This should not, however, lead us to think that heterophylly is a condition exclusive to, or even predominantly associated with, the aquatic habit, as emphasized long ago by Arber (1919). As summarized in table 2.1, heterophylly occurs also in at least 31 exclusively or predominantly terrestrial families, and there seems to be a trend for heterophylly to be particularly common in terrestrial plant communities of oceanic islands (Givnish et al. 1994). In some terrestrial heterophyllous plants, the magnitude of differences between leaf forms present on the same individual may be comparable to that exhibited by different leaf forms of aquatic taxa (fig. 2.1).

Structure and Function

Morphological, structural, and functional features of leaves tend to vary in unison (Gutschick 1999). In heterophyllous taxa, both aquatic and terrestrial, leaves of different forms in the same plant frequently differ also in other, less conspicuous but ecologically relevant features. These include microscopic structural features such as stomatal density, shape and arrangement of epidermal cells, mesophyll density, and cuticular thickness, along with features related to resource allocation such as carbon and

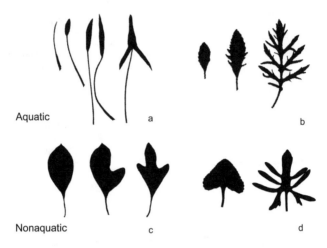

FIG. 2.1 Silhouettes of representative leaves of heterophyllous aquatic (*upper row*) and non-aquatic (*lower row*) plants. In each case, the different forms coexist on the same individual plant. *a, Sagittaria sagittifolia* (Alismataceae); *b, Hygrophila* (formerly *Synnema*) *triflorum* (Acanthaceae); *c, Sassafras albidum* (Lauraceae); *d, Viola septemloba* (Violaceae). Leaves from different species are not rendered at the same scale. *a* and *b* modified from Sculthorpe 1967; *c* from Ghent 1973; *d* from Winn 1999a.

nitrogen concentration, and pigment composition (Johnson 1926; Critch-field 1960; Bostrack and Millington 1962; Bodkin et al. 1980; Winn 1996a, 1999b; Robe and Griffiths 1998). In heterophyllous *Populus grandidentata*, the two leaf types (early- and late-season morphs) differ not only in shape and size, but also in the number and size of resin glands and extra-floral nectaries, these secretory structures being generally more prominent and active on late leaves (Curtis and Lersten 1978). "Biochemical heterophylly" is sometimes associated with morphological heterophylly. In the seasonally heterophyllous herb *Eryngium vesiculosum*, Palá-Paúl et al. (2003) found significant differences in the essential oil composition of summer prickly leaves and winter entire leaves. In heterophyllous species of *Potamogeton*, the floating ovate leaves generally contain an assortment of flavonoids, while the filiform submersed leaves tend to exhibit reduced flavonoid profiles (Les and Sheridan 1990). In *Hedera helix*, anthocyanins are abundant in the lobed juvenile leaves, and scarce or absent in the entire adult leaves (Stein and Fosket 1969). This difference between leaf morphs is due to the presence in young leaves, but not in adult ones, of dihydroflavonol reductase, an enzyme that plays a central role in the biosynthetic pathway leading to anthocyanin (Murray and Hackett 1991). All these structural and chemical correlates of morphological heterophylly contribute to within-plant variability in leaf features beyond mere variation in form.

Nevertheless, within-plant heterogeneity in subtle structural, chemical, or functional leaf features is by no means confined to heterophyllous plants exhibiting discrete variation in leaf features. A large number of traits may vary continuously among the leaves of individual plants in spite of overall morphological similarity. Probably the two longest-known examples of such cryptic within-plant variation in leaf features involve the differences in structure and volume of the different leaf tissues between sun and shade leaves of the same tree crown (Wylie 1951), and the steep stomatal frequency gradients from the lower leaves to the higher leaves in a plant (Smith 1941). Variations in leaf size and specific leaf area (area per unit mass) within the crown of individual trees have been recognized and thoroughly investigated for decades (Ford and Newbould 1971; Hutchison et al. 1986; Hollinger 1989; Casella and Ceulemans 2002). More recently, modern instrumentation and physiological procedures have revealed that leaves in the same plant may greatly differ in some cryptic features such as photosynthetic characteristics, water-use efficiency, and concentration of nutrients and secondary metabolites, even if they are otherwise similar in their gross external morphology. For example, the proportion of the heavy stable isotope ^{13}C in photosynthesis products, as reflected by the carbon isotope discrimination index $\delta^{13}C$, may exhibit tremendous variation within the crown of individual trees (Waring and Silvester 1994). Within-plant variation in leaf nitrogen concentration, chlorophyll content, and photosynthetic rates has been found in a wide variety of herbs, shrubs, and trees from tropical and temperate habitats (e.g., Young and Yavitt 1987; de Soyza et al. 1990; Bowers and Stamp 1992; Suomela and Ayres 1994; Traw and Ackerly 1995; Bassow and Bazzaz 1997; Torres Boeger and Poulson 2003), and similar variation is also well-known from cultivated plants (e.g., Bentz et al. 1995; Honěk and Martinková 2002). Nutrient concentrations in phloem sap, including amino acids and sugars, can also be highly variable among leaves of the same plant (Merritt 1996). A kind of cryptic physiological heterophylly, involving discrete within-plant variation in water-use efficiency of otherwise morphologically similar leaves, has also been reported. Plants of the tropical understory shrub *Psychotria marginata* produce two leaf types that greatly differ in specific mass, stomatal conductance, and water-use efficiency (Mulkey et al. 1992). Subtle variations of this kind will likely prove to be fairly common as more investigations are designed to look for them. In a study of the photosynthetic characteristics of canopy leaves of trees conducted in a seasonally dry tropical forest in Panama, Kitajima et al. (1997) found that, in four of the six species studied, leaves produced at the end of the rainy season (November–December) had higher photosynthetic capacities and higher water-use efficiencies than leaves produced by the same plants during the early rainy season (May).

Improved analytical techniques have also revealed with increasing frequency that the concentration and composition of secondary plant metabolites vary among leaves or branches of the same plant, and that individual host plants need to be considered as phytochemical mosaics when interpreting herbivore-plant interactions (Powell and Raffa 1999; Shelton 2005). This has been shown to apply, for example, to wax alkanes in cypress trees (Dyson and Herbin 1970), cardenolides in *Asclepias eriocarpa* (Nelson et al. 1981), phenolic compounds in *Populus* (Zucker 1982), monoterpenes in *Sequoia sempervirens* (Hall and Langenheim 1986), iridoid glycosides in *Plantago lanceolata* (Bowers and Stamp 1992), glucosinolates in *Brassica nigra* (Merritt 1996), foliar monoterpenes in *Larix laricina* (Powell and Raffa 1999), and cyanogenic activity in *Turnera ulmifolia* (Shore and Obrist 1992). In *Quercus ilex* trees, leaves located in different parts of the crown differ significantly in emission rate of volatile monoterpenes (Staudt et al. 2001).

Fine-Grained Variation and "Nontraits"

In addition to variation between distinct leaves of the same plant, a significant number of studies have also detected very fine-grained gradients of variation in chemical traits that take place at the reduced within-leaf level. For example, photoassimilates are not homogeneously distributed along the leaves of grasses (Williams, Collis, et al. 1993), the alkaloid nicotine is patchily distributed in the leaf blades of tobacco plants (Kester et al. 2002), concentration of phenolics increases from the base to the tip of individual leaves in *Populus angustifolia* (Zucker 1982), volatile terpenoid content varies significantly among the petiole, midrib, and leaf blade of cultivated carrot plants (Senalik and Simon 1987), and glucosinolates are randomly, patchily distributed within single leaves of *Raphanus sativus* (Shelton 2005). In needles of *Pinus banksiana*, the distribution of water, nutrients, and monoterpenes varies significantly between the distal and basal sections (Wallin and Raffa 1998). Individual mineral elements, like calcium and phosphorus, are also distributed heterogeneously within single leaf blades (Williams, Thomas, et al. 1993). Within-leaf variation of this kind has been relatively little explored so far, but the results obtained clearly demonstrate that it has the potential to represent a major source of within-plant heterogeneity in chemical traits (Shelton 2005).

Within-plant variation in leaf traits is not restricted to the morphological, structural, chemical, and functional features properly belonging to the plant's phenotype mentioned so far in this section. There are some leaf "nontraits" that, although not strictly forming part of a plant's phenotype, do exhibit within-plant variation that may be ecologically relevant from the viewpoint of the interaction of plants with animals or the abiotic environment. Prominent among these are endophytic fungal communities, made up of predominantly harmless fungi colonizing the interior of aerial plant tissues, which can play roles, among others, in the plants' phenotypic expression, physiology, and interaction with herbivores (Clay 1990; Cheplick 1997, 1998; Clay and Schardl 2002; Vicari et al. 2002). In *Sequoia sempervirens*, for example, the species composition of endophytic communities varies considerably over the tree's foliage, and patchiness is so extreme as to occur even at the very small spatial scale of individual branches (Espinosa-Garcia and Langenheim 1990). In several species of orchids of the genus *Lepanthes* in rain forests of Puerto Rico, heterogeneity of fungal endophytes in single plants and plant organs was found by Bayman et al. (1997) to be greater than differences between species. Likewise, the composition of endophytic fungal assemblages did not differ significantly among eight coexisting species of ericaceous shrubs in Japan, but did differ among different-aged leaves of individual plants (Okane et al. 1998). These fungal "nontraits" of leaves can vary not only among leaves of the same plant, but also among different parts of the same leaf, as found by Hata et al. (2002) for the endophytic fungi isolated from leaves of the temperate tree *Pasania edulis*. In this species, the distal, central, and basal region of leaf blades harbored taxonomically distinct fungal communities. Likewise, Deckert and Peterson (2000) found that both the frequency of infection and the taxonomic composition of endophytic communities varied significantly along single needles of *Pinus strobus*.

Floral Traits

In most species, flowers tend to be less variable than other reiterated structures. Nevertheless, examples abound where individual plants produce flowers that differ in structural or functional traits of either qualitative or quantitative nature.

Individual plants of many species produce alternative forms of flowers that differ in structure or sexual expression. Among the former are the cases described by Davis and Ramanujacharyulu (1971) where flowers on the same plants or the same stem may exhibit either clockwise or counterclockwise spiral estivations (the arrangement of perianth parts in the floral bud). Nevertheless, the best-known and commonest examples of discontinuous within-plant variability in floral features are provided by species where flowers differ in sexual expression or mating system. The former situation is exemplified by the many species in which one plant simultaneously bears separate unisexual male and female flowers. This monoecious condition is characteristic of whole families of temperate trees, for example, Pinaceae, Fagaceae, and Betulaceae, but it is also found in many other trees, shrubs, and herbs, sometimes as an imperfect monoecism where bisexual and unisexual flowers coexist on the same individual. This happens, for example, in andromonoecious species of *Solanum* that produce hermaphrodite and staminate flowers (Diggle 1991). In some mimosoid legumes of the genus *Neptunia*, up to three different floral morphs (hermaphrodite, male, and sterile) may coexist on the same inflorescence (Tucker 1988). Alternative floral variants characterized by contrasting mating systems are exemplified by hermaphroditic species that produce cleistogamous (closed, obligatorily self-pollinated) and chasmogamous (open, exposed to cross pollination) flowers on the same plant, often on the same inflorescence and according to a regular, fixed ontogenetic sequence (Lord 1980; Ellstrand et al. 1984). Instances of this dimorphic floral condition have been reported from 228 genera and 50 families (Culley and Klooster 2007). Species exhibiting monomorphic enantiostyly provide another example of discontinuous within-plant variation in floral features. Enantiostyly is a floral polymorphism where flowers exhibit a medial-lateral asymmetry, with the style of a flower being deflected either to the left (left-styled) or to the right (right-styled) of the floral axis (Barrett et al. 2000). In monomorphic enantiostylous plants, the two floral variants are found on the same individual.

But all these well-known forms of discontinuous variation related to sexual expression, mating system, and floral morphology are neither the sole nor, quite likely, the most frequent modalities of within-plant variation in floral features. Although not sufficiently investigated, significant within-plant variation in continuous and nearly continuous morphological and structural traits (e.g., number of flower parts, size and morphology of parts) and in functional attributes of flowers (e.g., nectar production) are probably the rule in nature.

Flower Parts

Flowers are reiterated structures themselves where each verticil (calyx, corolla, androecium, gynoecium) is in turn made up of a number of repeated parts. The number of parts in each verticil is often considered among the most constant angiosperm characters, and sets of these figures ("floral formulas") have been traditionally used as a discriminant trait that characterizes many taxonomic groups at the genus or family levels. Conspicuous exceptions to this generalization are the many angiosperm families belonging to or close to the magnoliid clade (Soltis et al. 2000). In species of these families (e.g., Lauraceae, Winteraceae, Magnoliaceae, Ranunculaceae), floral verticils are composed of a variable number of parts (Cronquist 1981), and at least in the few species for which detailed data are available, such variation commonly occurs also at the within-plant level. For instance, flowers on the same tree of *Drimys winteri* (Winteraceae) vary markedly in the number and arrangement of floral organs, and most of this variation occurs within individual plants (Doust 2001). Likewise, in *Actaea rubra* (Ranunculaceae) the number of petals and stamens vary widely among flowers of the same plant, and the extent of within-plant variation in petal number is similar or even exceeds the magnitude of variation found among species in the genus (Lehmann and Sattler 1994). In populations of *Helleborus foetidus* and *H. viridis* (Ranunculaceae), all individuals exhibit substantial within-plant variation in number of carpels, stamens, and nectaries (fig. 2.2).

Although it is rarely acknowledged, variability in number of floral parts is frequent even in species where the floral formula not only is considered constant, but also is used as a characteristic trait possessing systematic value. In *Paris quadrifolia*, the specific epithet makes reference to the fixed complement of "four leaves," or rather floral bracts, that supposedly characterizes this species. Lowndes (1931), however, found that flowers produced by the same clone had from three to seven bracts, and that the number of petals, sepals, stamens, and carpels was also variable. *Ipomopsis aggregata* has a 5–5–5–3 floral formula (5 sepals, 5 petals, 5 stamens, 3 carpels) that is remarkably constant throughout its genus and is almost invariant in its family, the Polemoniaceae (Cronquist 1981). In a study conducted on 13 populations of this perennial herb, Ellstrand (1983) found that about one-third of individual plants surveyed had at least one

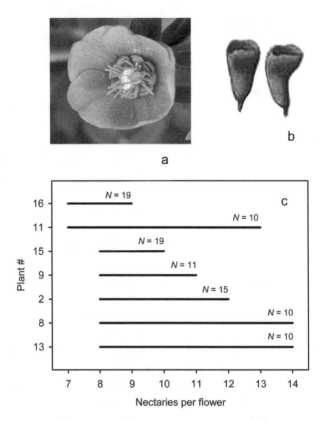

FIG. 2.2 As in other species of the genus *Helleborus*, nectaries of *Helleborus viridis* are discrete structures derived from modified petals that form a distinct verticil located between the sepals and the stamens (*a*). Each of these structures (*b*) produces large nectar volumes. The graph (*c*) illustrates variation in number of nectaries among flowers of the same plant for seven *Helleborus viridis* plants in a population from northwestern Spain. Each line in the graph corresponds to a different plant, and denotes the range of variation observed. N = number of flowers examined per plant. Based on unpublished data from J. Guitián.

deviant flower (i.e., departing from the supposedly constant, species- and family-specific merism denoted by the floral formula). In a later study of the same species, Ellstrand and Mitchell (1988) found that 28 out of 30 plants studied produced some deviant flowers over a five-week observation period. In these two studies, the four floral verticils were subject to variation around the supposedly constant number of parts. In a study of floral variation in several species of *Linanthus*, also in the Polemoniaceae, Huether (1969) found a nonnegligible frequency of flowers with corollas departing from the normal number of five corolla lobes, and part of that variation consistently occurred within individual plants.

Significant amounts of within-plant variation in the number of floral parts have also been reported for other species where a fixed number is a taxonomically diagnostic character at the generic or suprageneric level. Based on the examination of more than 20,000 flowers produced by four plants of *Jasminum multiflorum* over an entire flowering season, and more than 150,000 flowers produced by 20 trees of *Nyctanthes arbor-tristis* (both species in the Oleaceae), Roy (1963) provided what is possibly the most thorough description of within-plant variation in meristic floral characters ever published for any species. In his two species, within-plant variation in number of petal lobes occurred in all individuals studied. In *N. arbor-tristis*, the "normal" number of corolla lobes is six, but most individuals also produced a proportion of flowers having fewer and more lobes (fig. 2.3). As flowers differing in number of corolla lobes also differed markedly in the shape of the corolla in front view, within-plant variation in number of petal lobes causes highly visible within-plant heterogeneity in floral form (fig. 2.3). A similar situation was also reported by Huether (1968) for plants of *Linanthus androsaceus*, where the number of corolla parts varied among flowers of the same plant.

Although visually less conspicuous and more difficult to quantify than variations in the number of floral parts, flowers on the same plant frequently exhibit considerable continuous variation in size and shape. In *Lychnis dioica*, the shape of flowers and the form of petals vary within the

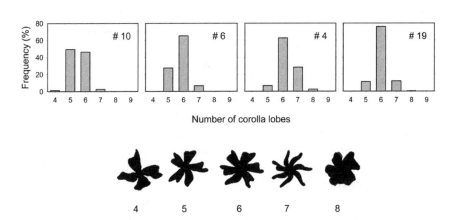

FIG. 2.3 Within-plant variation in the number of corolla lobes in four representative individuals of *Nyctanthes arbor-tristis* (Oleaceae), a small tree with tubular corollas where the "normal" number of petals is six. Each graph is based on $N > 4,000$ flowers. Frequency distributions are based on data from Roy 1963, table 7.

same plant. The terminal flower of a cyme has narrow toothed petals with distinct lateral lobes, bifurcate corona, and inflated calyx, while the second flower of the same cyme has narrow petals, both entire and toothed, some with lateral notches, a square corona, and cylindrical calyx (Curtis 1931). Although this sort of subindividual variation has long been known, there have been relatively few investigations measuring within-plant variation in continuous floral traits, a fact emphasized by Williams and Conner (2001). These studies, however, have shown that most species exhibit substantial amounts of continuous within-plant variation in structural floral features such as petal length and width, corolla length and diameter, and style length, among others (Goodspeed and Clausen 1915; Fenster 1991; Campbell 1992; Svensson 1992; Herrera 1996; Williams and Conner 2001). Quantitative information on continuous within-plant variation of flowers is reviewed in the next chapter.

Functional Traits

Empirical evidence, both observational and experimental, and theoretical analyses have consistently shown that the distribution and presentation of floral nectar rewards in insect-pollinated plants have a strong influence on the foraging patterns of their insect pollinators and, as a consequence, on plant mating system, gene flow, and male and female pollination success (e.g., Corbet 1978; Pleasants 1981; Pyke 1981; Ott et al. 1985; Zimmerman and Pyke 1986; Mitchell 1993; Cresswell 1999). For this reason, patterns of nectar composition, production, and availability have been thoroughly investigated at both the intra- and interspecific levels in recent decades, and it seems safe to state that nectar-related features are those functional floral traits whose variability has most frequently been examined.

At the local population level, the volume of nectar instantaneously available per individual flower at any one moment is extremely variable (Feinsinger 1978; Zimmerman and Pyke 1986; Zimmerman 1988). At a given site, most flowers contain little or no nectar, and a few flowers contain significant amounts. This characteristically patchy, roughly bimodal distribution of nectar availability among individual flowers is so common as to deserve distinct designations, and it has been termed a "blank-bonanza" pattern (Feinsinger 1978, 1983; Brink 1982), or as made up of "hot and cold" points (Pleasants and Zimmerman 1979) or "lucky hits" (Southwick 1982). Although relatively few investiga-

tions have carefully considered within-plant variation in nectar standing crops, there is little question that blank-bonanza patterns of nectar availability at the population level are due to differences in nectar
availability both between plants and between flowers on the same plant
(Feinsinger 1983).

Nectar standing crop depends not only on per-flower nectar secretion rate, but also on pollinator foraging (visitation frequency and nectar
extraction efficiency); hence blank-bonanza patterns may be promoted by
pollinator foraging rather than reflecting genuine variability in intrinsic
floral features. Although the "blurring" influence of pollinator foraging
on intrinsic patterns of nectar secretion cannot be negated (Zimmerman
1988), many studies have documented that different flowers on the same
plant frequently exhibit a remarkable variability in nectar secretion rates
(Steiner 1979; Feinsinger 1983; Pleasants 1983; Southwick 1983; Marden
1984a; Zimmerman and Pyke 1986; Real and Rathcke 1988; Boose 1997).
For example, in a large clone of the hummingbird-pollinated herb *Heliconia psittacorum* studied by Feinsinger (1983), individual flowers produced
from 0 to 202 µL of nectar during their one-day life span (fig. 2.4). In some
cases, within-plant variation in nectar standing crop and/or nectar produc-

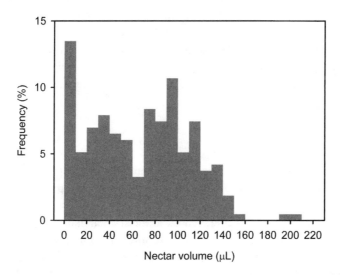

FIG. 2.4 Frequency distribution of the total volume of nectar secreted by each of 215 flowers of a single clone of *Heliconia psittacorum* during their one-day life span. Redrawn from
Feinsinger 1983.

tion rate per flower is mainly linked to differences between flowers in the total number of secretory structures, as is the case in *Helleborus* flowers with variable number of nectaries (Herrera and Soriguer 1983; fig. 2.2). Within-plant variation can also show some predictable spatial patterns. Vertical inflorescences frequently exhibit nectar gradients, either increasing or declining from top to bottom (Best and Bierzychudek 1982; Devlin and Stephenson 1985; Kudo et al. 2001).

Variation in nectar composition has been investigated far less frequently than variation in nectar abundance, and the vast majority of studies focusing on the variation in chemical composition of nectars have been concerned with interspecific variation (e.g., Baker and Baker 1982; Baker et al. 1998; Galetto et al. 1998; Torres and Galetto 2002; among many others). In one of the few detailed investigations of within-plant variation in nectar composition, Freeman and Wilken (1987) found that flowers on the same plant of field-grown *Ipomopsis longiflora* differed significantly in the relative proportions of fructose, glucose, and sucrose. They concluded that "samples from numerous flowers are needed to accurately characterize the nectar sugar composition of an individual under field conditions." Freeman and Wilken's results (1987) contrast with those of Lanza et al. (1995) in a study of variation in sugar and amino acid nectar composition of *Impatiens capensis* at the individual, plant, and population levels. They failed to detect statistically significant within-plant variation in any of the nectar constituents considered, but that negative result should be interpreted with caution in view of the limited sample on which the study was based (three flowers from each of three plants from each of three populations, or a grand total of only 27 flowers).

Very Fine-Grained Variation

As noted earlier for leaves, a very fine-grained level of within-flower variation in chemical characteristics has sometimes been reported. In *Turnera ulmifolia*, all flower parts exhibit cyanogenic activity, but this varies between floral verticils, being greatest at the stamen filaments, least at the ovary, and intermediate at other parts (Shore and Obrist 1992). In flowers of *Hypericum calycinum*, Gronquist et al. (2001) likewise found a heterogeneous distribution of defensive dearomatized isoprenylated phloroglucinols in different flower parts. These aspects of within-flower variation remain essentially unexplored so far, but may eventually prove to have some ecological significance, particularly in relation to the conflicting interaction of flowers with mutualistic pollinators and antagonistic flower visitors (Gronquist et al. 2001).

Fruit Traits

Discrete Variation

Discontinuous within-plant variation in fruit features occurs in many species, where one individual plant simultaneously produces two or more clearly different fruit types, often in the same infructescence. This heterocarpic condition occurs in some species of at least 18 plant families, but is particularly frequent among species of Asteraceae and Chenopodiaceae. In the former family, 138 species from 52 genera exhibit some form of heterocarpy, accounting collectively for 63% of heterocarpic species reported so far (Imbert 2002). Fruit differentiation in heterocarpic Asteraceae mainly occurs at the within-infructescence level, with central achenes differing in size, morphology, and/or presence of a dispersal structure, from those in the periphery of the capitulum (fig. 2.5). This differentiation often leads to discrete, clearly distinct fruit types, as in the case of dimorphic *Leontodon longirrostris* achenes depicted in figure 2.5 (see also, e.g., Venable and Levin 1985; Tanowitz et al. 1987). In the Balearic Islands, single plants of the endemic dwarf shrub *Thymelaea velutina* produce dry fruits (achenes) lacking any special dispersal mechanism along with fleshy fruits (drupes) that are eaten by small lizards, which act as the dispersal agents of the enclosed seeds (Tébar and Llorens 1993; de la Bandera and Traveset 2006). In other cases, there is a smooth gradation of fruit types within the same infructescence, and fruit variation can be considered closer to, if not indistinguishable from, a continuous phenomenon, as in the case of *Heterosperma pinnatum* (fig. 2.5; see Imbert 2002 for other examples). Heterocarpy can also be predictably associated with within-plant variation in the genetic constitution of seeds. In a study of the mating system of *Crepis sancta*, Cheptou et al. (2001) found that nondispersing achenes produced at the periphery of the capitulum were significantly more outcrossed, and pollinations involved a higher number of paternal parents, than the dispersing achenes produced by inner florets in the same capitulum.

Continuous Variation

Within-plant variation in fruit features is not restricted to heterocarpic species. Nonheterocarpic plants, although producing fruits not divisible

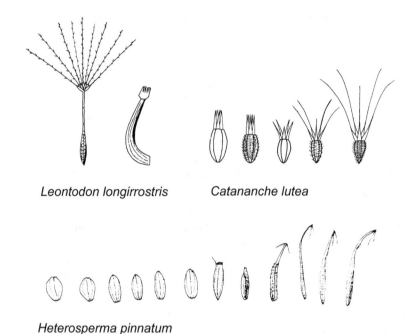

Leontodon longirrostris   *Catananche lutea*

Heterosperma pinnatum

FIG. 2.5 Within-plant variation in fruit characteristics in three representative heterocarpic species in the Asteraceae. In *Leontodon longirrostris* and *Heterosperma pinnatum*, the different
fruits pictured are produced in the same infructescence (capitulum). In *Catananche lutea*, the
two fruit types at the left are produced in subterranean capitula, and the other three types in
aerial capitula. Drawings from Venable et al. 1995; Ruiz de Clavijo and Jiménez 1998; Ruiz
de Clavijo 2001.

into types, often also exhibit considerable levels of continuous within-plant
variation in diverse fruit attributes. This kind of continuous variation and
its consequences, however, have been investigated far less frequently than
discontinuous variation associated with heterocarpy.

Size-related features of fruits (e.g., length, diameter, or mass of the
whole fruit or some of its component structures) often vary considerably
within individual plants. This occurs, for example, in fleshy-fruited species that produce berries, drupes, or functionally equivalent structures
(Obeso and Herrera 1994; Obeso 1998b). In *Helleborus foetidus*, an ant-
dispersed herb of the forest understory, Garrido et al. (2002) found extensive within-plant variation in the mass of individual elaiosomes, ancillary
structures attached to the seeds that provide a nutritious reward to dispersing ants. Within-plant variation in fruit characteristics is not restricted

to simple metric traits like length or width. As shown by Mazer and Wheelwright (1993) for the tropical tree *Ocotea tenera*, composite traits like fruit shape can also vary widely within the crops of individual plants.

Broad within-plant variation in the number of filled seeds contained in each ripe fruit seems to be the rule among species with multiovulate ovaries (Gorchov 1985; Jordano 1991; Obeso and Herrera 1994). Not surprisingly, within-plant variability in fruit seediness seems to be greatest in species that have many ovules per ovary, as exemplified in figure 2.6 by the frequency distributions of fruit seediness in representative individuals of four taxonomically and ecologically disparate species. These examples show that the number of seeds per ripe fruit can vary severalfold within the crop of a single plant, particularly in species with many ovules per ovary (e.g., *Guazuma ulmifolia* in fig. 2.6). Within-plant variation in fruit

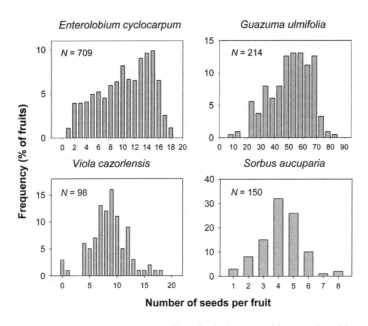

FIG. 2.6 Within-plant variation in fruit seediness in single crops of four species with multiovulate ovaries. Each panel show the frequency distribution of the number of filled seeds per fruit in the crop of one representative individual of each species. *Enterolobium cyclocarpum* (Fabaceae) and *Guazuma ulmifolia* (Sterculiaceae) are tropical dry forest trees producing dry indehiscent fruits. *Viola cazorlensis* (Violaceae) is a perennial suffruticose violet producing dehiscent capsules, and *Sorbus aucuparia* (Rosaceae) is a temperate tree producing fleshy berries. *N* = number of fruits examined per plant. Based on data from Janzen 1982b (*E. cyclocarpum*); B. Pías and M. Salvande, unpublished data (*S. aucuparia*); and C. M. Herrera 1993, and unpublished data (*V. cazorlensis* and *G. ulmifolia*).

seediness may generate concomitant variation in other fruit features. In *Vaccinium corymbosum* and *Amelanchier arborea*, Gorchov (1985) found that the ripening asynchrony of the fruits produced by the same plant was a consequence of variance in fruit developmental time (not flowering time). Fruit developmental time, in turn, was highly correlated with seed number in these species, with comparatively many-seeded fruits developing faster than few-seeded fruits.

In wind-dispersed species, both seed mass and the dimensions of ancillary structures aiding dispersal, such as wings or plumose structures, vary concurrently among the diaspores of the same fruit crop, which in turn translates into within-plant variation in wing loading and consequently in the dispersal ability of individual diaspores (Janzen 1978; Augspurger and Hogan 1983; Sipe and Linnerooth 1995; Sakai et al. 1997). Individuals of the tropical tree *Platypodium elegans*, for example, produce both single-seeded and double-seeded wind-dispersed, indehiscent legumes, and the two variants greatly differ in mass, area, wing loading, rate of descent, and dispersal distance (Augspurger 1986).

In fleshy-fruited plants, fruit pulp is the reward offered by plants to frugivorous animals that act as seed dispersers; hence its nutritional characteristics are a critical element in the plant-disperser interaction (Herrera 2002). Interspecific variation in nutritional fruit features, and its ecological and evolutionary consequences, have been documented in considerable detail by a large number of studies (Herrera 1987b; Jordano 1995a; and references therein). In contrast with the extensive attention paid to interspecific variation, only a handful of studies have investigated fruit nutritional variation among individual plants of the same species (Denslow 1987; Jordano 1987, 1989; Gargiullo and Stiles 1991). Still poorer is our knowledge on variation at the within-plant level, as I am not aware of any study of wild plants that has looked for possible within-plant variation in fruit nutritional properties. Important within-plant variation in the chemical characteristics of fruit pulp has been frequently reported for cultivated fleshy fruits (Fryer et al. 1954; Hopkirk et al. 1986; Barritt et al. 1987; Yamada et al. 1997; Broom et al. 1998); hence the absence of similar published information for wild fruits most likely reflects lack of interest from researchers in looking for that source of variation in wild plants along with the analytical difficulties involved in conducting chemical determinations on the small pulp samples obtainable from individual wild fruits. In the Mediterranean evergreen shrub *Osyris lanceolata*, refractometer readings of fruit-pulp juice, an approximate measurement of soluble-solid content (sugars, vitamins, and other solutes having refractive properties), vary considerably among drupes produced by the same plant (fig. 2.7). Within-plant variation in refractometer readings of fruit-pulp juice comparable to that of *O. lanceolata* has also been found for the berries of the vine *Smilax aspera* (C. M. Herrera, unpublished data). Refractometer readings of fruit-pulp juice have been considered a "quick and dirty" method of characterizing average fruit nutritional features in interspecific comparisons of wild plants (White and Stiles 1985). The method, however, is customarily used in the cultivated fruit industry to compare pulp properties between fruits, plants, or cultivars of the same species (e.g., Hopkirk et al. 1986). In these intraspecific contexts, measurements of soluble-solid content obtained from refractometer readings closely reflect variations in sugar and carboxylic acids content (Gurrieri et al. 2001), and are included among target variables in selective breeding programs (Yamada et al. 1997; Rodríguez-Burruezo et al. 2002, 2003).

Very Fine-Grained Variation

As noted earlier for leaves and flowers, a very fine-grained level of within-fruit variation in chemical characteristics has sometimes been reported.

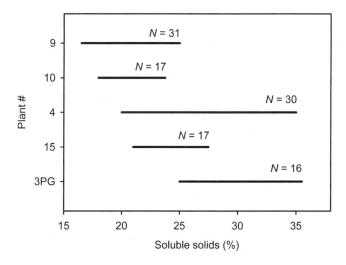

FIG. 2.7 Within-plant variation in solute concentration of fruit-pulp juice in ripe drupes of *Osyris lanceolata* (formerly *quadripartita*) from southern Spain. Each line in the graph corresponds to a different plant, and denotes the range of variation observed in percent dissolved solids in pulp juice, an inverse measurement of "fruit juiciness," as estimated using an approximate refractometry method. *N* = number of fruits measured per plant. C. M. Herrera, unpublished data.

In kiwifruit (*Actinidia deliciosa*), a marked gradation in soluble-solid concentration along the longitudinal axis of each individual fruit was reported
by Hopkirk et al. (1986). The distal end of the fruit exhibited consistently
higher soluble-solid concentrations than the basal end. Similar longitudinal patterns of soluble-solid concentration in individual fruits are also
known to occur in other cultivated fruits, including apples, melons, and citrus (Harding 1936; Scott and MacGillivray 1940; Ting 1969).

Seed Traits

Some examples of continuous and discrete within-plant variation in seed
traits that commonly occur in association with the heterocarpic condition
or, more generally, with variation in fruit features, were mentioned in the
preceding section. In other cases, within-plant variation in seed features
(size, dispersal ability, germination behavior) is related to the production
of cleistogamous and chasmogamous flowers on the same plant (Waller
1982; Antlfinger 1986; Baskin and Baskin 1998; Berg 2000). In this section
I consider only those seed traits that vary within individual plants without
such variation being correlated in obvious ways with concomitant variation in the characteristics of flowers or fruits.

Discrete Variation

As noted earlier, plants with heteromorphic fruits frequently also exhibit
discrete within-plant variation in seed features, which is known as heterospermy (Mandák 1997; Imbert 2002). One of the seed features that is
typically associated with the heteromorphic condition is the duration of
seed dormancy (Baskin and Baskin 1998; Imbert 2002; Matilla et al. 2005).
But discrete seed-dormancy polymorphisms are not necessarily restricted
to plants with heteromorphic fruits, as illustrated by the classic example
of *Xanthium pensylvanicum* seeds. In this species, each capsule bears two
seeds. The upper seed is much more dormant than the lower one, with
at least 12 months normally separating the germination of the two seeds
(Murdoch and Ellis 2000). A similar pattern was described by Marañón
(1987, 1989) in species of the grass genus *Aegilops*, where each spikelet
contains two seeds differing markedly in size. Within each pair of seeds,
the largest one consistently has the least dormancy. Discrete within-plant
variation in seed traits may sometimes involve seed coat color, chemical

composition, and dispersal mechanism. In the annual herb *Croton setiger*, single plants produce uniformly gray, chemically defended seeds unpalatable to avian seed predators along with variously mottled or striped, undefended palatable seeds (Cook et al. 1971). Individuals of the annual herb *Spergularia marina* may produce both winged and unwinged seeds (Mazer and Lowry 2003).

Many species regularly produce fruit crops that consistently contain fruits that are both normal (i.e., with filled seeds) and parthenocarpic (containing empty, unfertilized seeds) on the same plant. Whether this phenomenon is a genuine discrete polymorphism is open to question, but it undoubtedly exemplifies a situation of drastic within-plant variation in seed traits whereby normal seeds containing a viable embryo regularly coexist in the same crop with "pseudoseeds" consisting of just an empty coat. The phenomenon is particularly frequent among gymnosperms, such as species of *Juniperus* (Fuentes and Schupp 1998; García et al. 2000), but it occurs also in angiosperms, such as wild parsnip (*Pastinaca sativa*, Zangerl et al. 1991) and species of *Pistacia* (Zohary 1952; Jordano 1988; Traveset 1993) and *Ulmus* (López-Almansa and Gil 2003). In some species, multiseeded drupes commonly contain a mixture of filled seeds along with one or a few parthenocarpic "seeds," the two types being externally indistinguishable (*Ilex aquifolium*, Obeso 1996; *Crataegus laciniata*, C. M. Herrera, unpublished data). As discussed in chapter 9, studies of species where individual seed crops contain both filled and empty seeds inside otherwise normal fruits have provided some of the finest evidence of the influence of within-plant variation on the reproductive success of individual plants through its effect on the foraging behavior of animals.

Continuous Variation

Possibly most plants exhibit some sort of "cryptic heteromorphism" (Venable 1985) involving significant continuous within-plant variation in seed size, viability, germination behavior, or some combination of these (Matilla et al. 2005). Once thought to be one of the most intraspecifically constant plant characters (Harper et al. 1970; Harper 1977), seed size has eventually come to figure prominently among those reproductive traits exhibiting the greatest levels of intraspecific variation, with most of this variation taking place within individual plants' crops (Janzen 1977b, 1978; Michaels et al. 1988; Hendrix and Sun 1989; McGinley et al. 1990; Winn 1991; Mehlman 1993; Méndez 1997; among many others). Michaels et al.

(1988) found that within-plant variation exceeded among-plant variation in the vast majority of the 39 species considered in their survey, and was due in most instances to variation both among and within fruits. These and other quantitative data on seed-mass variation are reviewed in the next chapter.

Within-plant variation in seed size may translate into subsequent variation in seedling emergence and survival. This happens, for example, in the tropical tree *Virola surinamensis*, where variation in seed size is great enough within some tree crops as to give rise to significant differences in seedling vigor (height, shoot mass, leaf length) and survival (Howe and Richter 1982). Extensive within-plant variation in seed viability and germination rate has been documented for the shrub *Erica australis* (Cruz et al. 2003) and the tree *Alnus rubra* (Markham 2002), although a connection with variation in seed size was not proven in these studies.

Continuous within-plant variation in seed germinating behavior also seems to be an extremely common phenomenon, although it usually has been overlooked or ignored when not associated with a discrete polymorphism in seed size or shape. Silvertown (1984) provides a review and a model for the evolution of this sort of cryptic "somatic polymorphism," in which seeds from the same maternal crop differ widely in dormancy length and/or in their response to environmental germination cues. Differences in germination behavior are often, but not invariably, associated with differences in seed size, and can occur both within and between fruits of the same crop. Although most examples of somatic polymorphism considered by Silvertown (1984) involved variation in seed germination rates over relatively long time spans (e.g., over several germination seasons), within-crop heterogeneity in germination time may also occur over very short-term temporal scales. In the perennial herb *Phytolacca americana*, Armesto et al. (1983) found that seeds from different berries on the same raceme had widely different probabilities of germination over a trial period of only 9 days.

Relatively few investigations have so far examined the possibility of within-plant variation in the chemical composition of individual seeds. Nevertheless, as more-sophisticated analytical methods have become increasingly available in recent years, there is already some evidence revealing that individual seeds produced by the same maternal plant are far from homogeneous in their chemical composition, and that the within-plant variation involves both nutrients and potentially deterrent secondary metabolites. These studies have mostly been conducted on plants of economic interest. Using nondestructive methods based on nuclear magnetic resonance spectroscopy, Fick and Zimmerman (1973) demonstrated variation in total oil content of individual seeds within single heads of cultivated sunflowers (*Helianthus annuus*). More recently, Velasco and Möllers (2002) have been able to document significant within-plant differences in the protein content of the tiny individual seeds of rapeseed (*Brassica napus*) by means of near-infrared reflectance spectroscopy. In a study of 32 varieties of *Lupinus albus*, Velasco et al. (1998) found significant within-plant variation in every trait considered in their study, which included protein content, oil content, and oil composition. The oil from seeds of pods borne on the main stem, for example, was characterized by a higher percentage of saturated fatty acids and oleic acid, and a lower percentage of linoleic, linolenic, eicosenoic, and erucic acid than the seeds from pods located on the branches. In soybeans (*Glycine max*), Marchetti et al. (1995) found significant variation in the concentration of protease inhibitors, depending on the position along the main stem of individual plants.

Chemical variation among seeds of the same plants can take place at very small spatial scales. Within the capitulum of *Tragopogon dubius*, Maxwell et al. (1994) found a significant increase in phenolic content from the lighter central achenes to the heavier peripheral achenes. Calderini and Ortiz-Monasterio (2003) found that the concentration of macronutrients (Ca, Mg, K, P, and S) and micronutrients (Cu, Fe, Mn, and Zn) in individual seeds of wheat (*Triticum aestivum*) varied considerably according to grain position within the spike. Variation took place on both the vertical axis (spikelets on different positions along the rachis) and the horizontal axis (seeds at different positions within the same spikelet) of the spike, but grain variation within the same spikelet was the main source of variation. Although these studies refer to cultivated plants, it seems reasonable to expect that similar results would eventually be obtained if the same fine analytical procedures were applied to seed crops of wild plants. In fact, significant within-crop variation in the concentration of major mineral nutrients may be inferred from some investigations that have examined the relationship between nutrient concentration and seed size in tropical (Grubb and Burslem 1998) and temperate wild plants (Brookes and Wigston 1979).

CHAPTER THREE

Continuous Within-Plant Variation of Reiterated Structures

The extent of subindividual variation in continuously varying leaf, flower, fruit, and seed traits is assessed.

A s noted in the preceding chapter, instances of discrete within-plant variation in the form or function of reiterated structures such as leaves, flowers, and fruits have attracted so much attention from researchers, and for so long, as to eventually become classic research subjects in botany and ecology. Continuous variation, in contrast, has received only marginal attention. While there are hundreds of papers on heterophylly, cleistogamy, and fruit heteromorphism, only a handful of investigations focus careful attention on continuous variation within individuals.

Several reasons might be offered for the imbalanced interest in discrete and continuous within-plant variation. Chapter 2 lists a large number of disparate traits of reiterated structures that are widely known to exhibit continuous within-plant variation, so rarity of the phenomenon can hardly account for the eagerness to study discontinuous variation and the relative neglect of continuous variation. In fact, instances of discrete polymorphisms are probably rarer, in terms of the number of species showing them, than instances of marked continuous variation. Another simple explanation could be that continuous variation, although admittedly widespread and involving a long list of morphological, physiological, and functional traits of reiterated structures, is quantitatively unimportant. Dismissal of the phenomenon would then be explained because of some implicit judgment of quantitative irrelevance. No systematic assessment of the extent of within-plant variability shown by reiterated structures seems

to have been conducted apart from early biometricians' efforts noted earlier (which, by the way, suggested that continuous variability might be far
from negligible), yet the prevailing disregard for continuous within-plant
variation as a research topic might still stem from some generalized intuition among botanists and ecologists that it is a quantitatively minor phenomenon. A third explanation for neglect could be that no guiding theory
makes the phenomenon of continuous variation interesting to researchers, while alternative phenotypes (e.g., submerged and emergent leaves)
suggest adaptation to distinct functions.

    In this chapter, I address the question, how large is the within-plant
variability of continuous structural and morphological traits of reiterated
plant structures? First, I briefly describe some of the problems involved
in measuring and comparing variability of biological structures in general,
and suggest possible approaches to circumvent these difficulties in the particular case of within-plant variability in reiterated structures. In the rest
of the chapter I survey published and unpublished data for leaf, flower,
fruit, and seed traits, in an attempt to objectively evaluate the absolute
and relative magnitude of continuous within-plant variation. If, after such
an exercise, it turns out that within-plant variability is quantitatively negligible, then there would be sound reasons for the traditional neglect of
the phenomenon, and little justification would be left for this book. Having the book in hand, the reader can easily anticipate what the main conclusion of this chapter will be. Future chapters attempt to motivate more
interest in such variation.

Measuring and Comparing Within-Plant Variation

Measuring and comparing levels of variation bring up delicate issues in
biostatistics. This is evidenced by the long series of methodological publications that have addressed potential biases and pitfalls arising from inad-
equate statistical treatments of variation levels (Haldane 1955; Lewontin
1966; Lande 1977; Van Valen 1978; Sokal and Braumann 1980; McArdle
and Gaston 1992, 1995; among many others), and by the controversies
that have so frequently arisen around the interpretation of studies related
in one way or another to variability levels (Kluge and Kerfoot 1973; Sokal
1976; Rohlf et al. 1983). The concepts of variation and variability may have
simple, intuitive meanings in the vernacular, yet that simplicity proves
somewhat deceptive on closer examination, as the concepts are open to

multiple codifications and measurements. Problems may be exacerbated in the case of within-plant variation of reiterated structures. Before proceeding with the survey of data on within-plant continuous variability, and to justify the approach adopted in this chapter and in the rest of this book to analyzing within-plant variability, I first briefly review some problems involved in quantitatively evaluating and comparing within-plant varia-bilities. For a recent, synthetic treatment of the main issues involved in the study of relative levels of variation, see Lynch and Walsh 1998, chapter 11.

Coefficient of Variation

As the variance or standard deviation of a character is scale-dependent, that is, depends on the mean and the units of measurement, neither of these parameters can properly be used to compare variability levels. This was long ago noted by Pearson (1901, 360), who remarked, "Measures of the absolute variations as given by the standard deviation seem to me of no use when we are comparing different characters in different species." One way of circumventing this problem has been to apply to the original data some variance-stabilizing transformation, typically logarithms, which renders the variance on the transformed scale independent of the mean (Lewontin 1966). Another classic procedure for comparing relative amounts of variation for measurements involving different units or samples having different means involves using the ratio of the standard deviation to the mean. This statistic, the coefficient of variation (CV), expresses sample variability relative to the mean of the sample, and is most useful for variables that are always positive, because it is undefined when the mean is zero and frequent zero values can introduce important biases (Lande 1977). Pearson (1901, 360) noted that the coefficient of variation "seems to me the only satisfactory comparative measure we can find at present of variability," and he used the CV extensively to assess within-plant variability in continuously varying traits. Because the mean and the standard deviation are expressed in the same units, the CV is unitless, a fact emphasizing that it is a relative measure, divorced from the actual magnitude or units of measurement of the data (Zar 1999). These advantageous theoretical properties of the CV, along with its ease of interpretation, have made the CV the most well-known method of obtaining comparable variability measurements.

In practice, however, use of the CV is not exempt from problems, and artifacts may arise if one puts too much confidence on the theoretical scale independence of CV without actually checking the validity of the underlying premises. The most critical of these assumptions is the requirement that, for the CV to be truly scale-independent, the standard deviation (SD) must be directly proportional to the mean (X). This requires that the two magnitudes are linked by a relationship of the form SD = k · X, k being a constant. Departures from this requirement, however, can frequently occur caused by (1) the relationship between SD and X being extremely weak, or SD varying randomly with respect to the mean, that is, the two statistics being unrelated; or it may be caused by (2) the regression of SD on X having a nonzero intercept, that is, it is of the form SD = a + k · X. In this case the CV will be linked to the mean by an inverse, nonlinear relationship, as is clearly seen by dividing both sides of the equality by X. In both case 1 and 2, CV values are not truly scale-independent, and residual correlations with the mean are therefore to be expected. This problem can be consequential in analyses involving comparisons or correlations of CV values across distinct groups, and some patterns involving the variability of species and populations that were once thought to reflect genuine biological phenomena have eventually turned out to be caused by hidden CV-mean correlations. These include "allomeric variation," or the inverse relationship between the CV of a morphological trait and the square root of organ size (Soulé 1982), and the so-called Kluge-Kerfoot and Roginskii-Yablokov effects, involving relationships between population-level trait means and either between- or within-population CVs (Lynch and Walsh 1998). Verifying for particular data sets that the data do not suffer too seriously from condition 1 or 2 above is thus an important requisite to avoid artifacts when using the CV to look for ecological correlates of variability or to test particular hypotheses about variability. For descriptive purposes or rather crude comparisons like most of those performed in this book, however, the CV can still be useful as an index of variability provided that there is at least some reasonably linear relationship between SD and X.

Since Karl Pearson (1901) first introduced it for that purpose, the CV has often been used to measure intraspecific variability in structural or functional traits of reiterated structures, including flower corolla length (Fenster 1991; Herrera 1996), nectar production (Real and Rathcke 1988; Boose 1997), seed mass (Michaels et al. 1988; Krannitz 1997a), seed chemical constituents (Krannitz 1997b), and fruit dimensions (Obeso and Herrera 1994). Some of these investigations have explicitly considered the within-plant component of variability, but, to my knowledge, none of them seems to have verified whether the presumed direct relationship between variance and mean, which would justify use of the CV, does actually apply in the little-explored subindividual context. Are the within-plant variance and the corresponding individual mean for continuously varying metric traits directly related across individual plants? Figure 3.1 plots the relationship between within-plant variance and plant mean for span (corolla-spur length, fruit length) and mass (leaf fresh mass, seed mass) characters

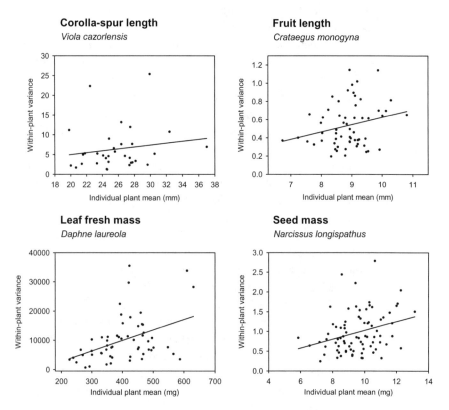

FIG. 3.1 The usual direct relationship between mean and variance holds in the case of within-plant variation in metric characters of reiterated structures, as illustrated here for two linear and two mass traits of different reiterated structures in four species of plants from southeastern Spain. Across individual plants, within-plant variance tends to increase with increasing individual mean, although the strength of the relationship differs among examples. In all graphs, each symbol corresponds to a different plant, and lines are fitted least-squares regressions. Sample sizes (N = number of plants sampled, number of structures measured) and R^2 and P-value for the regression in each graph are in parentheses after species names: *Viola cazorlensis* (N = 33, 969; R^2 = 0.026, P = 0.37), *Crataegus monogyna* (N = 60, 2,400; R^2 = 0.061, P = 0.06), *Daphne laureola* (N = 56, 895; R^2 = 0.18, P = 0.001), and *Narcissus longispathus* (N = 81, 2,944; R^2 = 0.064, P = 0.02). Based on data from C. M. Herrera and from C. Alonso, unpublished data for *D. laureola*.

of flowers, fruits, leaves, and seeds of four species of southeastern Spanish plants. Variance-mean relationships similar to those depicted in figure 3.1 have been reported, for example, by Real and Rathcke (1988) for the nectar volume available in individual flowers of *Kalmia latifolia*, and can also be inferred from raw data on seed mass presented by Janzen (1978) for the tropical legume tree *Ateleia herbert-smithii*.

Although it is not possible at present to know whether these rather limited examples depict general patterns of variation exhibited by these characters, they at least serve to suggest two tentative conclusions. For a variety of characters, within-plant variance is positively, linearly related to plant mean, which provides statistical justification for using the CV as a quantitative descriptor of the extent of within-plant variability, useful for comparative purposes. Nevertheless, the relationship between variance and mean is fairly weak, with R^2 values ranging between 0.026 and 0.18. This means that, sometimes, within-plant CVs may not be fully scale-independent, and CV-assessed variability data might generate some spurious patterns in comparisons involving groups with widely different means. Keeping this potential difficulty in mind, one could still reasonably use the CV as the main descriptor of within-plant variability in analyses that do not require strict scale independence of the variability measurement (e.g., when no hypothesis under test implies some direct or indirect relationship between variability and means).

Variance Partitioning

Another approach to quantitatively assessing within-plant variation in quantitative, continuously varying traits of reiterated structures consists of partitioning the total population-level variance of the character (Var_{total}) into its additive between-plant (Var_{among}) and within-plant (Var_{within}) components. As with use of the CV, Pearson (1901) also pioneered the use of variance partitioning to quantify within-plant variation, although in this case later thinkers created the terminology that we use today. Pearson's "homotypic correlation coefficient," a measurement of the degree of resemblance between "undifferentiated like organs" on the same plant individual, corresponds to the statistical parameter we currently know as the "intraclass correlation coefficient," a measure of the homogeneity of observations within the groups of a random factor relative to the variability of such observations among groups. The intraclass correlation coefficient equals the ratio of the variance accounted for by within-plant variance relative to total variance (Zar 1999).

Meaningful variance-partitioning analyses can be conducted even if the number of within-plant replicates is small, so the method can profitably be used with small plants or those producing just a few reiterated structures per individual. This contrasts with CV-based methods, which require more replicates per plant in order to obtain reliable plant-level estimates of variability. Another advantage of approaching the analysis of within-plant variability using character variance partitioning is that results will generally be amenable to straightforward biological interpretations. For example, the fraction of total population-level phenotypic variance in a metric character of a reiterated structure occurring within the confines of an individual plant may readily be interpreted in terms of limitations to the responses to natural selection, as discussed in chapter 10.

One drawback of variance partitioning is that the within-plant component of variance may be inflated by measurement error unless either large samples are measured per plant or individual reiterated structures are measured repeatedly, thus allowing proper estimation of measurement-error variance (rather than assuming that it is negligible, as is most often done; I return to this point in chapter 10). A second and more important drawback is that, if the within-plant component of variance is expressed in relative terms with respect to the total population-level variance (i.e., as an intraclass correlation coefficient), then it does not depend exclusively on the extent of within-plant variability, but is "contaminated" to a variable extent by between-individual variability. This is clearly seen by considering that

$$\%Var_{within} = Var_{within} / (Var_{within} + Var_{among}). \quad 3.1$$

By this definition, multiple populations with identical absolute Var_{within} values may yield widely different $\%Var_{within}$ figures if their differences in Var_{among} are large. This implies that $\%Var_{within}$ estimates may be nearly useless for comparative purposes in those situations where it reflects Var_{among} nearly as much as it reflects Var_{within}, a fact that does not seem to have been properly acknowledged in earlier investigations that have used the variance-partitioning approach to compare within-plant variability.

Within-plant variability in seed and fruit mass (e.g., Michaels et al. 1988; Hendrix and Sun 1989; Obeso and Herrera 1994) and floral morphology (Herrera 1996; Cresswell 1998; Williams and Conner 2001; Ehlers et al. 2002) has sometimes been examined by adopting a variance-partitioning perspective. Because of its inherent drawbacks, this approach should be seen as a complement, rather than as an alternative, to the one based on within-plant CVs. While the latter assesses within-plant variability at the per-plant level, the variance-partitioning approach provides a population perspective of the extent of within-plant variation *scaled to total population-level variance*. The CV-based approach allows us to investigate, for example, possible individual differences in within-plant variability levels (chapter 7), while the variance-partitioning approach may be used to compare species or populations (subject to the limitations and caveats noted in the preceding paragraph). As the two methods focus on different aspects of within-plant variability that may or may not be related (Pearson 1901), they should ideally be used in combination, and this is the approach followed in the next sections. All phenotypic variance components reported in this chapter and elsewhere in the book were computed using restricted maximum likelihood estimation (Searle et al. 1992; see Lessells and Boag 1987 for a readable account of problems associated with sum-of-squares estimation of variance components).

Leaf Variation

Of all reiterated structures, leaves have most frequently attracted the attention of botanists for assessing patterns and implications of continuous within-plant variation. Reasons for the continued interest in leaf variation are diverse. For example, the size, shape, and physiological characteristics of tree leaves, and the manner in which these traits vary within tree crowns, are of great importance to forest ecosystem function, since production and ecosystem-atmosphere exchange are influenced by architectural features of the canopy (Ford and Newbould 1971; Hutchison et al. 1986). Assessing vertical variation in leaf distribution and its relationship to patterns of photosynthesis among different canopy positions provides insight into how carbon, nitrogen, and nutrient resources in general are partitioned within forest canopies and individual plants, an aspect that may have important consequences for the fitness of individual plants, as discussed in chapter 9 (Field 1983; Ellsworth and Reich 1993; Hollinger 1996; Bassow and Bazzaz 1997). In the case of trees of economic interest, investigations of within-tree variation of leaf features have sometimes been motivated by attempts to characterize tree crowns for possible genetic selection directed at enhancing carbon fixation and timber production (Tucker et al. 1993). Analyses of within-plant variation in leaf features have also been prompted by taxonomic considerations (Kincaid et al. 1998).

Detailed quantitative information on within-plant variation in leaf traits is remarkably scarce in the literature, and good quantitative estimates of the magnitude of within-plant leaf variability are rare. This may seem ironic, given the interest elicited among ecologists by patterns of leaf variation. Nevertheless, the lack of rigorous analyses is probably just a reflection of the practical difficulties involved in adequately assessing the enormous within-plant variation ordinarily exhibited by continuously varying leaf traits, particularly in the case of large plants. For example, in a very detailed study of the morphology of leaves of *Quercus petraea*, Bruschi et al. (2003) found that the variation within trees was considerably greater than variation among trees for 19 out of the 32 leaf descriptive parameters investigated. In a study of olive trees (*Olea europaea*), Perica (2001) found that nitrogen content of individual leaves varied with canopy height, side (orientation) of the tree, and position along shoots, and concluded that adequate sampling protocols should necessarily incorporate these three concurrent levels of variation. Faced with such extensive within-plant variability, therefore, the characterization of leaf traits for individual plants is far from trivial, and ecologists and statisticians alike have long struggled to design optimal sampling strategies that adequately capture extensive within-plant variation in leaf traits while economizing time and effort (e.g., Wood 1972; Valentine and Hilton 1977; Gregoire et al. 1995; Luyssaert et al. 2001; Temesgen 2003; among many others). The effort required to sample variation at all levels (among individuals, among branches and subbranches, among leaves, among measurements) can be formidable, and this may explain the scarcity in the literature of reliable statistics of leaf variability at the plant population level. Leaf sampling problems become particularly serious in tall forests where, as summarized by Bassow and Bazzaz (1997), sampling leaves is "difficult, dangerous, and expensive." Not surprisingly, therefore, some of the most thorough investigations so far conducted on within-plant leaf variation of trees are based on data collected from single plants (Kincaid et al. 1998; Porté and Loustau 1998; Le Roux et al. 1999; Luyssaert et al. 2001), which precludes any assessment of within-plant variability at the population or species level. Difficulties inherent in leaf sampling, however, are not exclusive to large plants such as trees or shrubs, which bear many thousand, widely distributed leaves. Extensive variation and steep within-plant gradients in structural, chemical, and functional leaf traits are also known to occur even at the relatively reduced spatial scales of herbaceous plants, and properly sampling this variation can also be a formidable problem even in these apparently simple cases (Charles-Edwards et al. 1987; Hirose and Werger 1987; Lemaire et al. 1991; Williams, Collis, et al. 1993; Williams, Thomas, et al. 1993).

A compilation of available within-plant coefficients of variation and variance components for a variety of morphological, size-related, structural, and chemical leaf traits is shown in table 3.1. Although they refer to only a few tree and shrub species, these figures serve at least to illustrate that some leaf traits sometimes vary so extensively within-plants that nearly all population-level variance (>90% of total) takes place within single individuals. The proportion of total population-wide variance accounted for by within-plant variation ($\%Var_{within}$) ranged between 54 and 90% for size-related leaf traits (area, length, width, mass), between 52 and 92% for structural traits related to micromorphology, pubescence, thickness, and toughness, and between 60 and 87% for some leaf chemical properties such as amino acid, carbohydrate, and water content. $\%Var_{within}$ was lower for nitrogen and phenolics (36–49%). That leaf nitrogen concentration (on a per mass basis) tends to be among the least subindividually variable leaf traits is also supported by the results of, among others, Hollinger (1989) for *Nothofagus solandri*, Ellsworth and Reich (1993) for *Acer saccharum*, Le Roux et al. (1999) for *Juglans regia*, and Niinemets and Kull (1998) for *Populus tremula, Fraxinus excelsior, Corylus avellana*, and *Tilia cordata*. In Bassow and Bazzaz's study of variation in photosynthetic and structural leaf traits of ten tree species (1997), the variance among leaves within a single tree was greater than the variance among trees within a species for all parameters measured except nitrogen content. It must be noted, however, that their within-plant variance estimates are likely to underestimate actual variability levels, because only leaves from the top of the canopy ("sun leaves") were sampled and steep height-related gradients in leaf traits commonly occur in tall forest trees (Hollinger 1989; Ellsworth and Reich 1993; Tucker et al. 1993).

The ranges of within-plant variation in leaf traits furnished by some studies are also useful to supplement the scarce measurements of variability available based on CVs and variance components. These data also unequivocally point to extensive within-plant variation and often to greater within- than between-plant variability. In the single individual of the tropical tree *Pourouma tomentosa* sampled by Kincaid et al. (1998), leaf mass varied 9-fold (range = 0.6–5.5 g dry mass), leaf area varied 8.3-fold (53–443 cm²), and leaf specific mass varied 1.5-fold (98–143 g/m²). Within a single tree of *Juglans regia*, leaf specific mass varied 2.8-

TABLE 3.1 **Within-plant variation in continuously varying leaf traits in several tree (*Betula, Prunus, Quercus*) and shrub (*Daphne*) species.**

Leaf trait	Species[a]	$\%Var_{within}$ [b]	Mean $CV_{within}(\%)$ [c]	Reference
Size-related				
Length	*Daphne gnidium*	53.5	10.9	C. M. Herrera unpubl.
Width	*Daphne gnidium*	64.6	12.4	C. M. Herrera unpubl.
Area	*Daphne gnidium*	58.8	19.6	C. M. Herrera unpubl.
	Daphne laureola (2)	87.4	37.8	C. Alonso unpubl.
	Prunus mahaleb (5)	89.4	51.7	Alonso 1997b and unpubl.
Fresh mass	*Daphne laureola* (2)	78.1	37.5	C. Alonso unpubl.
	Prunus mahaleb (5)	89.7	52.5	Alonso 1997b and unpubl.
Structural				
Micromorphology[d]	*Quercus petraea*	80.5		Bruschi et al. 2003
Pubescence[e]	*Quercus petraea*	74.7		Bruschi et al. 2003
Specific weight[f]	*Betula pubescens*	59.0		Suomela and Ayres 1994
	Daphne laureola (2)	65.2	11.2	C. Alonso unpubl.
	Prunus mahaleb (5)	91.9	24.6	Alonso 1997b and unpubl.
	Mean for 10 tree species	52.4		Bassow and Bazzaz 1997
Thickness[g]	*Quercus petraea*	85.3		Bruschi et al. 2003
Toughness	*Betula pubescens*	88.0		Suomela and Ayres 1994
Chemical				
Amino acids[h]	*Betula pubescens*	73.1		Suomela, Ossipov, and Haukioja 1995
Carbohydrates	*Betula pubescens*	60.0		Suomela, Ossipov, and Haukioja 1995
Nitrogen content	*Betula pubescens*	36.0		Suomela and Ayres 1994
	Mean for 10 tree species	49.2		Bassow and Bazzaz 1997
Phenolics	*Betula pubescens*	36.0		Suomela, Ossipov, and Haukioja 1995
Water content	*Betula pubescens*	79.0		Suomela and Ayres 1994
	Daphne laureola (2)	75.4	2.7	C. Alonso unpubl.
	Prunus mahaleb (5)	87.4	7.2	Alonso 1997b and unpubl.

[a]For species with data from more than one locality, the number of populations sampled is shown in parentheses. In these instances, percent variance and CV figures are average values.
[b]$\%Var_{within}$ is the proportion of total population-level variance in a given leaf trait that is accounted for by differences between leaves of the same shrub or tree.
[c]Mean CV_{within} is the population mean of within-plant coefficients of variation.
[d]Average for ten micromorphological leaf traits, related to stomatal area and density, length and width of stomata, and trichome density.
[e]Average for five pubescence-related traits, related to the frequency of pubescence on different sections of leaf midrib and petiole.
[f]Leaf mass per area unit.
[g]Average for seven thickness-related leaf traits, related to thickness of lamina, epidermal cells, and palisade cells.
[h]Average for 25 amino acids.

fold (50–140 g/m^2) and total nonstructural carbohydrates varied 4.2-fold (4–17 g/m^2) (Le Roux et al. 1999). In a single willow tree (*Salix fragilis*) thoroughly sampled by Luyssaert et al. (2001), cadmium concentration in leaves varied 4.4-fold (2.4–10.6 mg/kg), and the CV of within-plant variation was 25.4%.

    Further support for greater within- than between-plant variation in leaf traits comes from investigations that failed to detect statistically significant differences between individual plant means yet found variation between branches of the same plant and between leaves of the same branch. This occurs, for example, in the New Caledonian nickel-hyperaccumulating shrub *Psychotria douarrei*. In this species, mean nickel content of leaves does not differ between shrubs, but there is considerable variation among different branches of the same plant (Boyd et al. 1999). In the temperate tree *Sassafras albidum*, de Soyza et al. (1990) failed to find statistically significant differences between plants in mean leaf chlorophyll content, yet the position of leaves along branches accounted for 30–50% of the total variance in chlorophyll content, which denotes considerable within-plant variation at the relatively small scale of individual shoots. This is a heterophyllous species with distinct leaf morphs (fig. 2.1), but within-plant variation in leaf chlorophyll content and photosynthetic activity is largely unrelated to differences in leaf shape (de Soyza and Kincaid 1991).

Floral Variation

In sharp contrast to leaves, flowers have been traditionally considered the least plastic, most intraspecifically constant of all reiterated plant structures (East 1913; Sinnott 1921). The small variability of flowers was nearly dogma among botanists because of the historical emphasis on reproductive traits by systematists and the observation that flowers are less phenotypically plastic than other plant features. De Candolle and Sprengel (1821, 85) stated that "the duration of plants, their stature, their taste and smell, even sometimes their colours, and also their situation and time of flowering, are all things and relations which we must consider variable; whilst, on the other hand, the forms and numerical proportions of the part of fructifications, are seldom subject to change. These, therefore, must constitute the principle of classification." This long-held belief applies particularly to the case of animal-pollinated flowers and is motivated by the consideration that, in these plants, the size and shape of flowers and flower parts have been precisely shaped by strong stabilizing selection exerted by pollinators in order to optimize pollen transfer (Stebbins 1970). Support for this interpretation was provided by early investigations of patterns of variation and covariation of floral and vegetative traits. These pioneering studies revealed that, within species of animal-pollinated plants, floral characters were more tightly integrated (i.e., phenotypically correlated to each other) than vegetative ones; that sets of floral and vegetative characters were only weakly correlated; and that dimensions of floral organs tended to be less variable than dimensions of vegetative parts (Stebbins 1950; Berg 1959, 1960). More recent studies have often provided support for these predictions (Conner and Via 1993; Conner and Sterling 1995; Armbruster et al. 1999), including direct or circumstantial evidence of stabilizing selection by animal pollinators on metric floral traits such as corolla dimensions or style length (Fenster 1991; Wolfe and Krstolic 1999; Cresswell 1998, 2000). Nevertheless, neither the evidence that animal pollinators may sometimes exert stabilizing selection on floral traits, nor the observation that these traits are often tightly integrated and vary little within species, should lead us to automatically embrace any sort of a priori truth about intraspecific floral invariability, a warning advanced nearly a century ago by Goodspeed and Clausen (1915, 371). These authors stated that "flower size, although very evidently not so markedly modified by environmental conditions as height of plant, leaf size, etc., still is not a stable character complex and is subject to marked modifications under the stress of both internal and external conditions attending development."

Two main lines of evidence run contrary to any claim of universality of the "floral invariability" paradigm for animal-pollinated plants. On one hand, some recent tests have failed to validate the canonical expectation that floral traits are always less variable than vegetative ones of the same species. In a study of nine species of wind- and animal-pollinated tropical plants, Armbruster et al. (1999) found that, while species with specialized pollination usually showed lower coefficients of variation for floral traits than vegetative traits (as predicted), the same was also true of species with unspecialized or wind pollination (unlike the prediction). Furthermore, there was no greater tendency toward small floral CVs in species with specialized animal pollination than in species with generalized animal pollination or wind-pollination, as would have been expected if precise matching of flowers and pollinators consistently reduced floral variability through the action of stabilizing selection. To quote Goodspeed and Clausen again (1915, 372), their careful greenhouse experiments on *Nicotiana* led them to conclude that "under favorable and unfavorable conditions of greenhouse culture flower size will vary distinctly and in the same direction as vegetative characters under such conditions," which supports the idea that, in some species, floral traits are not essentially different from vegetative ones insofar as variability is concerned.

On the other hand, there is considerable empirical evidence showing that flowers are not, by any means, exceptional among reiterated plant structures with regard to variability, and that some species exhibit considerable floral variation at both the within- and between-individual levels. This not only applies to meristic floral traits (e.g., part numbers), as already noted in chapter 2, but also to continuously varying floral dimensions. In a literature survey of intraspecific floral variation across a wide range of insect-pollinated taxa, Cresswell (1998) found that the mean CV for continuously varying floral traits related to advertising, pollen vector matching, gender, biomass, and sexual dimensions ranged between 14 and 29%. Variability was much larger for characters related to nectar production, with a mean CV of 54%. In 19 insect-pollinated species studied by Møller and Eriksson (1994), the CV for petal size varied between 7.7 and 24.8%. Kearns and Inouye (1993, table 9.1) presented a summary of means and standard deviations for continuous floral traits such as corolla width, spur length, petal length, and style length for a number of animal-pollinated plants. In that survey, population-level CVs range between 5.2 and 23.1%. Floral variability measurements reported in these studies represent a heterogeneous mixture of among-individual and population-level estimates that reflect the influence of variability at both the within- and among-plant levels, and thus are of little use to infer levels of within-plant variation. Despite this limitation, variability measurements presented by these authors should be kept in mind to dismiss any belief in the oft-repeated myth of floral invariability as a characteristic inherent to all animal-pollinated plants. Results of Cresswell's review are also particularly valuable in that they illustrate that levels of floral variability vary greatly among species and among floral traits that perform different functions (see also Ushimaru et al. 2003), a finding that should also caution us against simplistic generalizations on levels of intraspecific floral variability.

Within-plant variability in continuous or quasi-continuous structural floral traits is considerable in many species of animal-pollinated plants. In *Raphanus raphanistrum*, for example, within-plant variation in petal length, petal width, and corolla-tube length accounted, on average, for 71.5–82.5% of total variance in a common-garden experiment (Williams and Conner 2001). Similarly large within-plant variance components (range 44.3–72.0%) were also found by Campbell (1992) for corolla, calyx, stamens, and pistil mass in *Ipomopsis aggregata*. In *Rhizophora mangle*, within-plant variation accounted, on average, for 66% and 76%

of population-level variance in the first two principal components describing flower morphology (Domínguez et al. 1998). In midstyled plants of tristylous *Eichhornia paniculata*, within-clone variation accounted for between 43 and 98% of total population-level variance for 14 different floral traits (Seburn et al. 1990, table 3). In a common-garden study of 30 cultivars, each genetically homogeneous, of *Brassica rapa*, Syafaruddin et al. (2006) found that within-plant variance in several metric floral traits (e.g., petal area, stigma length and width, anther length) was often comparable to, and at times even exceeded, the combined variance among individuals and cultivars. Within-plant coefficients of variation (CV_{within}) in the range of 10–20% are not exceptional in the literature, as shown by, for example, Shull (1902) for number of bracts, disk florets, and ray florets in *Symphyotrichum puniceum*, and Sherry and Lord (1996a) for petal, sepal, style, and hypanthium dimensions in *Clarkia tembloriensis*. For two species of *Trillium*, Irwin (2000) reported very high coefficients of variation for floral characters (*T. erectum*, 17.3%; *T. grandiflorum*, 19.8%), which were similar to those for vegetative characters (*T. erectum*, 18.9%; *T. grandiflorum*, 20.0%).

An unambiguous picture of frequently large within-plant variabilities in continuously varying structural floral traits emerges from the taxonomically diverse sample of 97 species of animal-pollinated plants compiled in table 3.2. Traits considered are petal number, flower size, corolla-tube length, spur length, and petal and sepal length. In this sample of species, population means of within-plant CVs (mean CV_{within} hereafter) range between 1.7 and 16.3%, and % Var_{within} ranges between 5.8 and 100%. This latter magnitude was greater than 50% in 27% of species, which means that the variation between flowers of the same plant was the dominant source of variance in floral traits in a significant fraction of species.

Floral features related to nectar secretion are much more intraspecifically variable than structural ones (Cresswell 1998), and this applies to the within-plant scale of variation as well. The amount of nectar instantaneously available per flower is often extremely patchily distributed within individual plants. In *Helleborus foetidus*, 66.9% of population-wide variance in nectar volume per flower is accounted for by differences among flowers of the same plant (Herrera and Soriguer 1983). In a population of *Lavandula latifolia*, differences between flowers on the same shrub accounted for 97.7% of population variance in per-flower nectar volume, and CV_{within} for nectar volume per flower ranged between 175 and 363% (Herrera 1995a, and unpublished data).

TABLE 3.2 **Within-plant variation in continuously varying structural floral traits in 97 species of insect-pollinated plants.**

		Sample size				
Floral trait	Species[a]	Number of plants	Number of flowers	%Var_{within}[b]	Mean CV_{within} (%)[c]	Reference
Petal number	*Nyctanthes arbor-tristis* (T)	17	82,173	90.0	9.8	Roy 1963, table 7
Flower length or diameter	*Castilleja* sp. (S)	10	20	86.4	7.1	L. Navarro unpubl.
	Chelidonium majus (H) (2)	166	324	94.3		Kang and Primack 1991
	Gesneria cuneifolia (H)	10	20	20.7	5.1	L. Navarro unpubl.
	Helleborus foetidus (H) (13)	368	1,686	40.8	4.5	Herrera et al. 2002 and unpubl.
	Hormathophylla spinosa (S)	20	399	59.7	9.0	J. M. Gómez unpubl.
	Odontonema cuspidatum (H)	15	30	87.7	5.2	L. Navarro unpubl.
	Tabebuia chrysantha (T)	10	20	31.4	2.0	L. Navarro unpubl.
Corolla-tube length	*Acinos alpinus* (H)	20	100	42.2	4.4	Herrera 1996 and unpubl.
	Alloplectus tetragonoides (S)	10	20	64.3	1.8	L. Navarro unpubl.
	Anarrhinum bellidifolium (H)	20	100	42.2	5.2	Herrera 1996 and unpubl.
	Anchusa azurea (H)	20	93	18.4	4.2	Herrera 1996 and unpubl.
	Anchusa undulata (H)	20	98	37.9	4.3	Herrera 1996 and unpubl.
	Ballota hirsuta (S)	20	100	15.9	2.8	Herrera 1996 and unpubl.
	Calamintha sylvatica (H)	20	100	30.4	3.8	Herrera 1996 and unpubl.
	Capanea grandiflora (H)	10	20	41.4	2.5	L. Navarro unpubl.
	Cerinthe major (H)	21	93	19.8	1.9	Herrera 1996 and unpubl.
	Chaenorhinum serpyllifolium (H)	20	90	47.9	4.1	Herrera 1996 and unpubl.
	Cleonia lusitanica (H)	25	79	37.9	7.8	Herrera 1996 and unpubl.
	Columnea rubriacuta (S)	8	20	79.0	2.1	L. Navarro unpubl.
	Coris monspelliensis (H)	19	89	43.9	6.6	Herrera 1996 and unpubl.
	Dactylorhiza elata (H)	20	99	18.0	7.4	Herrera 1996 and unpubl.
	Daphne gnidium (S)	20	100	86.7	9.5	Herrera 1996 and unpubl.

(*continued*)

TABLE 3.2 *(continued)*

Floral trait	Species[a]	Sample size		%Var$_{within}$[b]	Mean CV$_{within}$ (%)[c]	Reference
		Number of plants	Number of flowers			
	Daphne laureola (S) (2)	18	270	40.9	12.5	Herrera 1996 and unpubl.; C. Alonso unpubl.
	Dipcadi serotinum (H)	20	77	60.6	8.7	Herrera 1996 and unpubl.
	Episcia cupreata (H)	9	20	39.4	4.5	L. Navarro unpubl.
	Fedia cornucopiae (H)	20	100	28.5	7.9	Herrera 1996 and unpubl.
	Ipomoea wolcottiana (T)	42	370	64.9	13.3	V. Parra-Tabla unpubl.
	Jasminum fruticans (S)	14	70	46.6	6.0	Herrera 1996 and unpubl.
	Lavandula latifolia (S) (20)	348	7,262	34.2	3.2	C. M. Herrera unpubl.
	Lavandula stoechas (S)	20	100	29.1	3.5	Herrera 1996 and unpubl.
	Lithodora fruticosa (S)	20	97	44.5	8.3	Herrera 1996 and unpubl.
	Lonicera arborea (S)	20	100	59.0	8.2	Herrera 1996 and unpubl.
	Lonicera etrusca (V)	20	100	34.6	6.6	Herrera 1996 and unpubl.
	Lonicera implexa (V)	27	138	23.8	4.6	Herrera 1996 and unpubl.
	Lonicera splendida (V)	17	101	28.2	3.8	Herrera 1996 and unpubl.
	Macleania rupestris (S)	22	107	21.8	3.7	P. Guitián unpubl.
	Marrubium supinum (H)	20	100	30.5	5.5	Herrera 1996 and unpubl.
	Mucizonia hispida (H)	20	96	39.4	5.1	Herrera 1996 and unpubl.
	Muscari comosum (H)	20	112	5.8	1.7	Herrera 1996 and unpubl.
	Origanum virens (H)	30	150	12.8	5.4	Herrera 1996 and unpubl.
	Pancratium maritimum (H) (9)	107	356	44.4	5.4	M. Medrano unpubl.
	Phlomis lychnitis (S)	20	100	36.3	2.8	Herrera 1996 and unpubl.
	Phlomis purpurea (S)	22	110	43.7	5.2	Herrera 1996 and unpubl.
	Pinguicula vallisneriifolia (H)	20	44	35.3	4.5	Herrera 1996 and unpubl.
	Pistorinia hispanica (H)	20	79	43.5	6.0	Herrera 1996 and unpubl.
	Polygala boissieri (H)	24	83	32.7	2.7	Herrera 1996 and unpubl.
	Polygonatum odoratum (H)	17	164	40.8	3.1	Herrera 1996 and unpubl.
	Primula vulgaris (H)	20	78	33.4	5.2	Herrera 1996 and unpubl.

	Species					Reference
	Prunella hyssopifolia (H)	18	81	63.7	5.0	Herrera 1996 and unpubl.
	Prunella laciniata (H)	19	113	23.8	2.7	Herrera 1996 and unpubl.
	Prunella vulgaris (H)	20	83	50.8	4.0	Herrera 1996 and unpubl.
	Rosmarinus officinalis (S)	20	100	43.7	6.1	Herrera 1996 and unpubl.
	Salvia argentea (S)	20	100	17.3	4.3	Herrera 1996 and unpubl.
	Salvia blancoana (S)	20	86	38.8	5.0	Herrera 1996 and unpubl.
	Salvia verbenaca (H)	12	58	33.0	5.3	Herrera 1996 and unpubl.
	Saponaria ocymoides (H)	20	100	67.6	5.0	Herrera 1996 and unpubl.
	Satureja intrincata (S)	20	100	43.5	6.5	Herrera 1996 and unpubl.
	Scabiosa turolensis (S)	20	100	35.6	6.0	Herrera 1996 and unpubl.
	Sideritis arborescens (S)	13	65	20.7	3.6	Herrera 1996 and unpubl.
	Sideritis incana (S)	20	100	37.0	5.3	Herrera 1996 and unpubl.
	Silene colorata (H)	20	89	27.5	2.7	Herrera 1996 and unpubl.
	Silene lasiostyla (H)	20	53	51.0	3.4	Herrera 1996 and unpubl.
	Silene legionensis (H)	20	83	15.8	3.8	Herrera 1996 and unpubl.
	Silene vulgaris (H)	16	45	19.7	4.2	Herrera 1996 and unpubl.
	Sphyrospermum sp. (S)	6	20	36.9	2.1	L. Navarro unpubl.
	Spigelia palmeri (H)	30	66	80.9	5.2	C. M. Herrera unpubl.
	Teucrium rotundifolium (S)	20	100	56.1	5.9	Herrera 1996 and unpubl.
	Teucrium webbianum (H)	20	99	55.5	4.5	Herrera 1996 and unpubl.
	Thymus orospedanus (S)	20	100	12.3	5.2	Herrera 1996 and unpubl.
	Trachelium caeruleum (H)	20	100	25.2	2.7	Herrera 1996 and unpubl.
Spur length	Impatiens balsamina (H)	30	90	73.2	8.9	C. M. Herrera unpubl.
	Linaria aeruginea (H)	28	124	27.6	3.1	Herrera 1996 and unpubl.
	Linaria anticaria (H)	20	78	62.7	5.9	Herrera 1996 and unpubl.
	Linaria viscosa (H)	20	93	48.2	5.8	Herrera 1996 and unpubl.
	Viola cazorlensis (H)	23	578	60.8	10.7	Herrera 1993 and unpubl.
Sepal length	Scleranthus annuus (H) (20)	400	4,000	26.5		Svensson 1992
Petal length	Adenocarpus telonensis (S)	20	60	12.8	1.9	J. Herrera unpubl.
	Calicotome villosa (S)	17	51	19.0	4.2	J. Herrera unpubl.
	Clarkia tembloriensis (H) (4)				9.0	Sherry and Lord 1996a
	Cytisus arboreus (S)	15	45	12.5	2.9	J. Herrera unpubl.

(continued)

TABLE 3.2 *(continued)*

		Sample size				
Floral trait	Species[a]	Number of plants	Number of flowers	%Var_{within}[b]	Mean CV_{within} (%)[c]	Reference
	Cytisus grandiflorus (S) (3)	62	177	19.4	3.1	J. Herrera unpubl.
	Cytisus scoparius (S)	20	60	17.1	2.0	J. Herrera unpubl.
	Genista hirsuta (S) (3)	61	174	34.4	4.9	J. Herrera unpubl.
	Genista triacanthos (S) (3)	62	177	24.0	3.9	J. Herrera unpubl.
	Lupinus angustifolius (H) (4)	69	195	30.8	2.5	J. Herrera unpubl.
	Lupinus hispanicus (H) (2)	38	108	64.8	2.2	J. Herrera unpubl.
	Petrocoptis glaucifolia (H)	10	30	100.0	10.8	L. Navarro unpubl.
	Petrocoptis grandiflora (H)	71	174	92.1	16.3	L. Navarro unpubl.
	Petrocoptis viscosa (H)	30	180	92.3	11.8	L. Navarro unpubl.
	Retama monosperma (S) (3)	63	179	32.8	4.0	J. Herrera unpubl.
	Silene acutifolia (H)	13	187	86.8	9.9	M. L. Buide unpubl.
	Stauracanthus genistoides (S) (3)	63	180	27.6	4.2	J. Herrera unpubl.
	Teline linifolia (S)	20	60	18.1	2.3	J. Herrera unpubl.
	Ulex argenteus (S)	20	60	13.1	2.7	J. Herrera unpubl.
	Ulex australis (S) (3)	58	187	25.7	4.3	J. Herrera unpubl.
	Ulex eriocladus (S)	12	24	11.3	3.6	J. Herrera unpubl.

[a]Growth form in parentheses: H, herb; S, shrub; T, tree; V, vine. For species with data from more than one locality, the number of populations sampled is shown in parentheses. In these instances, sample sizes refer to the sum over all sites, and percent variance and CV figures shown are average values.

[b]%Var_{within} is the proportion of total population-level variance in a given trait that is accounted for by differences between flowers of the same plant.

[c]Mean CV_{within} is the population mean of within-plant coefficients of variation.

Observed patterns of extensive within-plant variation in nectar standing crop are most likely related to concomitant variations in nectar secretion. Real and Rathcke (1988) studied patterns of within- and between-plant variation in nectar production for the shrub *Kalmia latifolia*, and this investigation is probably the most thorough one ever conducted on the subject. Although CV_{within} of per-flower nectar production varied widely among plants and days, in most cases it fell somewhere in the range 150–350%. The extent of within-plant variability in nectar production per flower, although still considerable, is not so extreme in other species studied. For *Epilobium canum*, Boose (1997) found that the CV_{within} of nectar production ranged from 6 to 117% in one study year, and from 15 to 101% in another. In *Echium vulgare*, the average CV_{within} of nectar production was only 37% (Klinkhamer and Van der Veen–Van Wijk 1999).

Much less is known about the extent of within-plant variation in nectar composition, but the limited evidence available suggests that such variation may often be important. In the plants of *Ipomopsis longiflora* studied by Freeman and Wilken (1987), within-plant coefficients of variation for glucose and fructose percent content in nectar fell in the 70–90% range at some of their study locations. Extensive within-plant variation in nectar sugar composition has recently been reported by Herrera, Pérez, and Alonso (2006) for *Helleborus foetidus* and Canto et al. (2007) for two species of *Aquilegia*.

Fruit Variation

There have been few attempts at measuring the extent of within-plant variation in fruit characteristics of wild plants, and most of these refer to fleshy-fruited plants whose seeds are dispersed by frugivorous vertebrates. It was shown earlier that in the evergreen shrub *Osyris lanceolata* the concentration of soluble solids in the juice of ripe fruits varies considerably among drupes produced by the same plant (fig. 2.7). Further emphasizing the quantitative importance of within-plant variation are the observations that $\%Var_{within}$ in that trait was 41.7%, and that the ranges of within-plant variation exhibited by some of the *O. lanceolata* plants shown in figure 2.7 are comparable or even broader than differences between species means reported by White and Stiles (1985) for a set of North American wild plants. In eight fleshy-fruited southern Spanish species studied by Jordano (1991) and Obeso and Herrera (1994), within-plant variance for fruit length and width, pulp and seed mass, and percent pulp mass, was of roughly the same order of magnitude as, and often much larger than, variance among plants. Extensive within-plant variability in total fruit mass, pulp mass, and seed mass has been also reported by Obeso (1998b) for the fruits of *Ilex aquifolium*. In *Juniperus phoenicea*, Jordano (1991) found that the within-plant variance for the number of seeds per berry nearly doubled the variance between individual plants. This situation is probably widespread among species with multiovulate ovaries, regardless of whether fruits are fleshy or not. Similarly high $\%Var_{within}$ for fruit seediness have also been reported by Acosta et al. (1993) and Jacquemyn et al. (2001) for the capsules of the shrub *Cistus ladanifer* and the perennial herb *Primula elatior*, respectively; by Markham (2002) for the cones of *Alnus rubra*; and by Cipollini and Stiles (1991) for the pods of two *Phaseolus* species. In all these cases, $\%Var_{within}$ was greater than 50%, meaning that the within-plant component of variance in fruit seediness was larger than the among-plant component.

Available estimates of within-plant variability in fruit width for 25 species of vertebrate-dispersed plants listed in table 3.3 indicate that extensive within-plant variation in this fruit trait seems to be the rule. In this sample of species, $\%Var_{within}$ ranges between 12 and 87%, and is larger than 50% in 20 out of the 25 species. Mean CV_{within} ranges between 4.1 and 13.5%.

Studies of cultivated fruits have likewise documented that within-plant variation in different fruit characteristics is generally extensive. For example, in a group of apple trees studied by Broom et al. (1998), tree means for calcium concentration in fruits varied only 1.3-fold (range = 30–39 µg/g). In contrast, variation between individual fruits of the same tree was nearly 4-fold (range = 16–60 µg/g), thus showing that between-tree variation was comparatively unimportant in relation to within-tree variation. This situation seems to be the rule among cultivated fruit plants, and de Silva and Ball (1997, 411) noted that "in a well managed orchard block the within-plant component generally forms the predominant source of variation for most fruit attributes."

Seed Variation

Seed size can affect seed dispersal and predation, as well as the likelihood of seedling emergence, establishment, growth, and survival (Willson 1983). It is probably because of these important ecological consequences that seed mass has become, by far, the trait for which patterns of varia-

TABLE 3.3. **Within-plant variation in fruit width (transversal diameter) in 25 species of vertebrate-dispersed plants producing berries, drupes, or other types of fleshy diaspores.**

Species[a]	Sample size		%Var$_{within}$[b]	Mean CV$_{within}$ (%)[c]	Reference
	Number of plants	Number of fruits			
Arum italicum (H)	10	130	82.0	9.7	C. M. Herrera unpubl.
Berberis hispanica (2) (S)	41	1,001	60.2	11.1	Obeso 1986; C. M. Herrera unpubl.
Corema album (13) (S)	195	5,850	56.6	5.4	A. Rodríguez Larrinaga unpubl.
Crataegus laciniata (T)	8	306	76.9	7.5	C. M. Herrera unpubl.
Crataegus monogyna (4) (T)	60	2,400	60.6	8.9	C. M. Herrera unpubl.
Daphne gnidium (S)	36	488	56.8	8.0	C. M. Herrera unpubl.
Daphne laureola (4) (S)	45	913	55.2	6.3	Obeso 1986; C. M. Herrera unpubl.
Gonzalagunia hirsuta (S)	10	100	55.3	12.8	L. Navarro unpubl.
Guaiacum officinale (T)	7	105	73.2	6.8	L. Navarro unpubl.
Guazuma ulmifolia (T)	40	998	64.2	10.4	C. M. Herrera unpubl.
Hedera helix (V)	5	100	20.0	8.6	Obeso 1986
Juniperus communis (S)	10	164	52.7	9.5	Obeso 1986
Juniperus phoenicea (T)	12	240	55.8	7.9	Obeso 1986
Juniperus sabina (S)	6	172	33.8	6.9	Obeso 1986
Miconia prasina (T)	11	110	82.4	8.5	L. Navarro unpubl.
Olea europaea (2) (T)	50	1,481	12.0	5.6	C. M. Herrera unpubl.
Osyris lanceolata (S)	37	3,201	43.9	7.2	C. M. Herrera unpubl.
Palicourea crocea (S)	8	101	76.5	13.5	L. Navarro unpubl.
Parathesis crenulata (S)	15	150	65.3	6.3	L. Navarro unpubl.
Phillyrea latifolia (2) (T)	51	1,227	35.2	6.7	Herrera et al. 1994 and unpubl.
Prestoea montana (T)	7	175	65.6	4.1	L. Navarro unpubl.
Rosa canina (S)	17	345	50.4	10.4	Obeso 1986
Sorbus aucuparia (T)	15	750	58.2	7.5	B. Pías and M. Salvande unpubl.
Syzygium jambos (T)	11	55	76.2	13.1	L. Navarro unpubl.
Viburnum lantana (S)	7	251	86.9	9.9	C. M. Herrera unpubl.

[a]Growth form in parentheses: H, herb; S, shrub; T, tree; V, vine. For species with data from more than one locality, the number of populations sampled is shown in parentheses. In these instances, sample sizes refer to the sum over all sites, and the percent variance and CV figures shown are average values.
[b]%Var$_{within}$ is the proportion of total population-level variance that is accounted for by differences between fruits of the same plant.
[c]Mean CV$_{within}$ is the population mean of the within-plant coefficients of variation.

tion have been most thoroughly investigated for a long time. Although the vast majority of studies have focused on interspecific or individual differences and their environmental correlates (Salisbury 1942; Harper et al. 1970; Leishman et al. 2000), there is still a sizeable number of quantitative data available in the seed ecology literature bearing on patterns of seed-mass variation within single plants (Michaels et al. 1988; Hendrix and Sun 1989). These studies have almost invariably documented extensive levels of within-plant variability in seed size. For example, variation in seed mass within individual plants can be as high as 16-fold and 8-fold in the herbs *Lomatium salmoniflorum* (Thompson and Pellmyr 1989) and *Lomatium grayi* (Thompson 1984), respectively, and 4-fold in *Mucuna andreana*, a tropical woody vine (Janzen 1977b). McGinley et al. (1990) found that seed mass varied 2- to 3-fold within trees of *Pinus contorta*, and in 18 *Ateleia herbert-smithii* individuals studied by Janzen (1978) within-tree variation in seed mass was 1.7–2.6-fold. A summary of quantitative data on the magnitude of within-plant variability in seed mass is shown in table 3.4 for 86 species of trees, shrubs, herbs, and vines. The magnitude of within-plant variation in seed mass varies extraordinarily among species, with $\%Var_{within}$ figures virtually matching the whole 0–100% range of possible values (observed range = 10–98%). Despite broad interspecific differences, however, extensive within-plant variation in seed mass was the rule in this sample of species. The within-plant component of population-level variance in seed mass exceeds the between-plant component in the majority of species (71%), and mean CV_{within} varies between 8.1 and 39.2%.

There is little quantitative information on the degree of within-plant variability in seed traits other than seed mass. These scanty data, however, are also unambiguously indicative of extensive within-plant variation in several structural and functional seed features. Variation within *Alnus rubra* trees is responsible, on average, for 61% of total population-wide variance in percent seed viability per catkin (Markham 2002). In the shrub *Erica australis*, within-plant variation in percent seed germination accounts for 42.1% of total population variance in this magnitude (Cruz et al. 2003). In a study of eight species of wind-dispersed Asteraceae, Andersen (1992, table 3) found that within-plant variability in seed-settling velocities (an inverse surrogate for dispersal ability) largely exceeded variation among plants of the same species. Greene and Johnson (1992) also reported large within-plant CVs for seed-settling velocities and dispersal distances for a taxonomically diverse sample of temperate wind-dispersed trees, which also suggests broad within-plant variability in diaspore dispersal ability. In the perennial herb *Phytolacca americana*, the CV of percent germination of different seed lots originating from a single plant may be as high as 86% (Armesto et al. 1983). I reanalyzed some of the data obtained in a field experiment consisting of sowing *Lavandula latifolia* seeds of known maternal parentage in the species' natural habitat (Herrera 2000). Within-plant variation accounted for nearly all (98.9%) observed variance in seed germination date, while

TABLE 3.4 **Within-plant variation in individual seed mass for 86 species of trees, shrubs, vines, and herbs.**

Species[a]	Sample size		%Var_{within}[b]	Mean CV_{within} (%)[c]	Reference
	Number of plants	Number of seeds			
Acer saccharum (T)	10	472	70.0		Michaels et al. 1988
Aesculus glabra (T)	13	252	44.0		Michaels et al. 1988
Aesculus pavia (T) (2)	18	282	55.0		Michaels et al. 1988
Alliaria petiolata (H)	127	9,979	77.0		Susko and Lovett-Doust 2000
Arum italicum (H) (3)	28	3,027	74.4		Méndez 1997
Asarum canadense (H)	10	1,559	85.0		Michaels et al. 1988
Asclepias syriaca (H)	20	500	39.0		Michaels et al. 1988
Asclepias verticillata (H)	10	701	75.0		Michaels et al. 1988
Asimina triloba (T)	27	1,599	53.0		Michaels et al. 1988
Asphodelus albus (H) (2)	62	3,677	55.7	15.8	Obeso 1993 and unpubl.
Ateleia herbert-smithii (T)	20	1,793		13.4	Janzen 1978
Banksia marginata (S)	10	1,800	69.0		Vaughton and Ramsey 1998
Banksia spinulosa (S)	19	1,722	67.7		Vaughton and Ramsey 1997
Baptisia lanceolata (S)	19	5,347	53.5		Mehlman 1993
Cassia fasciculata (H)	10	705	72.0		Michaels et al. 1988
Cassia grandis	10	621		9.7	Janzen 1977c
Celtis occidentalis (T)	11	529	38.0		Michaels et al. 1988
Cercis canadensis (T)	4	380	51.0		Michaels et al. 1988
Chelidonium majus (H)	163	324	82.2		Kang and Primack 1991
Cicuta maculata (H)	14	1,046	98.4		Hendrix and Sun 1989
Claytonia virginica (H) (3)	54	792	67.0		Michaels et al. 1988
Cornus racemosa (T)	17	853	64.0		Michaels et al. 1988
Crataegus monogyna (T) (2)	27	610	38.7	18.1	Obeso and Herrera 1994 and unpubl.; A. Rodríguez Larrinaga unpubl.
Cryptotaenia canadensis (H)	18	912	80.8		Hendrix and Sun 1989

(continued)

TABLE 3.4 (continued)

Species[a]	Sample size		%Var$_{within}$[b]	Mean CV$_{within}$ (%)[c]	Reference
	Number of plants	Number of seeds			
Daphne gnidium (S)	27	346	52.9	14.1	C. M. Herrera unpubl.
Daphne laureola (S) (2)	17	590	45.5	10.9	Obeso and Herrera 1994 and unpubl.; C. M. Herrera unpubl.
Daucus carota (H) (2)	20	1,735	67.0		Michaels et al. 1988
Dentaria laciniata (H) (2)	32	1,528	65.5		Michaels et al. 1988
Dicentra canadensis (H)	18	154	69.0		Michaels et al. 1988
Dicentra cucullaria (H)	17	430	61.0		Michaels et al. 1988
Epilobium dodonaei (H)	23	1,156	82.0		Stöcklin and Favre 1994
Epilobium fleischeri (H)	26	952	77.7		Stöcklin and Favre 1994
Fraxinus americana (T)	2	184	23.0		Michaels et al. 1988
Fraxinus pennsylvanica (T)	4	174	100.0		Michaels et al. 1988
Geranium maculatum (H)	15	140	70.0		Michaels et al. 1988
Gleditsia triacanthos (T)	5	634	59.0		Michaels et al. 1988
Gymnocladus dioica (T)	5	612	61.0		Michaels et al. 1988
Helleborus foetidus (H) (2)	60	4,425	62.7	18.0	Garrido 2003 and unpubl.
Hepatica nobilis (H)	13	726	57.0		Michaels et al. 1988
Ilex aquifolium (T) (2)	16	2,248	55.7	26.5	J. R. Obeso unpubl.
Impatiens biflora (H)	12	321	84.0		Michaels et al. 1988
Ipomopsis aggregata (H)	13	2,085	96.0	22.7	Wolf et al. 1986
Juglans nigra (T)	8	351	70.0		Michaels et al. 1988
Juniperus communis (T)	5	169	87.7	26.4	J. R. Obeso unpubl.
Lavandula latifolia (S)	15	1,500	93.6	26.9	C. M. Herrera unpubl.
Lindera benzoin (S)	6	64	10.0		Michaels et al. 1988
Lomatium grayi (H)	17	2,273	59.5	26.4	Thompson 1984
Maianthemum racemosum (H)	19	397	79.0		Michaels et al. 1988
Menispermum canadense (V)	16	797	39.0		Michaels et al. 1988
Narcissus longispathus (H)	103	2,828	28.1	10.1	C. M. Herrera unpubl.

Species				Reference
Olea europaea (T)	10	356	29.9	23.1	J. Alcántara unpubl.
Osmorhiza claytoni (H) (2)	37	3,129	68.2		Michaels et al. 1988; Hendrix and Sun 1989
Osmorhiza longistylis (H)	22	636	53.3		Hendrix and Sun 1989
Osyris lanceolata (S)	23	528	39.6	16.3	C. M. Herrera unpubl.
Oxypolis rigidior (H)	12	567	86.5		Hendrix and Sun 1989
Paeonia broteroi (H)	25	219	29.4	22.5	A. M. Sánchez-Lafuente unpubl.
Pancratium maritimum (H) (3)	29	604	43.4	15.0	M. Medrano unpubl.
Phillyrea latifolia (T) (2)	38	1,227	55.8	21.2	Herrera et al. 1994 and unpubl.
Phlox divaricata (H)	10	512	67.0		Michaels et al. 1988
Pinus nigra (T) (3)	24	2,576	94.9	16.2	P. A. Tíscar unpubl.
Pinus sylvestris (T) (2)	16	1,041	35.4	14.5	Castro 1999 and unpubl.
Polygonatum commutatum (H) (2)	17	2,999	71.5		Michaels et al. 1988
Primula elatior (H)	34	330	65.3		Jacquemyn et al. 2001
Prunella vulgaris (H)	10	2,283	87.0	39.2	Winn 1991
Prunus mahaleb (T)	6	85	98.0	13.4	J. Guitián unpubl.
Prunus serotina (T)	10	554	48.0		Michaels et al. 1988
Purshia tridentata (S)	241	8,040	68.4	17.9	Krannitz 1997a
Quercus coccifera (T)	22	334	34.4	18.8	C. M. Herrera unpubl.
Quercus rotundifolia (T)	28	420	18.8	17.0	R. Bonal unpubl.
Rhamnus alpinus (S)	10	72	59.4	17.0	Bañuelos and Obeso 2003; M. J. Bañuelos unpubl.
Sanguinaria canadensis (H)	11	934	80.0		Michaels et al. 1988
Sanicula gregaria (H)	19	684	80.5		Hendrix and Sun 1989
Smilax aspera (V)	13	678	67.0	18.7	C. M. Herrera unpubl.
Smilax hispida (V)	13	646	20.0		Michaels et al. 1988
Smilax lasioneura (V)	5	711	70.0		Michaels et al. 1988
Staphylea trifolia (T)	6	1,008	75.0		Michaels et al. 1988
Tamus communis (V) (3)	31	2,451	52.6	18.7	J. R. Obeso unpubl.
Taxus baccata (T) (5)	66	1,756	33.5	8.1	J. R. Obeso unpubl.
Tilia americana (T)	10	1,118	55.0		Michaels et al. 1988

(continued)

TABLE 3.4 (*continued*)

Species[a]	Sample size		%Var_within[b]	Mean CV_within (%)[c]	Reference
	Number of plants	Number of seeds			
Ulex europaeus (S) (4)	12	817	37.1	12.0	J. R. Obeso unpubl.
Ulex gallii (S) (4)	10	252	46.3	11.8	J. R. Obeso unpubl.
Uvularia grandiflora (H)	22	819	49.0		Michaels et al. 1988
Viola pubescens (H)	10	187	76.0		Michaels et al. 1988
Virola surinamensis (T)	46	460	21.6		Howe and Richter 1982
Vitis vulpina (V)	3	222	43.0		Michaels et al. 1988
Zizia aurea (H)	12	1,273	94.9		Hendrix and Sun 1989

[a]Growth forms in parentheses: H, herb; S, shrub; T, tree; V, vine. Data from more than one locality were available for some species. In these cases, the number of populations sampled is shown in parentheses, sample sizes denote the sum over all sites, and percent variance and CV figures represent average values.
[b]%Var_within is the proportion of total population-level variance that is accounted for by differences between seeds produced by the same plant.
[c]Mean CV_within is the population mean of within-plant coefficients of variation.

differences among mother plants accounted for a negligible fraction (1.1%).

Information from cultivated plants reveals that within-plant variation in chemical composition of seeds can also be extensive. In soybean (*Glycine max*) seeds, for example, within-plant variation was responsible for 88% of total variance in the oil content of individual seeds (Brim et al. 1967), and in three different cultivars of rapeseed (*Brassica napus*) the CV_{within} figures for seed protein content of individual seeds were 8.9, 9.5, and 11.7% (Velasco and Möllers 2002).

Summary and Comparison

As shown in the preceding sections, all types of reiterated structures may exhibit considerable levels of continuous within-plant variation, and, for any given structure, variability varies greatly among species. Is it possible to draw a broad conclusion about the comparative levels of variability exhibited by different types of reiterated structures? Comparisons between different structures should ideally be performed on data obtained from the same species and individuals by means of some paired comparison method, in order to avoid possible artifacts due to some combination of phylogenetic correlations and differences in taxonomic composition of species samples. An explicit test of this sort has never been attempted for any group of species, and cannot be performed on the data listed in tables 3.1 to 3.4 either, since variability data for different structures come from largely nonoverlapping species sets. Taken with the necessary caution, however, comparisons of the central tendency and range of CV_{within} and $\%Var_{within}$ for the four types of structures considered are still useful in revealing some broad patterns.

Results of the comparison between structures differ depending on the metric used to measure variability (fig. 3.2). When CV_{within} is used, there is a steadily declining trend in *both* the mean and the range of within-plant variability, running in the direction leaves-seeds-fruits-flowers. Variability of leaves is greatest and has the broadest range, while at the opposite extreme flowers exhibit the smallest variability and narrowest range. Fruits and seeds are intermediate. This clear gradient vanishes when $\%Var_{within}$ is used to measure variability, and within-plant variance is thus scaled to total population-wide variance. On this metric, flowers are still, on average, less variable than leaves, seeds, and fruits, but the range of

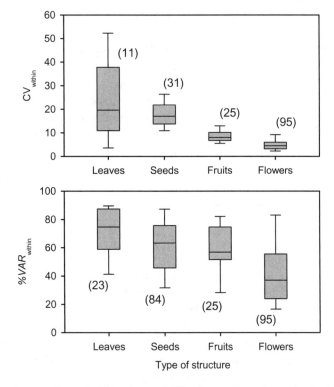

FIG. 3.2 A comparison of within-plant variability for continuously varying leaf, flower, fruit, and seed features using two variability metrics. $\%Var_{within}$ is the proportion of total population-level variance accounted for by differences between structures produced by the same plant. CV_{within} refers to within-plant coefficients of variation. Box plots show the 10%, 25%, 50%, 75%, and 90% percentiles of the distributions. Figures in parentheses beside boxes denote the number of distinct data sets included in each case. Based on the data in tables 3.1 through 3.4.

$\%Var_{within}$ values is roughly similar for the four types of structures. Leaves, seeds, and fruits have extensively overlapping variability distributions and roughly similar mean variabilities.

The contrast between the conclusions reached by the two methods exemplifies the difficulties associated with measuring variability, noted at the beginning of this chapter. It is important to emphasize, however, that figure 3.2 shows average trends in the comparative magnitude of within-plant variability of the different reiterated structures, and that patterns exhibited by individual species may depart significantly from that overall pattern. This is exemplified by two congeneric shrub species,

Daphne gnidium and *D. laureola*, the only two species listed in tables 3.1 to 3.4 with data simultaneously available for within-plant variability of leaves, seeds, fruits, and flowers. In *D. gnidium*, CV_{within} values for leaf area, seed mass, fruit width, and corolla length are 19.6, 14.1, 8.0 and 9.5, respectively. In *D. laureola*, the corresponding figures are 37.8, 10.9, 6.3, and 12.5. In both species, therefore, fruits rather than flowers are the least subindividually variable structures.

CHAPTER FOUR

Distribution of Subindividual Variability in Time and Space

How are variants of reiterated structures organized along temporal, spatial, and architectural axes?

Within-plant variation is considered in the preceding chapters from a very abstract perspective, as if the morphological, structural, or functional differences between reiterated structures of the same plant lacked organization and occurred in some sort of vacuum or undifferentiated matrix. The total variance of a character within a plant is a value that does not depend on whether it is predictably distributed over the plant in relation to some spatial gradient or with some discernible directionality. The crude figures presented in chapter 3 are useful in performing comparisons and in convincing skeptics that continuous within-plant variation is sufficiently widespread and extensive to deserve our attention. Nevertheless, crude measurements of within-plant variability are just the bare bones of a very complex phenomenon involving multiple mechanisms (see chapters 5 and 6). For example, subindividual variability may have a temporal component, as when flowers or fruits are continuously produced by single plants over extended intervals of the annual cycle or, in the case of long-lived perennial plants, in different years. Also, within-plant variability may be closely tied to the particularities of a plant's architecture and to environmental gradients extrinsic to the plants. As shown below, within-plant variation in fruit features or morphological and functional leaf traits of trees is often associated with vertical and horizontal light gradients in the crown, and flower features often vary with position on the inflorescence. But variation can also be quite fine-grained, occurring over the restricted spatial scale of the same twig or the interior of one fruit, without bearing any obvious relationship to external environmental gradients. Different patterns of distribution in time and space may be attributed to different proximate mechanisms, and will also presumably have different ecological consequences or even evolutionary implications. Therefore, in order to understand the variety of mechanisms that generate within-plant variation, and to gain insight into its ecological and evolutionary significance, it is essential to consider how such variation is organized in time and space. This chapter considers the ways in which subindividual variability is distributed in time and among different parts and spatial locations within a plant. The next section deals with the issues of the temporal component of within-plant variation and the relative importance of simultaneous and sequential components. After that, patterns of spatial organization are considered.

Simultaneous and Sequential Components of Variation

The different variants of the same organ produced by an individual plant can either be formed simultaneously on the plant or be separated temporally, hence giving rise to simultaneous and sequential components of within-plant variation. This distinction applies regardless of whether differences between organs are of the discontinuous type (e.g., heterophylly, heterocarpy) or the continuous type (e.g., continuous variation in size or shape).

A sequential component arises whenever individual plants produce variants of the same organ over the same season or, in the case or perennials, in different years. The different leaf morphs that characterize heterophyllous plants are often produced at different times of the growing season (Critchfield 1960; Mulkey et al. 1992; Winn 1999b). This phenomenon is clearly seen, for example, in some heterophyllous carnivorous plants, where the leaves with secretory glands that act as trapping and digestive organs are produced only during relatively short periods of the seasonal cycle (Green et al. 1979; Zamora 1995). In the case of flowers, it has long been known that their size and characteristics often vary regularly in the course of the same reproductive episode. This occurs, for example, in species of the Ranunculaceae, a family where the number of parts of floral verticils often vary widely within species. In *Ficaria ranunculoides*, for example, the number of pistils, the number of stamens, and their correlation change from early to late flowers during the same flowering season (Weldon 1901), and in *Aquilegia canadensis* the number of ovules and anthers per flower, and the number of pollen grains per anther, decline across the flowering sequence of individual plants (Kliber and Eckert 2004). In species of *Nicotiana*, Goodspeed and Clausen (1915) showed that, on the same plants, mean corolla length and width declined steadily from the beginning to the end of the flowering period. In *Narcissus dubius*, a steady decline in flower size occurs from early- to late-opening flowers in the same inflorescences (Worley et al. 2000). Similar seasonal declines in the size of flowers or floral parts on individual plants or inflorescences have also been reported for numerous species from a broad variety of plant families, including *Anemonopsis macrophylla* (Pellmyr 1987), *Mimulus guttatus* (Macnair and Cumbes 1990), *Raphanus sativus* (Young and Stanton 1990), *Chelidonium majus* (Kang and Primack 1991), *Hydrophyllum appendiculatum* (Wolfe 1992), *Sidalcea oregana* (Ashman 1992), *Chamaecrista fasciculata* (Frazee and Marquis 1994), *Solanum hirtum* (Diggle 1995), and *Raphanus raphanistrum* (Williams and Conner 2001). This prevailing pattern of seasonal decline is sometimes reversed, as in the tropical orchid *Myrosmodes cochleare*, where flower length increases through the flowering season (Berry and Calvo 1991). In a given species, different floral traits may follow different within-plant seasonal courses, thus giving rise to complex patterns of within-plant variation in floral morphology. In *Oenothera macrocarpa*, corolla diameter decreased and floral-tube length increased from early- to late-produced flowers within the same plant (Mothershead and Marquis 2000).

The sizes of fruits and seeds, and aspects of their chemical composition, also vary frequently in the course of the same reproductive episode. In *Celtis durandii*, a tropical tree that produces fruit year-round, Worman and Chapman (2005) found that the lipid content of ripe fruit pulp varied between 0.3 and 30.8% depending on time of year. This is the most extreme example of intraspecific variation in fruit chemical composition reported to date, although it is not possible to ascertain if such variation mainly reflects within-tree seasonal variation in fruit chemistry or if it arises because individual trees differing in fruit composition ripen fruit at different times of year. In cultivated safflowers (*Carthamus tinctorius*), the oil content of seeds from the earlier flowering heads is lower than that of the late ones (Williams 1962). Cavers and Steele (1984) found a consistent seasonal component in seed size in eight species of herbaceous annuals. Seeds from individual plants exhibited the same general pattern over a period of time: the largest seeds were produced first, and then progressively smaller seeds ripened as the plants aged. After reviewing the literature available at the time, these authors concluded that the phenomenon of declining mean seed weight over the fruiting season is common, if not universal. Although exceptions sometimes occur and the pattern is reversed in some species (e.g., *Lobelia inflata*; Simons and Johnston 2000), Cavers and Steele's conclusion has generally been upheld by subsequent studies (Levy 1988; Winn 1991; Wolfe 1992, 1995; Obeso 1993). Although the phenomenon has received less attention, the size of fruits produced by individual plants or infructescences also tends to decline gradually throughout a single fruiting season (Macnair and Cumbes 1990; Kang and Primack 1991; Herrera 1988; Wolfe and Denton 2001).

The simultaneous component is generally due to plants producing multiple variants of the same organ simultaneously or over a relatively short period of time, in such a way that organ variants are borne at the same time by individual plants. This frequently happens, for example, in the case of plants that produce at about the same time cleistogamous and chasmogamous flowers (Culley and Klooster 2007), mixtures of different fruit types (Imbert 2002), or leaves of different sizes or shapes (Sculthorpe 1967). Sometimes variation that appears simultaneous is actually due to the protracted persistence on the plant of different cohorts of the same organ produced at different times of the seasonal cycle or, in the case of perennials, in different years. For example, nectar amino acid composition may vary with the age of flowers (Gottsberger et al. 1990); hence coexistence of flowers of slightly different age on the plant will give rise to snapshot heterogeneity within a plant in the amino acid composition of nectar. In *Narcissus dubius* inflorescences, where early-opening flowers are the largest and late-opening flowers are the smallest, within-plant heterogeneity in flower size is mainly due to the simultaneous display of flowers of different ages (Worley et al. 2000).

Similar effects also occur frequently in leaves (both on deciduous and evergreen plants), because the longevity of individual leaves often varies considerably with position in the crown (Miyaji et al. 1997; Harlow et al. 2005) and so coexisting leaves are not all the same age. Chemical and physical properties of leaves vary with age, with new leaves tending to be higher in nitrogen, water, and secondary metabolites than mature leaves (e.g., Feeny 1970; Young and Yavitt 1987; Bowers and Stamp 1992; van Dam et al. 1994; de Boer 1999; Ikonen 2002; among many others). Therefore, age heterogeneity of the leaves on a plant at a given time will normally enhance the simultaneous component of within-plant variance in these, and possibly other, age-dependent leaf features. Some deciduous species produce new leaves over relatively prolonged periods of time, which leads to "early" and "late" leaves simultaneously possessing different morphological and structural characteristics. This occurs, for example, in *Liquidambar* (Smith 1967), *Populus* (Critchfield 1960), *Betula* (Clausen and Kozlowski 1965), and *Salix* (Ikonen 2002). The contribution of age heterogeneity of leaves to the simultaneous component of within-plant variation is considerable among evergreen trees and shrubs, where individual leaves may last for up to 12–16 years (Chabot and Hicks 1982; Rogers and Clifford 1993; Niinemets and Lukjanova 2003; Reich et al. 2004; Harlow et al. 2005). Leaves of some evergreen Mediterranean plants produce large amounts of monoterpenes, and in *Quercus ilex* trees, leaves of the same plant differ greatly in total terpene emission. These differences are largely explained by heterogeneity of leaf ages, since emissions decline steadily from 1-year-old through 3-year-old leaves (Staudt et al. 2001). One would generally expect that, everything else being equal, the greater the average longevity of individual organs of the same kind on a plant, the greater the age snapshot variance of organs on the plant and hence the simultaneous component of within-plant variation due solely to this effect.

Rogers and Clifford (1993) analyzed the taxonomic and evolutionary correlates of leaf longevity in a large sample of 202 taxa of vascular plants. They found a strong correlation between leaf longevity and taxonomical affiliation, with 60% of total variance in leaf longevity represented in the sample being due to differences at the superorder or higher taxonomic levels. They also found an inverse relationship between leaf longevity and Sporne's advancement index, a rough index of evolutionary advancement (Chapman 1987), with more primitive vascular plants tending to have longer-lived leaves than the more advanced. Although Rogers and Clifford's conclusions should be corroborated by more refined techniques of evolutionary analysis, their findings lead one to predict that, at least among evergreen plants, the extent of within-plant variation in leaf features due to age heterogeneity should exhibit strong taxonomic correlates.

Ascertaining the relative quantitative importance of simultaneous and sequential components of within-plant variance is important because it is the simultaneous occurrence on the plant of different forms of the same organ that opens the way for the animals that interact with these organs (e.g., pollinators, seed predators, leaf eaters) to detect such variations and make choices among them. This topic is considered at length in chapter 8, and it is central to the argument in chapter 10 that subindividual variability can have some evolutionary implications for plants. In addition, assessing the relative importance of the sequential and simultaneous components of within-plant variation has some important methodological implications (Cavers and Steel 1984). If seasonal or annual variation in the characteristics of the organs produced by a plant is comparatively large in relation to simultaneous variation, then estimates of within-plant variation will be artificially small when organs are sampled only once, as is usually done. In the next section I examine some cases that allow for objectively dissecting the within-plant variance in flower, seed, leaf, and fruit features into its simultaneous and sequential (seasonal and annual) components. As there is little published information amenable to an analysis of that kind, I mainly rely on ad hoc reanalyses of data from my studies and those of my associates.

Relative Magnitude of Simultaneous and Sequential Components

In the tree *Nyctanthes arbor-tristis*, flowers of the same plant produced at different times differ in mean petal number, which first increases, then decreases, and finally increases again during the course of a 3.5-month flowering season (Roy 1959). As the raw petal-number data were tabulated separately by tree and observation date in table 9 of Roy's study (1963), it was possible to partition the total variance in petal number exhibited by each tree during its entire flowering season into sequential (i.e., due to seasonal variation) and simultaneous (i.e., due to variation within date) components. Despite the seasonal trend in mean number of petals, the sequential component accounted for less than 1% of total within-plant variance, which was almost entirely due to the simultaneous component. In other words, virtually all the season-long variance in petal number was instantaneously present on plants.

The fruiting phenology of the evergreen Mediterranean shrub *Osyris lanceolata* (= *O. quadripartita* of earlier publications) is quite unusual. Ripe fruits are continuously produced on individual plants throughout the year. The stock of developing ovaries resulting from the single annual flowering episode ripens slowly and gradually, taking over a year after flowering (Herrera 1984a, 1985). Fruit-diameter measurements were collected regularly on a large sample of shrubs during two reproductive seasons (Herrera 1988), and there was considerable seasonal variation within-plants in the diameter of ripe fruits (fig. 4.1). Despite this variation, and despite the unusually long fruiting season of the species, in *O. lanceolata* the simultaneous (i.e., within-date) component still accounts for 70% of total within-plant variance in fruit diameter. This is apparent in figure 4.1, where the diameter of individual fruits is plotted against ripening date for three representative shrubs. Although there is a discernible seasonal trend in fruit size in all shrubs, most diversity of fruit sizes is simultaneously represented on individual plants at any given time.

The relative importance of annual variation as a component of sequential within-plant variation can be most easily examined for leaf characteristics on deciduous plants. Leaves are renewed once per year in these species, and after completion of leaf flushing, all mature leaves simultaneously borne on a plant have roughly the same age. Data on leaf characteristics of individual plants from several years can thus be used to partition within-plant variance into its simultaneous (i.e., within-year) and sequential (i.e., between years) components. Alonso (1997b and unpublished data) studied during two consecutive years five southeastern Spanish populations of the deciduous tree *Prunus mahaleb*. Her data allow for a decomposition of total within-plant variation in leaf fresh mass, area, and percent water content into within-year and between-year components. In this species, the simultaneous component of leaf-trait variance

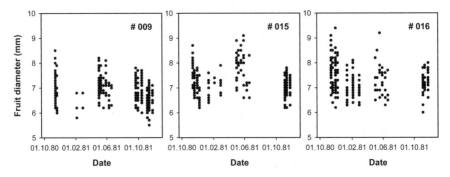

FIG. 4.1 Simultaneous and sequential components of within-plant variation in a fruit's size in the southern Spanish shrub *Osyris lanceolata*. Graphs show the variation in the diameter of ripe fruits produced by three representative plants during two reproductive episodes (the spring flowering periods of 1980 and 1981). Each symbol corresponds to a single fruit. Despite distinct seasonal trends in fruit size on each plant, most fruit size variants produced by each shrub are simultaneously present on the plant at any given time. Based on data from Herrera 1988.

(85.1%, 69.6%, and 68.9%, for leaf fresh mass, area, and percent water content, respectively) was considerably greater than the sequential component (14.9%, 30.4%, 31.1%, respectively).

Annual seed production by perennial plants often fluctuates considerably from year to year (Herrera et al. 1998; Kelly and Sork 2002). A trade-off between seed size and seed number frequently exists across a variety of nested scales, ranging from species and populations down to individual plants, and fruits within individuals (Mazer et al. 1986; Simons and Johnston 2000; Abrahamson and Layne 2002; Parciak 2002; Karrenberg and Suter 2003). One would therefore expect annual fluctuations in seed production by species and individuals to run parallel to variation in seed size, which would give rise to a long-term sequential component of within-plant variation in seed size, although this aspect does not seem to have received much attention in the otherwise abundant literature on the ecology of seeds. In the evergreen Mediterranean shrub *Lavandula latifolia*, the mean mass of single seeds produced by individual plants does vary significantly between years, but this component of within-plant variation is quantitatively negligible. In a group of 15 plants studied during two consecutive years, the sequential (i.e., between-year) component accounted for only 10%, while the simultaneous component predominated and was responsible for the remaining 90% of total within-plant variation in seed mass (C. M. Herrera, unpublished data). In six individuals of the temperate tree *Sorbus aucuparia* studied during three consecutive years in northwestern Spain by B. Pías and M. Salvande (unpublished data), the sequential and simultaneous components of within-plant variation in seed mass were 31% and 69%, respectively. *Lavandula latifolia* shrubs are characterized by extremely low annual variability in seed production (Kelly and Sork 2002), while *S. aucuparia* is well-known for its intense fluctuations in fruit crop size (Sperens 1997; Raspé et al. 2000). For *Quercus petraea* and *Quercus robur*, two oak species characterized by strong annual fluctuations in crop sizes, Brookes and Wigston (1979) found marked variation in the size and shape of acorns produced by individual trees on different years. For some of the trees studied by these authors, the ranges of acorn weight and shape index showed little or no overlap on consecutive years. These data although limited are consistent with the expectation advanced above that, across species, the relative importance of the sequential component of within-plant variation in seed size would increase with increasing annual fluctuations in crop size. From this it would also follow, as a practical corollary, that tree and shrub species exhibiting marked inter-annual fluctuations in seed production, as do many of those listed in table 3.4, will be particularly prone to underestimation of actual levels of within-plant variability in seed size when single-sample estimates are used.

The last example in this section illustrates both the seasonal and year-to-year components of within-plant variation. In the southern Spanish endemic hawk moth–pollinated perennial violet *Viola cazorlensis*, each plant produces a relatively small number of flowers during its May–June flowering period. In the population studied by Herrera (1993, and unpublished data), corolla size (mean petal length) tended to increase on the same plant from early- to late-opening flowers, and there was also some annual variation. Flowers open at a very slow pace, but as each of them lasts for up to two weeks, there are always several flowers simultaneously open on each plant. I collected detailed flower measurements for individual plants of *V. cazorlensis* during two different flowering seasons, which allows for a partition of total within-plant variation in corolla size occurring over a two-year period into its simultaneous (within date) and sequential (both seasonal and annual) components. On average, the simultaneous component accounted for 64% of the total, the sequential component due to seasonal variation for 35%, and the sequential component due to annual variation for less than 1%. These figures indicate that, despite seasonal and annual variation in the average characteristics of the flowers produced by individual plants, a predominant fraction of total within-plant variation is simultaneously present on plants at any given moment.

The main conclusion emerging from the preceding examples is that, regardless of the organ considered (leaves, flowers, fruits, seeds), the simultaneous component tends to be the predominant source of within-plant variation. This means that most of the variants of any reiterated structure produced by an individual plant in the same reproductive season, or over several years, are simultaneously or nearly simultaneously borne on the plant. Sampling plants only once, therefore, although obviously missing some variability and leading to variance underestimation, will generally not lead to any fatally flawed estimate of within-plant variability. Furthermore, a substantial portion of the phenotypic range of organs produced by an individual plant are simultaneously available to interacting animals, namely leaf eaters, seed predators, frugivores, and flower visitors. This is an important aspect, as it means that foraging animals are exposed to organs possessing different features, which in turn make possible discrimination and choice at the within-plant scale, as discussed in chapter 8.

Spatial Structuring of Simultaneous Variation

It has been shown in the preceding section that the sequential component of within-plant variation can sometimes be substantial but that the simultaneous component generally predominates. How is this instantaneous or nearly instantaneous variability spatially organized on the plant? What are the patterns of intraplant variability in one, two, and three dimensions?

The study of the spatial organization of within-plant variability can be approached from two perspectives. First, one may examine whether spatial variation in the trait of interest bears some predictable relationship to some reference system of coordinates, which may be either inherent or external to the plant itself. This approach leads to the identification of spatial gradients of intraplant variation. The second approach is mainly concerned with elucidating the magnitude and extent of distance-dependent similarities in organ features, regardless in principle of their position in relation to coordinate systems. Under this latter approach, what is operationally the focus is the issue of spatial autocorrelations, or "spatial continuity" (Rossi et al. 1992), in organ features at the within-plant scale. This may be explicitly addressed by means of spatial models of three-dimensional variation or, in a much simpler but biologically straightforward way, by taking advantage of the nested structural organization of most plants. I consider the two main approaches to studying the spatial organization of within-plant variation in the next two sections.

Gradients

The phenotypic characteristics of reiterated structures produced by a plant often vary more or less predictably in relation to some spatial reference system, giving rise to discernible intraplant phenotypic gradients. Two major types of intraplant gradient may be distinguished, depending on whether the reference system that underlies it is extrinsic or intrinsic to the plant.

Extrinsic gradients of within-plant variation are related to variation in some environmental parameter that is largely independent of the plant's architecture, such as height above the ground, compass direction, or light intensity. A substantial fraction of within-plant variance in structural and functional leaf features is often explained in terms of one or more of these major gradients. Within trees, the longevity (Miyaji et al. 1997; Harlow et al. 2005), size (Hollinger 1989; Tucker et al. 1993; Cowart and Graham 1999), mass per unit area (Ford and Newbould 1971; Ellsworth and Reich 1993; Kull and Niinemets 1993; Tucker et al. 1993; Casella and Ceulemans 2002; Leal and Thomas 2003), and photosynthetic capacity of leaves (Ellsworth and Reich 1993; Porté and Loustau 1998), as well as the concentration of nitrogen and phosphorus (Lemaire et al. 1991; Le Roux et al. 1999; Casella and Ceulemans 2002; Leal and Thomas 2003) and heavy metals in leaves and stems (Sanders and Ericsson 1998; Luyssaert et al. 2001), all tend to vary regularly with height above the ground. Some of these height-related gradients can be quite steep, as shown by Harlow et al. (2005) for leaf longevity within the crown of *Thuja plicata* trees, where maximum duration of individual leaves decreased at an average rate of 0.3 year per each 1-m increase in height above the ground, declining from a mean of 10.6 years in the lower third of the crown to 6.8 years in the upper third.

Diverse gradients in leaf photosynthetic capacity, nitrogen concentration, and specific leaf area occurring in the interior of tree crowns are generally related to variation in light intensity (Klein et al. 1991; Leuning et al. 1991; Kull and Niinemets 1993; Traw and Ackerly 1995; Casella and Ceulemans 2002). In isolated trees, leaf dry weight per area and nitrogen concentration may also vary regularly along horizontal transects drawn across the tree crown (Le Roux et al. 1999). Compass direction also accounts for patterns of within-plant variation in leaf water (fig. 4.2) and nitrogen content, particularly in isolated shrubs or trees where differences in canopy orientation are the main source of within-plant variation in incident radiation and transpiration (Infante et al. 2001). In olive trees (*Olea europaea*) the largest single source of within-plant variation in leaf nitrogen content was found to be canopy orientation, with leaf nitrogen concentration declining in both the north-south and the east-west directions. Leaves from the south side of the tree had 12% more nitrogen than north-side leaves (Perica 2001). Abrupt environmental transitions, such as the water-air discontinuity faced by amphibious plants, can be considered as merely extreme cases consisting of particularly steep extrinsic gradients. The discontinuous variation in leaf form commonly exhibited by heterophyllous aquatic plants depends closely on the leaf position along the extrinsic water-air gradient, with submerged leaves finely divided and the floating or aerial ones relatively simple (Sculthorpe 1967).

Information on extrinsic gradients in features of reiterated structures other than leaves is very scanty for wild plants. Such gradients, however,

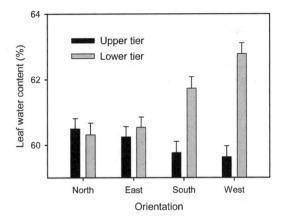

FIG. 4.2 The water content of individual leaves of *Prunus mahaleb*, a deciduous tree, varies within plants according to two extrinsically defined gradients, orientation (compass direction) and height above the ground (upper or lower tier). The two gradients, however, are not independent of each other. The height-related gradient occurs only in the southern and western parts of the crown, while the orientation-related gradient holds only in the lower tier of the tree. *Bars* represent average values for eight trees from one population in the Sierra de Cazorla, southeastern Spain (vertical segments are standard errors). Based on data from Alonso 1997a, and unpublished.

must be frequent in fruits, flowers, and seeds if one judges from the frequent reports found in the literature on cultivated plants. For *Hamelia patens*, a small tropical tree, Levey (1987) reported that berries on exposed portions of the tree had pulps with significantly higher sugar content than those shaded within the same tree crown. In cultivated apple trees (*Malus domestica*), Broom et al. (1998) found that fruit calcium concentration decreased an average of 2.7 µg/g with every 1-m increase in height above the ground, and fruit weight increased with height in the canopy and decreased toward the outside of the trees. Increases in fruit size, weight, and soluble-solid concentration, and decreases in fruit starch, nitrogen, phosphorus, potassium, and mineral-elements concentration with height above the ground, have been also reported by other investigations of a variety of apple cultivars (e.g., Jackson et al. 1971; Barritt et al. 1987). Within a single Valencia orange tree studied by Sites and Reitz (1950), vitamin C concentration in the juice of individual fruits was found to vary predictably with compass direction, and increased regularly with height at an average rate of 2.4 mg/100 mL of juice per meter. In lychee (*Litchi chinensis*), average soluble-solid concentration and acid content of fruits vary predictably within tree crowns, depending on both orientation and height above the ground (Tyas et al. 1998). For peach (*Prunus persica*), Génard and Bruchou (1992) reported that fruits in the upper parts of the tree were more purple, less orange red, and less firm, and had a higher sucrose content, lower citric acid content, and higher pH than fruits in the lower parts. Although it has been investigated less frequently, compass direction can also set an extrinsic gradient for within-plant variation in the characteristics of reiterated structures other than fruits. In *Annona cherimola* trees, for example, mean petal length of flowers located at north and south exposures of the same plant is significantly different (Perfectti and Camacho 1999).

Intrinsic gradients of within-plant variation are associated with some coordinate system that is closely linked to the plant's own architecture, and are thus inherent to the plant itself. Regular longitudinal variation of organ features along nodal positions on basically linear supporting structures such as main stems, lateral branches, and inflorescence or infructescence axes is probably the commonest, or at least the most frequently acknowledged, form of intrinsic intraplant gradients. This type of variation has been extensively documented since the classic studies on the morphogenesis of leaves in species with heteroblastic development (i.e., successive leaves produced on the same stem are not all alike), which provide some of the best-described examples of intrinsic gradients of within-plant variation (Ashby 1948; Critchfield 1960, 1970; Kozlowski and Clausen 1966; Steingraeber 1982). In species of *Delphinium* and *Gossypium*, for example, the degree of segmentation of leaves depends predictably on nodal position along the main stem (Ashby 1948). In *Ipomoea caerulea* there is a marked change in leaf shape from node to node, from entire leaves to three-lobed leaves in which the lobing is more and more pronounced as one moves up the main axis (Njoku 1956). In *Nicotiana rustica*, leaves are larger and more ovate at the base and become successively smaller and more lanceolate toward the inflorescence (Paxman 1956). Not only the size and shape of leaves, but also their chemical composition, can vary predictably along stems. In olive trees (*Olea europaea*), leaf nitrogen content declines from basal through central to distal leaves on the same stem, and this pattern remains invariant regardless of height or orientation in the canopy (Perica 2001). In *Mentha piperita*, monoterpene composition of essential oil from leaves exhibits a distinct pattern of longitudinal variation along stems. The concentration of menthol, menthyl acetate, and neomenthol decline, and that of menthone and isomenthone increases, from the base to the tip (Maffei et al. 1989; Rohloff 1999).

Within *Turnera ulmifolia* plants, there is a regular increase in the extent of cyanogenesis from basal to distal leaves along stems (Shore and Obrist 1992).

Although intrinsic intraplant gradients have been most thoroughly studied for leaf features, they are by no means exclusive to these organs. For example, in species of the Ranunculaceae, intraplant variation in number of floral parts (sepals, carpels, stamens) often depends on the position of the flower in the inflorescence, and in many species of Compositae the proportion of disk and ray florets in flower heads depends on the position of the flower head on the plant, as reviewed long ago by Baten (1936). In *Collomia grandiflora*, the production of cleistogamous and chasmogamous flowers is strongly dependent on nodal position along the inflorescence. Plants of this species grown in an uniform environment produce only cleistogamous flowers in the basal nodes of the inflorescence, both cleistogamous and chasmogamous ones in the central positions, and only chasmogamous ones in the distal positions (Ellstrand et al. 1984). Predictable linear variation along nodal positions of flowering shoots or inflorescences also frequently involves continuous or quasi-continuous floral traits (Ellstrand et al. 1984; Berry and Calvo 1991; Diggle 1995; Vallius 2000). Acropetal decline in flower size has even been considered a diagnostic feature of raceme inflorescences (Weberling 1989). In *Collomia grandiflora*, anther length and corolla length increase from basal to distal nodal nodes (Ellstrand et al. 1984), while in the orchid *Myrosmodes cochleare* perianth length declines from an average of 9.2 mm at bottom positions of the inflorescence to 6.2 mm at the top positions (Berry and Calvo 1991). In *Polygonatum odoratum*, a hermaphroditic perennial herb, the number of ovules and pollen grains per flower declined regularly from basal to apical nodal positions along the flowering shoot (Guitián et al. 2004). As the rate of decline was much steeper for ovules than for pollen grains, the functional gender of flowers shifted toward greater relative maleness with increasing distance from the base of the shoot. Numerous additional examples of intrinsic gradients of within-plant variation in morphological and functional floral traits can be found in Diggle (2003).

Seed and fruit features also frequently exhibit intrinsic gradients of continuous variation within-plants. The size and shape of reproductive structures of gymnosperms often vary regularly with relative position along the length of branches, as illustrated for example by Niklas (1994, fig. 4.6) for a series of ovulate cones from a single branch of *Larix*. In *Impatiens capensis*, the mean weight of seeds (measuring only cleistogamous flowers) varies regularly from the base to the top of the main stem, increasing significantly and roughly linearly with increasing node number (Waller 1982), and in *Asphodelus albus*, seeds from fruits located at the bottom of the long linear infructescence are heavier than those from fruits at the distal positions (Obeso 1993). *Pastinaca sativa* produces seeds on primary, secondary, and tertiary umbels of the flowering stalk. Within plants, variation in seed weight is about twofold, with secondary and tertiary seeds weighing 73% and 50% of primary seed weight, respectively (Hendrix 1984). Within the inflorescence of *Lolium perenne*, mean seed weight declines sharply from the proximal seed (1.86 mg) to the distal (0.71 mg) within individual spikelets, and 89% of the total variation in seed weight is accounted for by differences between seeds at different positions within the spikelet (Warringa, de Visser, and Kreuzer 1998; Warringa, Struik, et al. 1998). Similar declines in individual seed mass from basal to distal positions occur in infructescences of *Arum italicum* (Méndez 1997) and *Banksia spinulosa* (Vaughton and Ramsey 1997). In both cases, seeds in the apical positions weigh about 15% less than those from the basal positions. The inverse situation is exemplified by *Lobelia inflata*, where seed size increases with nodal position along the main stem (Simons and Johnston 2000). In kiwifruit vines (*Actinidia deliciosa*), apple trees, and hawthorns (*Crataegus monogyna*), the size of individual fruits varies regularly according to their position along stems (de Silva and Ball 1997; de Silva et al. 2000; Rodríguez Larrinaga 2004).

Intrinsic gradients of intraplant variation can also involve detailed

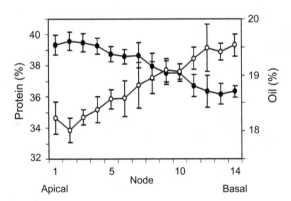

FIG. 4.3 Variation in mean protein (*filled circles*) and oil (*open circles*) content of individual soybean (*Glycine max*) seeds produced at different nodal positions along plants. Nodes are numbered from apical (left) to basal (right) positions. Redrawn from Bennett et al. 2003.

aspects of the chemical composition of fruits and seeds, although the information available on this type of variation is almost exclusively restricted to cultivated plants. In tomato (*Lycopersicon esculentum*) plants grown under controlled greenhouse conditions, the vitamin C content of fruits increases linearly from those located in the basal nodes of the plant to those in the uppermost positions (Fryer et al. 1954). Seeds from the lower part of the soybean (*Glycine max*) plant are higher in oil and lower in protein than those from the upper part (Collins and Carter 1956; Brim et al. 1967), and the patterns of variation of these two components along nodal positions of the plant are essentially linear and mirror each other (fig. 4.3). There is thus a clear seed compositional gradient within soybean plants, with seeds near the plant apex being high in protein and low in oil content, and those around the basal region exhibiting the opposite pattern. In addition, the concentration of protease inhibitors in soybean seeds (Marchetti et al. 1995) and the fatty acid composition of seed oil (Bennett et al. 2003) vary predictably with nodal position along the stem, further contributing to the position-dependent compositional gradient. The percentage of total fatty acid made up by oleic acid decreases with increasing nodal position along the stem, and the proportion of linoleic acid increases. Similar chemical changes take place along corn (*Zea mays*) ears, where the content of amylose starch and oleic acid decreases, and that of palmitic and linoleic acids increases, from seeds located at the base to those at the tip (Fergason et al. 1966; Jellum 1967). Within sunflower heads (*Helianthus annuus*), fatty acid composition of seed oil varies depending on seed position, although patterns of variation seem to be somewhat cultivar-specific. Zimmerman and Fick (1973) reported that the proportion of palmitic and linoleic acids tended to increase, and that of stearic and oleic acids to decrease, from the central to the peripheral positions in the capitulum. In the cultivars studied by Munshi et al. (2003), in contrast, there was a decline in the proportion of palmitic and stearic acids from the peripheral to the central whorls, and no discernible within-capitulum variation was found for oleic and linoleic acids. The fatty acid composition and the concentration and composition of glucosinolates in seeds of *Brassica napus* and *B. campestris* grown under uniform controlled conditions all depend on the position of pods within the plants (Bechyne and Kondra 1970; Kondra and Downey 1970). In mustard seeds (*Brassica juncea*), Munshi and Kumari (1994) found an increase in free fatty acids, phospholipid, glycolipid, and sterol content toward the apical positions of the inflorescence. In sesame (*Sesamum indicum*) seed oil, the percentage of palmitic and linoleic acids declines, and total oil content and the percentage of stearic, oleic, and linolenic acids increase, from basal to distal nodal positions along the plant (Mosjidis and Yermanos 1985; Tashiro et al. 1991). Within spikes of wheat (*Triticum aestivum*), protein content of seeds decreases from the basal to distal positions (Bramble et al. 2002), and parallel declines occur also for some macronutrients, including Ca, Mg, P, and S (Calderini and Ortiz-Monasterio 2003). Although the preceding selection of examples is far from comprehensive, it leaves little doubt about the frequent occurrence of intrinsic within-plant gradients in seed composition in cultivated plants from a broad variety of plant families and growth forms. There is no reason to suspect that it is a peculiarity restricted to cultivated species. If sought for, one would expect to find similar patterns in their wild counterparts.

Intrinsic gradients of intraplant variation may also take place over very small spatial ranges, as when seed size varies regularly within the confines of individual fruits. This happens frequently in fruits where seeds are arranged linearly, as in legume pods and crucifer siliques. In these cases, the size of a seed depends on its location within the fruit relative to the fruit pedicel (Schaal 1980; Stanton 1984; Mazer et al. 1986; Rocha and Stephenson 1990; Gutiérrez et al. 1996; Susko and Lovett-Doust 1999). In the legumes of *Lupinus texensis*, for example, mean seed mass increases from 27.8 mg for seeds closest to the fruit peduncle to 30.2 mg for seeds farthest away (Schaal 1980). In those of *Phaseolus coccineus*, mean seed mass likewise increases from the basalmost (549 mg) to the distalmost (628 mg) positions (Rocha and Stephenson 1990), and a similar pattern was reported for the congeneric *Phaseolus vulgaris* (Nakamura 1988). In *Raphanus raphanistrum*, in contrast, the largest seeds tend to occur at positions nearest the pedicel or in the middle of the fruit, and individual seed mass may vary up to sixfold within single fruits (Stanton 1984). Within the cones of *Pinus contorta*, seed size declines regularly from the basal to the apical positions (McGinley et al. 1990).

Gradients in organ features that take place along stems or branches are not necessarily linear. In shoots of *Sassafras albidum*, unlobed leaves predominate at proximal and distal nodes of shoots, while lobed leaves are most common at intermediate nodes. In this species, the highest values of leaf area, nitrogen content, chlorophyll content, and net photosynthetic rates occur at intermediate nodal positions, and decline toward both the basal and distant positions (de Soyza et al. 1990; de Soyza and Kincaid 1991). A similarly nonlinear pattern was found by Traw and Ackerly

(1995) in the tropical pioneer tree *Ficus insipida*, where leaf mass per area was highest in the central positions of shoots and declined toward both the apex and the base. Intrinsic gradients can become extraordinarily complex, as in Neotropical rain-forest herbaceous vines of the genus *Syngonium* (Ray 1987). The growth habit of these unbranched vines consists of a succession of terrestrial and arboreal growth phases, in which the plant either grows horizontally on the forest floor or climbs and descends from trees. Within each growth stage, leaf size and shape fluctuate cyclically, which gives rise to a tremendously complex pattern of longitudinal variation in leaf features along the shoot of the vine. Nonlinear gradients can also involve structures other than leaves. In fruiting heads of cultivated sunflower (*Helianthus annuus*), the oil content of seeds was found to be highest in the middle whorl positions, and to decline toward both the central and peripheral whorl positions (Fick and Zimmerman 1973).

The distinction between different types of gradients provides a useful general framework to differentiate the main axes along which within-plant variation is organized. Nevertheless, it is important to realize that directions of change can interact with each other, and that the pattern of variation in some feature of a reiterated structure along one axis frequently depends on the structure's position relative to another reference axis. This happens, for example, when there is an interaction between two different, independent extrinsic gradients, such as height above ground and compass direction. In trees of *Prunus mahaleb*, leaves from the upper and lower tiers of the crown may differ or not, depending on orientation, as illustrated in figure 4.2 for leaf water content. In this example, an orientation-dependent extrinsic gradient exists in the lower branches of the tree but not in the upper ones, and a height-dependent gradient exists in southern and western exposures but not in others. Similar interactions between extrinsic gradients are probably common in trees, although they have been reported in detail only infrequently. In olive trees (*Olea europaea*), for example, the increase in leaf nitrogen content (percent dry mass) from lower through middle to higher position in the canopy is most marked among leaves in north-facing and west-facing crown locations, intermediate in south-facing locations, and does not exist in east-facing orientations (Perica 2001).

In addition to interactions between extrinsic gradients, a tight link is generally to be expected between extrinsic and intrinsic gradients. Clear examples of the intersection between the two main types of gradients are furnished by situations where the sign and magnitude of an intrinsic gradient associated with some plant structure (node number on a shoot) depend critically on the position of the structure in relation to some relevant extrinsic gradient (light, orientation, height). In apple trees of the 'Royal Gala' cultivar studied by de Silva et al. (2000), fruit weight increases from the base to the tip in the lowermost tree branches, but that intrinsic trend is reversed in the uppermost branches, where fruit weight declines from base to tip. The nature of intrinsic gradients may also depend on higher-level architectural characteristics of the plants. In cultivated kiwifruits (*Actinidia deliciosa*), fruit weight may either decline or remain constant with increasing distance along shoots, depending on the method used to train plants in supporting structures (de Silva and Ball 1997).

There is little information on the relative importance of extrinsic and intrinsic gradients as predictors of within-plant variation in features of re-iterated structures. As a broad generalization, it would be tempting to pre-dict that external gradients would be the prevailing source of intraplant variation in those cases where reiterated structures are widely distributed in space and thus most likely face quite a broad range of external environ-mental conditions (height above the ground, light intensity). This would apply, for instance, to tall trees in dense forests whose crowns encompass broad ranges of a steep vertical gradient of light intensity, often span-ning several orders of magnitude. Nevertheless, this intuitively appeal-ing notion is probably simplistic, and is contradicted by some empirical data. Hall and Langenheim (1986) studied within-tree variation in the leaf monoterpenes of *Sequoia sempervirens*, focusing on their possible varia-tion with height above the ground. As this tree is one of the largest of liv-ing organisms, it seemed to make a particularly favorable study system for documenting patterns of spatial variation within an individual. Despite this, Hall and Langenheim found no significant differences in composition or concentration of leaf monoterpenes between the three canopy levels considered, 2–6 m, 8–15 m, and 20–25 m above the ground. In contrast to this failure to detect height-dependent spatial variation at the very large scale of whole *Sequoia* trees, Espinosa-Garcia and Langenheim (1990) did find significant within-branch longitudinal gradients in the composition of endophytic fungal communities in the same species, which might be related to fine-scale variations in leaf age and structural or chemical char-acteristics. This example illustrates well that simple consideration of plant size alone, and therefore the potential range of environmental conditions faced by different parts of the same individual, may have little predictive power in ascertaining the relative quantitative importance of intrinsic and extrinsic gradients as sources of within-plant variation. Traw and Ackerly's study of five species of rain forest pioneer trees (1995) also supports this conclusion. These authors applied path analyses to ask whether leaf nitrogen concentration was better explained by leaf nodal position along branches than by measured light level. Leaf position on the shoot (intrinsic gradient) explained significantly more of the within-plant variation in leaf nitrogen concentration (up to 60–70% of within-plant variance) than did leaf light level (extrinsic gradient) in four of the five species. Further evidence suggesting that intrinsic gradients can be responsible for most within-plant variance in leaf features is provided by the finding that in *Sassafras albidum* leaf nodal position alone accounts for 30–50% of total within-plant variance in leaf chlorophyll content (de Soyza et al. 1990).

Spatial Analyses

Spatial analyses of within-plant variation have attempted to elucidate, at the whole plant level, spatial dependencies of organ characteristics. They have addressed questions such as whether structures that are closer within a plant are more likely to be similar than structures that are farther apart. These analyses allow for determining the "grain size" of within-plant spatial variation, a shortcut used here rather informally to designate the scale of spatial domains within which organ similarities remain predictably higher than the overall average value for all structures on the plant (for a rigorous definition of the grain concept and related notions in the context of spatial statistical analysis, see, e.g., Dungan et al. 2002). Leaving aside purely descriptive methods based on the visual inspection of computer-abstracted three-dimensional representations of organ-trait distribution on plants (e.g., Smith et al. 1992), three main analytical approaches have been adopted to address the problem of effectively describing within-plant variation in the attributes of reiterated structures. These are geostatistical techniques, mixed-model analyses of variance, and hierarchical variance partitioning of organ traits into nested architectural levels. The first two are statistically more powerful and have the capacity to provide objective and rigorous spatial analyses, yet they have the important practical drawback of requiring extensive spatial data that are hard or impossible to collect on wild plants in the field. The third method takes advantage of the hierarchically nested distribution of reiterated structures within-plants, and fully random nested analyses of variance are used to partition total within-plant variance in organ attributes into hierarchical components due to different structural levels (e.g., branch, infructescence, fruit). This method provides poorer spatial resolution than the geostatistical or mixed-model ones, since spatial proximity is considered only at an ordinal level, yet it has the important advantage of being much less stringent as to the quality and amount of spatial data required, and thus easier to use with wild plants.

When geostatistical methods are applied to within-plant variation, the plant is considered as a spatial set of points (i.e., branches, shoots) on which a random function can be defined in the form of measurements made on some reiterated organ (Monestiez et al. 1990). Conventional geostatistical methods of the kind ordinarily used in geological surveys or ecological investigations of spatial patterns (e.g., Webster 1985; Rossi et al. 1992) are then applied to the data, and variogram functions are typically obtained that depict the relationship between the degree of spatial dependence of two structures, that is, the relationship between their similarity in measured attributes and the distance that separates them. Due to the particularities of the spatial structure of plants, different types of "distances" between organs may be defined and used in geostatistical analyses of within-plant variation, including the natural metric distance obtained following the branch system, the number of intervening bifurcations between them (Monestiez et al. 1990), or the euclidean distance on a straight line (Luyssaert et al. 2001). So far, geostatistical methods have been implemented mainly in studies of within-plant variation of cultivated fruit plants (but see Luyssaert et al. 2001; Shelton 2005). Although cultivated and wild plants may differ in detailed aspects of the spatial patterns of within-plant variation, there is no reason to expect drastic differences; thus I will briefly summarize here some results obtained by these investigations.

Habib et al. (1991) used geostatistical techniques to study the spatial pattern of within-plant variation in weight, percent dry matter, soluble-solid content of pulp, and acidity of kiwifruits. The distance between fruits was taken as the number of branching points (forks) to go from one fruit to another. Regardless of the fruit characteristic considered, they found a significant spatial dependence between fruits taken at short distances. The variogram function (see the legend for fig. 4.4 for a summary of concepts related to the variogram function) first increased with increasing distance, indicating declining spatial correlation of fruit attributes with increasing separation, and then stabilized at distances equal to or greater than two forks, indicating that fruit characteristics were essentially uncorrelated when they were borne by canes separated by two or more forks. Results of Monestiez et al. (1990) and Audergon et al. (1993) for peach trees (*Prunus persica*) are similar in showing that fruit features (size, soluble-solid concentration) are spatially autocorrelated only at very short distances within the plant (<1 m), and are uncorrelated beyond this very local threshold (fig. 4.4a). This contrasts with the pattern of spatial autocorrelation exhibited by leaf nitrogen content on the same peach tree. For this trait, the variogram function exhibits a nearly monotonic increase over most of the range of distance between leaves (fig. 4.4b), which denotes that the spatial dependence of this leaf trait takes place over a broader spatial range. Shelton (2005) also used geostatistical methods based on the construction of variograms to dissect the spatial pattern of the concentration of glucosinolates in plants of *Raphanus sativus* down to the within-leaf scale. One striking result of her study was the extremely fine-grained distribution

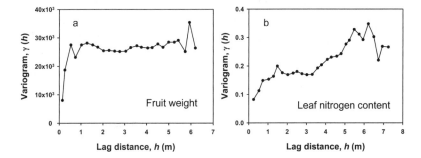

FIG. 4.4 Variogram functions for the spatial variation of (*a*) fruit size and (*b*) leaf nitrogen content in a peach tree. The variogram function for lag distance *h*, γ(*h*), is defined as the average squared difference of values separated by *h*, expressed in meters and measured using the natural distance following the branch system. Three main elements may be distinguished in variogram functions (Habib et al. 1991, Rossi et al. 1992): (1) There is a discontinuity at the origin. Although the variogram function is by definition null at the origin, there is often a significant value observed for a distance close to zero, which is known as the nugget. The nugget represents all unaccounted-for spatial variability at distances smaller than the smallest sampling distance. Nonnegligible nuggets in the graphs indicate that leaf and fruit variability still occurs at very short distances within the tree. (2) The variogram function increases with increasing separation of structures, which means that the farther apart the structures, the more independent they are. (3) The function eventually reaches a steady value, the sill, which is close to the variance of the whole sample. The lag distance at which the variogram levels off is known as the range, which defines the average distance within which the structures remain correlated spatially. In the graphs shown, the range is considerably smaller for fruits than for leaves, meaning that in the single peach tree studied, the threshold for spatial independence is smaller for fruits than for leaves. Redrawn from Monestiez et al. 1990, figs. 4 and 6.

of glucosinolates within-plants. The smallest scale examined, variation among samples within a specific region of a leaf, accounted for the greatest variation of any of the spatial scales examined (plants, leaves of the same plant, and regions of the same leaf), and there was no spatial autocorrelation at scales as small as 1–2 cm. Such small-grained subindividual distribution of plant defenses has a number of important implications for the interaction between plants and small herbivores, as discussed in detail by Shelton (2004, 2005) and in chapter 8.

Geostatistical methods can lead to inaccurate descriptions of the spatial structure of within-plant variation if the observed features of two organs on the plant depend on their location, thus violating the assumption of stationarity underlying these methods (de Silva and Ball 1997; stationarity assumes constant means and variances throughout the sampling space). As illustrated by the intrinsic and extrinsic gradients described in the preceding section, organ attributes are generally dependent on their position relative to the plant architecture or to some external coordinates; hence the stationarity assumption is unlikely to be generally valid.

As an alternative to geostatistical methods, de Silva and Ball (1997) proposed the application of mixed-model analyses of variance that incorporate position effects as known fixed and random effects (i.e., intrinsic and extrinsic gradients), and that model the spatial dependence of the residual variation by choosing an appropriate covariance structure allowing for spatial autocorrelations. Technical details of the application of mixed-model analyses of variance to the study of spatial patterns may be found in Littell et al. (1996). As with geostatistical methods, these procedures seem to have been used so far only in connection with cultivated plants. Results of studies on within-plant variation in fruit weight conducted on apple trees and kiwifruit vines are similar in showing that fruit size tends to be spatially autocorrelated only at short nodal distances within the same shoot; that the correlation between fruits in different shoots is generally negligible; and that by far the greatest variance component is accounted for by differences between fruits borne on the same shoot (de Silva and Ball 1997; de Silva et al. 2000). Miles et al. (1996) also found that variance occurring at the smallest spatial scales was the main source of within-plant variation in size and flesh firmness of kiwifruits.

The third method used to investigate the spatial organization of within-plant variation is based on partitioning the total within-plant variance of a given organ attribute into different architectural levels that can be envisaged as forming a hierarchically nested structure. Such data are

thus amenable to analysis through purely random models of nested analysis of variance. Under this approach one might dissect, for example, the
total within-plant variance in seed mass into components due to differences between branches, between infructescences borne by the same
branch, between fruits of the same infructescence, and in the case of multiseeded fruits, between seeds of the same fruit. From a strictly statistical
viewpoint, the spatial information provided by this technique is of poorer
quality than that obtained using either of the two preceding, more sophis-
ticated methods. From a biological viewpoint, however, it has some advantages. First, it is of interest in assessing the extent to which membership in
the same supporting structure (branch, inflorescence, infructescence, fruit)
influences the similarity of organs as they might be perceived by herbivores, pollinators, and fruit consumers. Second, information on the apportionment of within-plant variance in organ traits among different-level
supporting structures may provide insight into the relative importance of
the diverse mechanisms that underlie the variation, as discussed in chapter 6. In addition, partitioning total within-plant variance into a series of
nested structural levels is relatively easy to apply to data from wild plants
obtained for other purposes whenever sufficiently detailed records were
kept of the structural level each measured structure belonged to.

Available quantitative information for wild plants on the components of within-plant variance attributable to different structural levels
is summarized in table 4.1 for leaf, flower, fruit, and seed features. Variance partitions involving the among-branch versus within-branch levels
yield remarkably similar conclusions, regardless of the organ and character considered. The vast majority of within-plant variance in organ attributes is due to variation occurring within individual branches, and only
secondarily to differences between branches or ramets (75% vs. 25% of
total variance on average, respectively, for the data in table 4.1). Variance
partitions at the within and between inflorescence or infructescence levels
also reveal that most within-plant variance take place at a very local scale
(table 4.1). On average, 77% of all within-plant variance in flower, fruit,
or seed traits is accounted for by variation among organs of the same
inflorescence or infructescence. Although the data listed in table 4.1 are
very limited, they suggest a declining trend in the relative importance of
variance due to within-inflorescence or within-infructescence variation as
a source of within-plant variation, in the direction of flowers (96%) to
fruits (86%) to seeds (67%). I have not included in table 4.1 information on cultivated plants, but these provide similar examples documenting

TABLE 4.1 **Hierarchical distribution of the within-plant variance in flower, fruit, and seed traits between and within different structural levels.**

| | | Structural level | | | | | | |
| | | Branch or ramet | | Inflorescence or infructescence | | Fruit[b] | | |
Species[a]	Trait	Between	Within	Between	Within	Between	Within	Reference
Leaf traits								
Betula pubescens	% water content	34.2	65.8					Suomela and Ayres 1994
	Specific mass	33.9	66.1					Suomela and Ayres 1994
	% nitrogen	0	100					Suomela and Ayres 1994
	% amino acids	18.7	81.3					Suomela, Ossipov, and Haukioja 1995
	% carbohydrate	16.7	83.3					Suomela, Ossipov, and Haukioja 1995
	% phenolics	50.3	41.7					Suomela, Ossipov, and Haukioja 1995
Floral traits								
Heliconia psittacorum	Nectar secretion	38.9	61.1					Feinsinger 1983
Helleborus foetidus	Corolla length			1.9	98.1			C. M. Herrera unpubl.
	Corolla width			5.0	95.0			C. M. Herrera unpubl.
Ipomoea wolcottiana	Corolla length	43.7	56.3					V. Parra-Tabla unpubl.
Lavandula latifolia	Calyx length			2.7	97.3			C. M. Herrera unpubl.
	Corolla length			8.4	91.6			C. M. Herrera unpubl.
Silene acutifolia	Petal length			0	100			M. L. Buide unpubl.
Fruit traits								
Cultivated apple	Fruit mass	5.3	94.7					de Silva et al. 2000
Crataegus laciniata	Fruit width	0	100					C. M. Herrera unpubl.
Crataegus monogyna (4)	Fruit width	11.5	88.5					C. M. Herrera unpubl.
Sorbus aucuparia	Fruit width			12.8	87.2			B. Pías and M. Salvande unpubl.
	Seediness[c]			14.9	85.1			B. Pías and M. Salvande unpubl.

Seed traits

Species	Trait							Reference
Alnus rubra	Viability[d]	8.8	91.2					Markham 2002
Erica australis	Viability[d]	62.5	37.5					Cruz et al. 2003
Phillyrea latifolia (2)	Seed mass	23.0	77.0					Herrera et al. 1994 and unpubl.
Mean for 4 Compositae species	Settling velocity			52.0	48.0			Andersen 1992
Anthoxanthum odoratum	Seed mass			21.5	78.5			Roach 1987
Banksia marginata	Seed mass			16.7	83.3			Vaughton and Ramsey 1998
Banksia spinulosa	Seed mass			18.1	81.9			Vaughton and Ramsey 1997
Cicuta maculata	Seed mass			46.8	53.2			Hendrix and Sun 1989
Cryptotaenia canadensis	Seed mass			44.7	55.3			Hendrix and Sun 1989
Lomatium grayi	Seed mass			26.2	73.8			Thompson 1984
Osmorhiza claytoni	Seed mass			49.0	51.0			Hendrix and Sun 1989
Osmorhiza longistylis	Seed mass			14.9	85.1			Hendrix and Sun 1989
Oxypolis rigidior	Seed mass			18.3	81.7			Hendrix and Sun 1989
Prunella vulgaris	Seed mass			29.5	70.5			Winn 1991
Sanicula gregaria	Seed mass			44.8	55.2			Hendrix and Sun 1989
Zizia aurea	Seed mass			40.5	59.5			Hendrix and Sun 1989
Aesculus glabra	Seed mass					54.6	45.4	Michaels et al. 1988
Aesculus pavia (2)	Seed mass					48.8	51.2	Michaels et al. 1988
Arum italicum (2)	Seed mass					13.0	87.0	Méndez 1997
Asarum canadense	Seed mass					78.8	21.2	Michaels et al. 1988
Asclepias verticillata	Seed mass					24.0	76.0	Michaels et al. 1988
Asimina triloba	Seed mass					41.5	58.5	Michaels et al. 1988
Asphodelus albus (3)	Seed mass					37.5	62.5	Obeso 1993
Baptisia lanceolata	Seed mass					33.2	66.8	Mehlman 1993
Cassia fasciculata	Seed mass					69.4	30.6	Michaels et al. 1988
Cercis canadensis	Seed mass					72.6	27.4	Michaels et al. 1988
Corema album	Seed mass					65.0	35.0	A. Rodríguez Larrinaga unpubl.
Dentaria laciniata (2)	Seed mass					42.8	57.2	Michaels et al. 1988
Dicentra canadensis	Seed mass					27.5	72.5	Michaels et al. 1988
Dicentra cucullaria	Seed mass					26.2	73.8	Michaels et al. 1988

(continued)

TABLE 4.1 (*continued*)

Species[a]	Trait	Branch or ramet		Inflorescence or infructescence		Fruit[b]		Reference
		Between	Within	Between	Within	Between	Within	
Gleditsia triacanthos	Seed mass					39.0	61.0	Michaels et al. 1988
Gymnocladus dioica	Seed mass					19.7	80.3	Michaels et al. 1988
Helleborus foetidus (2)	Seed mass					52.8	47.2	Garrido 2003 and unpubl.
	Elaiosome mass					2.2	97.8	Garrido 2003 and unpubl.
Impatiens biflora	Seed mass					61.9	38.1	Michaels et al. 1988
Maianthemum racemosum	Seed mass					79.8	20.2	Michaels et al. 1988
Pancratium maritimum (3)	Seed mass					60.1	39.9	M. Medrano unpubl.
Phaseolus vulgaris	Seed mass					15.5	84.5	Cipollini and Stiles 1991
Phaseolus lunatus	Seed mass					18.7	81.3	Cipollini and Stiles 1991
Polygonatum commutatum (2)	Seed mass					26.8	73.2	Michaels et al. 1988
Rhamnus alpinus	Seed mass					9.5	90.5	Bañuelos and Obeso 2003; M. J. Bañuelos unpubl.
Sanguinaria canadensis	Seed mass					41.2	58.8	Michaels et al. 1988
Smilax aspera	Seed mass					38.4	61.6	C. M. Herrera unpubl.
Smilax hispida	Seed mass					0	100	Michaels et al. 1988
Smilax lasioneura	Seed mass					37.1	62.9	Michaels et al. 1988
Sorbus aucuparia (2)	Seed mass					52.5	47.5	B. Pías and M. Salvande unpubl.
Staphylea trifolia	Seed mass					40.0	60.0	Michaels et al. 1988
Uvularia grandiflora	Seed mass					16.0	84.0	Michaels et al. 1988
Vitis vulpina	Seed mass					30.2	69.8	Michaels et al. 1988

Note: Entries represent the mean proportion of total within-plant variance that is accounted for by variation at a particular level.

[a] Number of distinct populations and/or years sampled is shown in parentheses.

[b] Applicable only to species with multiseeded fruits.

[c] Number of seeds per fruit.

[d] Percent seeds germinating.

the predominant role of within-inflorescence variation as a source of within-plant variation. For example, in a particularly thorough study of the spatial structure of variance in single-seed protein content of wheat (*Triticum aestivum*), Bramble et al. (2002) found that most within-plant variance in this trait was accounted for by variation occurring within individual spikes, while variance among the different spikes of the same plant was much less important.

Plants producing multiseeded fruits provide an opportunity to further partition within-plant variance of seed traits into among- and within-fruit levels, and data for 33 such instances are listed in table 4.1. In this sample, within-fruit variation in seed traits (mainly seed mass) accounts on average for 61% of total within-plant variance, or in other words, about two-thirds of the total variance in seed features exhibited by an individual plant occurs within the spatially very constrained limits of a single capsule, legume, or follicle.

In species with apocarpous ovaries made up of several separate, multiovulate carpels (e.g., Ranunculaceae), a single fruit comprises several distinct follicles, and the within-plant variance in seed traits can thus be partitioned one step beyond the within-fruit level, that is, between and within follicles of the same fruit. *Helleborus foetidus* is a perennial herb with apocarpous flowers, and its fruits are generally made up of two or three independent follicles. The seeds are dispersed by ants and bear an elaiosome. Within-plant variance in elaiosome mass is almost entirely due to variation between seeds of the same fruit (97.8%; table 4.1). When this predominant source of variance is further partitioned into components due to variation between and within follicles of the same fruit, all variance is accounted for by differences between seeds belonging to the same follicle (Garrido 2003). Therefore, virtually all the variation in elaiosome mass that occurs within a plant of *H. foetidus* occurs at the very small spatial scale of the individual follicle. If one adds to this that within-plant variation is responsible for an estimated 82.2% of all observed variance in elaiosome mass in this species (Garrido 2003), then the striking conclusion follows that about 80% ($= 97.8 \times 0.822$) of the total variance in elaiosome mass exhibited by *Helleborus foetidus* seeds over the whole Iberian Peninsula is to be found within the few centimeters of a single follicle.

Further examples of extremely small-scale within-plant variation in chemical composition of reiterated structures are provided by Shelton's investigation of the distribution of glucosinolates in *Raphanus sativus* plants (2005) mentioned above, and by Herrera, Pérez, and Alonso's study of nectar sugar composition in *Helleborus foetidus* (2006). Flowers of this species generally have five separate, independent nectaries. Nectar sugar composition at the level of individual nectaries was determined using HPLC analytical techniques, and population-level variance in sugar composition was then dissected into hierarchical components due to variation among plants, flowers of the same plant, and nectaries of the same flower. Nectar sugar composition varied extensively among nectaries, and nearly all major combinations of individual sugars were recorded in the population. This large population-wide variance was mainly accounted for by variation among nectaries of the same plant, and only minimally by differences among plants (14%). On average, 35% of total within-plant variance in nectar sugar composition was due to differences among nectaries in the same flower. As these are only a few millimeters apart, these results reveal extremely fine-grained spatial variation in nectar composition in this species. Similarly fine-grained variation in nectar sugar composition was also demonstrated by Canto et al. (2007) for wild-growing plants of two species of *Aquilegia*.

Spatial Organization of Variation: Summing Up

The two main approaches adopted to study patterns of within-plant variation in the features of reiterated structures, namely gradient-oriented and explicitly spatial ones, furnish complementary views. Combining the results from these two views leads to some general insights on the overall spatial organization of within-plant variation.

On one hand, gradient-oriented analyses reveal the existence of *predictable trends* whereby the value acquired by a given phenotypic trait of an organ can be partly anticipated by virtue of its location along either an external environmental gradient (e.g., height, light intensity) or one linked to the plant's own architecture (e.g., nodal position along a stem or flowering axis). As the variation in organ attributes will in most instances be simultaneously subject to the influence of both intrinsic and extrinsic gradients, the superimposition of the two types of gradients produces a relatively broad-scale coarse-grained three-dimensional pattern of relatively predictable within-plant variation in organ features, as illustrated in figure 4.5 for the variation in leaf cadmium concentration in the crown of a single *Salix fragilis* tree. The ultimate template defining the details of this type of broad-scale within-plant variation is, of course, the plant's own spatial structure, organized around the hierarchical repetition of supporting structures (branches, stems, inflorescences, infructescences, fruits). The prevalence of the extrinsic gradients over the intrinsic, or vice versa, as major determinants of the scale and main trends of variation will therefore depend closely on the architecture of the plant, and distinct plant architectural models (sensu Hallé et al. 1978) should be expected to lead eventually to grossly different "maps" of spatial within-plant variation. In long-lived plants, these maps of variation will vary with age, insofar as ontogenetic changes modify both the size and the general architecture of the plant.

On the other hand, spatial analyses of within-plant variation in organ

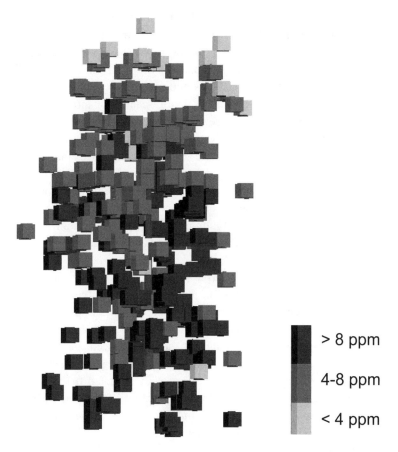

> 8 ppm

4-8 ppm

< 4 ppm

FIG. 4.5 Three-dimensional representation of variation in cadmium concentration in leaves of a single *Salix fragilis* tree sampled at 292 different locations regularly distributed over the crown. Modified from Luyssaert et al. 2001.

features reveal that, after the broad-scale variation disclosed by gradient-oriented analyses has been statistically accounted for, much intraplant variation remains, occurring over very restricted spatial scales. Methods based on spatial autocorrelations, mixed-model analyses of variance, or the partition of organ-trait variance into different plant structural levels, all reveal that extremely fine-grained variation can sometimes be the main, or at times almost the sole, source of within-plant variation in organ features. This very small-scale variation, exemplified by variation in seed size within the same pod, nectar composition within the same flower, secondary-metabolite concentration within the same leaf, and fruit or flower size within the same inflorescence or infructescence, is responsible for the fraction of within-plant variance that is left unaccounted for by gradient-based explanations. The overall spatial organization of within-plant variance in organ traits may therefore be seen as the combined result of one or more predictable trends of variation occurring along intrinsic or extrinsic gradients and potentially involving different spatial scales (whole plant, branches, inflorescences, or infructescences), *plus* a certain amount of variance that is unexplained by such gradients. As shown by geostatistical analyses described above, this unexplained variation largely occurs at very local scales. One further example is provided by the spatial pattern of variation in leaf tannin content within *Acer saccharum* trees described by Schultz (1983). Adjacent leaves on single branches may differ by a factor of 2 or more on any given date for this trait. Leaves of widely different tannin content appear randomly arrayed on branch segments, the position of a leaf of a given value being unpredictable on individual branches.

Luyssaert et al.'s study of intratree variation of cadmium concentration in leaves of *Salix fragilis* (2001), in addition to exemplifying a powerful analytical approach, illustrates well a dissection of within-plant variation into components that are explained and unexplained by gradients. These authors found a trend of declining leaf cadmium concentration from the lower to the upper parts of the tree (fig. 4.5). After statistically removing the effect of this broader-scale trend, they modeled the spatial distribution of residuals using geostatistical methods. The fitted variogram revealed a clear spatial structure, with most of the total variance occurring at lag distances less than 0.8 m, which means that residuals are spatially autocorrelated only at relatively small scales. The overall pattern of within-tree variation in leaf cadmium concentration in the studied individual of *Salix fragilis* resulted therefore from the combination of a gradient-based trend plus very local variations unrelated to the gradient. These two main sources of within-plant variation differ with regard to the most likely proximate mechanisms involved, as described in detail in chapter 6. Furthermore, recognizing the existence of both spatially predictable (related to gradients) and unpredictable (small scale, largely random) components of subindividual variation is also important for a better understanding of the interactions between variable plants and animal consumers (e.g., leaf eaters, frugivores, seed predators), as will be shown in chapters 8 and 9.

CHAPTER FIVE

Causes of Subindividual Variability

Mutations within individuals and organ-level responses to environmental cues are the main classes of remote causes of within-plant variability in reiterated structures.

The preceding chapters have shown within-plant variation in virtually every conceivable morphological, structural, compositional, and functional feature of reiterated structures that has been ever examined. Furthermore, subindividual variation is often very large, and takes place at a variety of nested spatial scales and in relation to environmental gradients that are both extrinsic and intrinsic to the plant. In view of the heterogeneity of the traits affected and the multiplicity of spatial scales at which the phenomenon occurs, many causes are likely to be involved. Understanding the variety of causes underlying within-plant variation and the possible interactions among them is an essential step before undertaking any examination of the possible functional, ecological, and evolutionary significance of subindividual variability.

Causes come in several forms, from remote to immediate. In the next chapter, I tackle immediate causes (called mechanisms) that involve the organizing influences of plant physiology, architecture, and morphogenesis. The present chapter gives an account of remote causes of subindividual variation. These fall into two main classes, namely genetic heterogeneity within an individual, and semiautonomous, programmed responses of organs within an individual to environmental cues, that is, developmental phenotypic plasticity of individual organs as governed by more or less rigid organ-level reaction norms. Admittedly, mutations and developmental phenotypic plasticity are not causes themselves—rather they are "inducers" of responses that involve physiology and morphogenesis—but it is nevertheless useful to review them as remote causes of subindividual phenotypic variation.

Genes, the first class of inducers, can vary within an individual because of mutation in meristems. Think of a growing shrub with many apical meristems all subject to mutation. In principle, the branches can acquire a different genetic makeup. There are also several ways other than conventional mutations for branches or leaves or flowers to have different active genetic elements. At any rate, one can recognize a class of variation that is not the result of changes in the environmental milieu of the plant, and I will call these "hard-wired" inducers. The second major class of inducers involves programmed (i.e., genetically based) organ-level responses to microenvironmental variation, for example, variation through the growing season in temperature, day length, and soil moisture, or variation in how much sun hits this or that leaf. There is an interplay between microenvironmental variation and the inducible systems of individual organs just as there is an interplay between hard-wired variation and the inducible systems of the organism. For that matter, genetic variation and environmental variation among modules within an individual surely would interact to the extent that both kinds of variation exist. This chapter reviews these two classes of "inducers," and it is followed by a chapter on how the organism offers an ontogenetic contingency plan that responds to the action of various inducers as well as simply generating subindividual variation without induction.

Genetic Mosaicism

When cells with different genotypes coexist in one individual plant (i.e., a genetic individual originating from one zygote), the plant is a genetic mosaic. Most of the phenotypic traits shown in the preceding chapters to be subindividually variable, such as leaf size and shape, floral traits, and size and chemical composition of seeds, are known to have a genetic basis. In *Arabidopsis thaliana*, for example, up to 94 different genes are known to yield mutations causing abnormal leaf morphologies (Berná et al. 1999), and dozens of quantitative trait loci (QTL) have been identified that control quantitative variation in leaf, flower, and seed features (Alonso-Blanco et al. 1999; Juenger et al. 2000; Pérez-Pérez et al. 2002). The heterogeneity in genetic makeup that is characteristic of genetic mosaics thus has the potential for inducing genetically based subindividual variation in phenotypic attributes of reiterated structures, a possibility that received considerable attention in the ecological literature around two decades ago (Whitham and Slobodchikoff 1981; Whitham 1981, 1983; Whitham et al. 1984; Gill 1986; Gill et al. 1995).

Because of their contrasting conceptual implications, it is important to distinguish intraplant variation in organ phenotypes caused by genetic mosaicism (i.e., differential cell genotypes) from that due to intraplant variation in gene expression. Even if all the cells of a plant have identical genotypes, gene expression can still vary among different organs or among parts of the same organ, sometimes giving rise to striking within-plant phenotypic heterogeneity. This happens, for example, in some variegated plants (having patches of two or more colors in leaves or flowers) whose cells all have the same genotype, yet the genes responsible for the synthesis or destruction of pigments are expressed in only some of the cells (Marcotrigiano 1997). Further examples are provided by the so-called bud sports (D'Amato 1997) or bud variations (Masters 1869; Darwin 1883), which are sudden genetic variations of somatic cells in a shoot apical meristem producing phenotypically altered leaves, flowers, or fruits in many ornamental plants (de Schepper et al. 2004). These changes can be caused by mutations, but also by abrupt changes in gene expression. I ignore for the moment situations where intraplant phenotypic variation is related to variations in gene expression, an aspect that will be considered in chapter 6. In this section I focus exclusively on sensu stricto genetic mosaicism, that is, situations where cells of different genotypes coexist in a single individual.

Multiple Genotypes in Individuals

Individual plants are built up by successive cellular divisions of a single zygote and thus in principle are genotypically homogeneous. On occasion, however, permanent genetic changes take place in somatic cells that lead to within-plant multiplicity of genotypes. Such changes can be caused, for example, by alterations in the normal diploid chromosome number (aneusomaty) or by somatic mutations (Klekowski 1988a; Gill et al. 1995; D'Amato 1997). Whitham et al. (1984) presented a detailed account of cases of cytogenetic within-plant heterogeneity known at the time (see also D'Amato 1997). High frequencies of aneusomatic plants occur in natural populations of, for example, *Orobanche gracilis*, *Poa pratensis*, and *Claytonia virginica* (D'Amato 1997). In some populations of

Claytonia virginica, up to 68% of individuals are internally heterogeneous with regard to the number of chromosomes, which are significantly more numerous in the aerial parts of the plant than in the roots (Lewis et al. 1971). Similar within-plant variation in chromosome numbers has been also reported in *Xanthisma texanum* (Berger and Witkus 1954), *Haplopappus spinulosus* (Li and Jackson 1961), and *Pilosella officinarum* (Chapman et al. 2000). The frequency of the phenomenon led Lewis et al. (1971) to consider that "constancy of chromosome number within all individuals [of a species or population] is merely a convenient fiction." Within-plant variation in ploidy levels is also frequent among cultivated plants. In some azaleas (*Rhododendron simsii*) with two-colored flowers, cells in the petal margins are tetraploid, while the rest of the corolla, as well as the entire plant, are diploid (de Schepper et al. 2004).

Somatic mutations, however, have generally received far more attention than cytogenetic variations or any other possible source of genetic mosaicism in plants (Klekowski 1988a; Gill et al. 1995). At least four reasons, both theoretical and applied, have fueled a persistent interest in the study of somatic mutations in plants. First, numerous asexually reproducing and clonal plant populations exhibit considerable genetic diversity (for reviews see Ellstrand and Roose 1987; Khudamrongsawat et al. 2004), and somatic mutations might be one major source contributing to standing genetic variation in these species (King and Schaal 1990; Fernando and Cass 1996; Corradini et al. 2002). Second, somatic mutations may be subjected to selection within the individual and inherited by naturally occurring mechanisms, and theoretical models confer profound evolutionary implications when this is so (Antolin and Strobeck 1985; Klekowski et al. 1985; Otto and Orive 1995; D'Amato 1997; Otto and Hastings 1998; Pineda-Krch and Poore 2004). Third, the combination of phenotypically discernible somatic mutations (e.g., chlorophyll deficiency) and the peculiar organization of plant meristems into developmentally quasi-autonomous layers (described below) has provided a powerful tool for investigating organ morphogenesis in plants (Poethig 1989; Szymkowiak and Sussex 1996; Korn 2001). And fourth, evaluating the occurrence of somatic mutations in clonally propagated plants of economic importance has become in recent years an important tool for either assessing the genetic integrity of lineages possessing desirable features or identifying additional sources of selectable genetic variability (Albani and Wilkinson 1998; Bregitzer et al. 1998; Al-Zahim et al. 1999; Rahman and Rajora 2001; Palombi and Damiano 2002; Ravindra et al. 2004).

Although none of these aspects is directly relevant to the objectives of this chapter, I will take advantage of the empirical information generated by these four lines of enquiry to assess the possible significance of genetic mosaicism originated by somatic mutations as a cause of within-plant variation in phenotypic traits of reiterated structures in wild plants. I first present a short overview of evidence on somatic mutations and associated genetic mosaics in cultivated and wild plants. Then I turn to the issue of the frequency with which genetic mosaics are expected to translate into within-plant phenotypic mosaics giving rise to subindividual variation in reiterated structures. This course of reasoning differs slightly from some earlier treatments of genetic mosaicism in relation to within-plant variation (e.g., Gill et al. 1995) in that it makes explicit the crucial issue that *genetic* mosaicism should not be automatically assumed to directly translate into associated *phenotypic* mosaicism. The space devoted in this section to deal with genetic mosaicism may perhaps seem excessive for its presumably minor role as a cause of within-plant phenotypic variation in natural plant populations. This assessment is necessary, however, in view of the significant role attributed by some earlier studies to genetic mosaicism as a determinant of within-plant phenotypic variation, and because within-plant phenotypic variation has sometimes been improperly used as an indicator of genetic mosaicism (Gill and Halverson 1984; Gill 1986; Gill et al. 1995).

FREQUENCY OF GENETIC MOSAICS   Point mutations occurring in somatic cells (= somatic mutations) are possibly the most frequent source of genetic mosaics in plants (Klekowski 1988a). The occurrence of somatic mutations and genetic mosaics in plants was thoroughly reviewed in a series of publications by Whitham and Slobodchikoff (1981), Whitham (1981), Whitham et al. (1984), Gill (1986), and Gill et al. (1995), in the context of the then-emerging hypothesis that genetic mosaicism played a crucial role in the evolution of defensive strategies of long-lived plants against herbivores. These publications summarized the abundant evidence documenting genetic mosaicism among cultivated plants. Somatic mutations, and resulting genetic mosaics, arise naturally in many crops and ornamentals, and have long been used in developing new cultivars that are superior to parent stock or that have ornamental appeal (e.g., variegated plants). Mutations uncovered by cloning research conducted on cultivated plants were also interpreted as indicative of somatic mutations. In contrast with the wealth of information from cultivated plants, the earlier reviews mentioned above were able to gather very little evidence of genetic mosaicism from wild plants, and even that evidence was mostly indirect. The latter included the occurrence of somatic novelties in populations of some apomictic plants, and the genetic heterogeneity found among the stems (ramets) of heavily cloning, asexually propagating plants. In retrospect, the main conclusions emerging from these earlier reviews of genetic mosaicism in plants conducted from the perspective of evolutionary ecology were that somatic mutations and ensuing genetic mosaicism doubtless occur in both cultivated and wild plants; that their frequency is probably high among crops and ornamentals; and that there was little direct empirical information at the time on their actual prevalence among wild plants.

Recent increased availability of molecular genetic markers with high discriminatory ability has considerably enhanced the possibilities of detecting genetic mosaicism in plants. Random amplified polymorphic DNA (RAPD), amplified fragment length polymorphism (AFLP), inter simple sequence repeat (ISSR), and single sequence repeat (SSR, or microsatellite) markers have all been used to assess the occurrence of genetic mosaicism in wild and cultivated plants, and have provided considerable empirical evidence of somatic mutations. As shown below, the evidence available is still very unequally distributed among natural conditions (wild plants in the field) and artificial ones (crop plants, transgenic, or in vitro cultures of plants of economic interest).

Investigations of crop plants and ornamentals using molecular genetic markers have often revealed genetic mosaics presumably derived from somatic mutations, most often in the form of stable periclinal chimeras (see next section). This has been found, for example, in several grapevine cultivars where the existence in some individual plants of more than two alleles at the same microsatellite locus revealed a genetic mosaic due to periclinal chimerism (Franks et al. 2002; Riaz et al. 2002; Crespan 2004; Hocquigny et al. 2004). Likewise, many of the spontaneously occurring variegated foliage plants are periclinal chimeras resulting from plastid mutations (Marcotrigiano 1997; Korn 2001). Further evidence on somatic mutations comes from the observation that clonally micropropagated plants originating from genetically homogeneous stocks often become phenotypically and/or genetically heterogeneous, a phenomenon known as "somaclonal variation" (Larkin and Scowcroft 1981). This may reflect either the induced expression of genetic differentiation that preexisted in somatic cells or the occurrence of somatic mutations in the cell lineages involved. By means of molecular genetic markers, the second possibility has been documented for dozens of species of herbaceous and woody plants of economic interest, including *Picea abies* (Fourré et al. 1997), *Hordeum vulgare* (Bregitzer et al. 1998), *Solanum tuberosum* (Albani and Wilkinson 1998), *Allium sativum* (Al-Zahim et al. 1999), *Picea glauca* (de Verno et al. 1999), *Populus tremuloides* (Rahman and Rajora 2001), *Chrysanthemum* varieties (Martín et al. 2002), *Actinidia deliciosa* (Palombi and Damiano 2002), *Humulus lupulus* (Patzak 2003), and *Pelargonium graveolens* (Ravindra et al. 2004). These and many other investigations conducted on plant material originating from the culture of plant cells, tissues, and organs provide unambiguous confirmation of the proclivity of certain plant genomes to experience high rates of somatic mutations under some circumstances. Nevertheless, they are of little value to draw inferences on the frequency of occurrence of somatic mutations in wild plants under natural conditions. Tissue-culture environment often causes a general disruption of cellular controls, leading to numerous genomic changes in the derived progeny (McClintock 1984; Phillips et al. 1994). In fact, it is precisely the general recognition that in vitro culture techniques used in plant micropropagation induce genetic variability, that ultimately explains the proliferation of investigations aimed at evaluating the genetic integrity of artificially obtained somaclones. For related reasons, the very high frequencies of spontaneous somatic mutations found by recent studies using mutation-monitoring systems based on transgenic plants (Kovalchuk et al. 2000) may represent gross overestimates of the somatic mutation frequencies actually experienced by wild plants. Increased somaclonal variation in transgenic plants versus nontransgenic plants that are similarly derived from tissue cultures has been documented for barley (*Hordeum vulgare*, Bregitzer et al. 1998), sugarcane (*Saccharum* hybrid, Arencibia et al. 1999), and rice (*Oryza sativa*, Labra et al. 2001), among others. This suggests that some of the biotechnological steps customarily used to integrate foreign genes into the genome of crop plants to produce transgenic clones may enhance the likelihood of somatic mutations over the already high background levels generated by tissue culturing itself.

Investigations of intraplant genotypic variation in wild plants using molecular markers continue to be relatively scarce in recent years, and only a few of these provide unambiguous evidence of somatic mutations and genetic mosaicism under natural, unmanipulated conditions. Possibly the most conclusive example of a genetic mosaic in a wild plant comes from the study by O'Connell and Ritland (2004) of western red cedar (*Thuja plicata*). By sampling haploid megagametophytes from two or

three different crown positions in each of 80 different trees, and geno-
typing them at eight microsatellite loci, they detected a single mutation
corresponding to a stepwise increase in one dinucleotide repeat occurring in the upper crown of a red cedar tree. From their data, mutation
rate was estimated as 6.3×10^{-4} per microsatellite locus per generation (or
3.1×10^{-4} per allele per generation), which falls within the range of 10^{-3}
to 10^{-4} mutations per generation reported for microsatellites in animals
(Ellegren 2000). Further recent evidence of genetic mosaicism in wild
plants comes from genetic analyses based on AFLP and RAPD markers conducted on colonies of *Taxus canadensis*, each presumably originating from a single genotype. These analyses demonstrated the existence
of small genotypic differences between pairs of physically connected ramets that could reasonably be explained as somatic mutations (Corradini
et al. 2002). In *Butomus umbellatus*, an aquatic plant forming large clones
by vegetative propagation of rhizomes, Fernando and Cass (1996) found
that a single RAPD fragment (out of 137 RAPD fragments scored) differentiated one ramet from all the other ramets in the clone, and interpreted this small difference as due to somatic mutation at that particular
RAPD locus.

Somatic mutations have also been customarily implied among the pos-
sible sources of observed genetic diversity (assessed with molecular markers) in clonal stands or vegetatively spread patches of wild trees (*Populus
tremuloides*, Tuskan et al. 1996; *Populus euphratica*, Rottenberg et al. 2000;
Ilex leucoclada, Torimaru et al. 2003), shrubs (*Haloragodendron lucasii*,
Sydes and Peakall 1998; *Vaccinium stamineum*, Kreher et al. 2000), herbs
(*Calamagrostis porteri*, Esselman et al. 1999), and reeds (*Arundo donax*,
Khudamrongsawat et al. 2004). Although some of these studies are particularly compelling in their support of the occurrence of somatic mutations (e.g., Tuskan et al. 1996; Sydes and Peakall 1998), none of them provided the critical evidence necessary to prove that somatic mutations are
actually the cause of within-patch genetic diversity, namely demonstrating genetic heterogeneity in sets of physically interconnected ramets, and
alternative explanations (e.g., multiple founders, seedling recruitment
during patch formation) could not be discarded. In one of these investigations, Kreher et al. (2000) concluded that it seemed unlikely that somatic
mutations accounted for the high genetic variation observed within *Vaccinium stamineum* patches, and that recruitment of genets from seed was
a more likely explanation. One further source of seemingly intraclonal
genetic heterogeneity may be the occurrence of "pseudomosaics," or

apparently unitary plants originating by close anatomical and even physiological fusion of several individuals initially originating from different zygotes. This phenomenon is considered in some detail later in this chapter.

PHENOTYPIC EXPRESSION OF GENETIC MOSAICS   Even if a plant individual is a proven genetic mosaic, this circumstance will not necessarily translate into enhanced phenotypic variability of its reiterated structures. Two main reasons account for this suggested decoupling between genetic and phenotypic mosaicism. The first has to do with the well-known mechanisms that limit the phenotypic expression of genetic variation. For example, genetic mosaics are not expected to have any discernible phenotypic consequences on the individual plants bearing them when mosaicism involves only genetic markers that are supposed to evolve neutrally (or nearly so), such as allozymes or microsatellites, or that do not necessarily have an immediate regulatory or structural functionality, such as AFLP or RAPD markers. Furthermore, even in cases where the mutation giving rise to the genetic mosaic is phenotypically consequential and produces large effects (e.g., chlorophyll deficiencies), it will still remain unexpressed in the phenotype if dominance or epistasis are involved. In culture-derived plants, somatic mutations underlying somaclonal variation involving major phenotypic changes are most frequently recessive alleles, which are inherited as single Mendelian factors (Phillips et al. 1994). Under these circumstances, any phenotypic consequences of the somatic mutations will be phenotypically expressed in part of the progeny but not in either the parent plant where it first originated or their clonally propagated descendants (e.g., Franks et al. 2002; Hocquigny et al. 2004). The second major reason that genetic mosaics do not always result in increased within-plant variability in the characteristics of reiterated organs such as leaves, flowers, and fruits is that, most often, genetic mosaics are expected to generate *within-organ* variation in cell characteristics, rather than *among-organ* phenotypic variance. This fact is related to certain anatomical and developmental peculiarities of plants that, although long known to plant morphologists (Tilney-Bassett 1963, 1969), do not seem to have been sufficiently appreciated in some earlier treatments of plant genetic mosaicism as a source of intraplant phenotypic heterogeneity. To provide some necessary background, the following paragraphs summarize relevant anatomical and developmental information presented in detail by Esau (1977), Poethig (1989), and Marcotrigiano (1997, 2000, 2001).

All the tissues in the aerial parts of a plant are ultimately derived from specialized regions at the tip of the shoot, termed the shoot apical meristem. The primary shoot meristem arises during embryogenesis and gives rise to the primary axis of the shoot. Lateral shoot meristems arise regularly at the base of each leaf and can differentiate as vegetative shoots (branches) or as flowering shoots. It is this continuously branching structure, in combination with the anatomical and developmental peculiarities of the meristem, that allows a mutant cell lineage to remain developmentally isolated from other apical lineages in the plant, leading eventually to a persistent genetic mosaic. In the angiosperms and some gymnosperms, independent cell lineages develop in shoot meristems, which normally appear as discrete cell layers. This stratification is the consequence of the orientation of cell division, which is almost always anticlinal (perpendicular to the surface) in the outer cell layers of the meristem (fig. 5.1). The peripheral layers in which cell divisions are anticlinal are called the *tunica* layers, while the body of subtending cells not displaying these restricted divisions is termed the *corpus*. This *tunica-corpus* organization of apical meristems is a characteristic feature of seed plants. Within each cell layer there are cells called *shoot apical initials*. Each of these cells by division gives rise to two cells, one of which remains in the meristem; the other is added to the plant body.

Following the appearance of a somatic mutation in one actively dividing

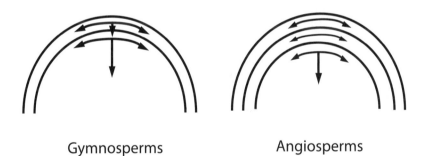

Gymnosperms   Angiosperms

FIG. 5.1 Schematic representation of the two patterns of cell division in the shoot apical meristems of seed plants. Gymnosperms have two fairly distinct cell lineages. The outermost cells in the shoot divide predominantly in an anticlinal plane, thereby giving rise to a single discrete cell layer. The subepidermal layer of cells divides in an irregular fashion. In angiosperms, two or three discrete cell layers surround a core of irregularly arranged cells. This tunica-corpus organization arises because cell division is restricted to an anticlinal plane near the surface of the meristem, but occurs in many different orientations in the center of the meristem. From Poethig 1989.

cell of a growing meristem, propagation within the plant may generate stable genetic mosaics whereby different plant parts have cells with different genotypes. There are two main types of genetic mosaics. In one, distinct clones of genetically different cells are scattered in the body of the plant and are not exclusively derived from the division of shoot apical initials. They are typically generated experimentally by grafting, radiation treatment, or mutagenic action of transposable genetic elements (Tilney-Bassett 1963, 1969; Chaparro et al. 1995; Marcotrigiano 1997). The most frequent type of mosaic, however, is the intrapical mosaic, in which shoot apical initials are genetically dissimilar and produce a persistent unique cell lineage whose origin can be traced back to particular shoot meristems. These mosaics are generally termed *chimeras* (Tilney-Bassett 1963). If a genetic change occurs in a shoot apical initial, its daughter cells can eventually populate an entire apical cell layer, which may subsequently remain developmentally independent from adjacent layers and will continue to give rise to the cells that form the body of the plant. This condition results in a periclinal chimera, a specific type of genetic mosaic in which one or more entire apical cell layers are genetically distinct from adjacent layer(s). Once a mutant lineage populates one of the three cell layers in a shoot, the cell-division pattern in the shoot ensures that this condition is perpetuated and is inherited by all the lateral shoots produced by the original chimeral meristem. Periclinal chimeras are therefore the most stable form of genetic mosaicism in angiosperms, where the apical meristem is stratified with cell layers remaining independent and where the axillary buds possess the same apical organization as the terminal bud from which they were generated.

Chimeras have been thoroughly studied because their anatomical peculiarities make them particularly well suited to unravel plant morphogenetic patterns when used in conjunction with cell-lineage analysis (Klekowski 1988a; Poethig 1989, 1997; Szymkowiak and Sussex 1996; Marcotrigiano 1997, 2001). The study of periclinal chimeras has demonstrated that the different cell layers that form tunica-corpus meristems generally contribute predictably to the formation of different parts of leaves and homologous organs (Marcotrigiano 2001). In *Datura stramonium*, for example, the outermost meristematic layer (L1) forms the epidermis of leaves and flowers. The initiation and development of the leaves, sepals, and petals depend primarily on the activity of the second layer (L2), while the initiation of the carpel and early development of its wall, septa, and placentas depend primarily on the activity of the innermost layer (L3)

(Satina and Blakeslee 1941, 1943). In *Nicotiana*, the L1 layer produces the epidermis of leaves and flowers, the L2 layer gives rise to the subepidermal layer of cells over the entire leaf lamina, as well as pollen and egg cells, and tissue derived from L3 occupies the central, internal portion of the leaf (Stewart and Burk 1970; Marcotrigiano 1986). In peach (*Prunus persica*), the L1 layer produces some parts of the ovary and the epidermal tissue of flowers, the L2 layer gives rise to internal tissues in the calyx, corolla tube, anther, and ovules, and the L3 layer contributes only to the central region of the calyx and corolla tubes and the ovary wall (Dermen and Stewart 1973). In flowers of *Arabidopsis thaliana*, the L1 contributes to the epidermis, the stigma, part of the transmitting tract, and the integument of the ovules, while the L2 and L3 contribute, to different degrees, to the mesophyll and other internal tissues (fig. 5.2). The "fate map" shown in figure 5.2 for *Arabidopsis* differs in some respects from those of other species. For instance, in *Arabidopsis* the perianth organs are composed almost exclusively of L1 and L2 cells. Species-specific variations presumably reflect differences in the developmental programs that give rise to the floral organs in each species (Jenik and Irish 2000).

Taken together, these and other studies suggest that even dominant somatic mutations may not be phenotypically expressed if they are sequestered in a meristematic cell layer where they are not developmentally expressed, given that some genes are expressed only in certain cell layers (e.g., Marcotrigiano 2000; Hocquigny et al. 2004). Furthermore, genetic mosaics of a chimeral nature are unlikely to generate, by themselves, within-plant variation in whole-organ features. If all organs in a genetic mosaic plant are identical with regard to the precise layering of the different genotypes represented in the chimera, then genetic mosaicism in itself will not enhance within-plant variance in whole-organ phenotypic features. This is illustrated by the results of an experimental investigation of artificial periclinal chimeras conducted by Szymkowiak and Sussex (1992). These authors generated periclinal chimeras between tomato (*Lycopersicon esculentum*) plants expressing the mutation *fasciated*, which increases the number of floral organs per whorl, and wild type tomato. When the *fasciated* mutation occupied the L3 meristematic layer, the mean number of carpels in flowers of chimeral plants (11.9 carpels/flowers) was three times that in flowers of the wild type (4.0 carpels/flowers), yet the relative variability of this magnitude remained unchanged (CV = 34% and 35% in chimeral and wild type, respectively).

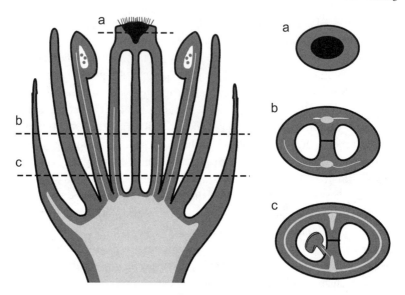

FIG. 5.2 The shoot apical meristem of *Arabidopsis thaliana*, like that of most angiosperms, consists of three cell layers (L1, L2, L3; shown in the graph as black, dark gray, and pale gray areas, respectively) that proliferate to give rise to the aerial organs of the plant. By labeling cells in each layer using a transposable element, Jenik and Irish (2000) mapped their contribution to the floral organs, as shown schematically in this figure reproduced from their study. *a, b,* and *c* are transverse sections of the pistil at the level of (*a*) the style, (*b*) the middle of the ovary, and (*c*) the base of the ovary.

Genetic mosaicism will most likely enhance within-plant variance in whole-organ phenotypic features only in situations where some sectors of the plant are chimeral for the genes concerned while others are not, or when the distribution of genetic variants among the different meristematic layers in the chimera varies among different parts of the plant. Within-plant variation in the distribution of genetic variants among meristematic layers can result, for example, from the invasion of the outer L1 layer by cells from the inner layers (a process known as displacement), or vice versa, that is, L1 cells invading the inner layer (replacement). The frequency of these spontaneous layer rearrangements in periclinal chimeras is, however, very low (Stewart and Burk 1970; Marcotrigiano 1986; D'Amato 1997). This is illustrated by the results of Hocquigny et al.'s study of one chimeral grapevine cultivar (2004) where L1 cells have a gray-berry genotype and L2 cells a white-berry genotype. Despite this, the vast majority of chimeral plants produce homogeneously gray berries. White-berry clusters, single white berries, or white-berry sectors are only sporadically observed, and this happens only if anomalous periclinal divisions of L2 cells and their invasion of the L1 layer lead to the expression of the white-berry genotype, which under conditions of normal cell divisions remains permanently locked and unexpressed in L2. Working also with chimeral grapevine cultivars involving hairy (L1 layer) and hairless (L2 layer) leaf genotypes, Franks et al. (2002) found that only on rare occasions was a leaf observed on some plants that displayed a sectored appearance, with part of the leaf lacking the normal hairy phenotype as a result of the L2 cell layer replacing the L1 cell layer. In artificial interspecific chimeras between *Nicotiana glauca* and *Nicotiana tabacum* synthesized by Marcotrigiano (1986), both quantitative (e.g., corolla, calyx, style, and pedicel length) and qualitative (corolla and calyx color, calyx pubescence) floral traits were closely dependent on the particular genotypic combination of chimeral layers, each histogenic arrangement resulting in a unique floral phenotype (Marcotrigiano 1986, fig. 3). Nevertheless, only very rarely did any flower appear in a plant that departed from the particular floral phenotype expected from the plant's specific chimeral arrangement, which generally reflected occasional cell displacements from the L2 to the L1 cell layer.

External agents may increase the probability of within-plant chimeral layer rearrangements, thus inducing the "unlocking" of genotypes that lay hidden in a chimeral layer and enhancing within-plant phenotypic variability. This was experimentally demonstrated by Marcotrigiano (2000) in plants of a variegated form of *Nicotiana sylvestris*. Control plants were a homogeneous green-white-green periclinal chimera (L1-L2-L3 meristematic layers, respectively) with all leaves similarly variegated. When the terminal and primary axillary meristems were removed to simulate herbivory, secondary shoots became phenotypically different from control or primary axillary shoots (e.g., through the appearance of entirely white leaves), as a consequence of repositioning of cell layers in the secondary meristems induced by simulated herbivory. Within-plant variation in degree of stem spinescence has a similar origin in some thornless horticultural varieties of blackberries (*Rubus laciniatus*). These are periclinal chimeras in which the epidermis (originating from the L1 layer) has mutated to a thornless phenotype while the internal portions of the plant (originating from L2 and L3 layers) possess the wild thorny genotype. If shoots emerge from a inner tissue layer, as often happens with adventitious shoots, then their epidermis is genetically thorny, prickles develop, and within-plant heterogeneity in spinescence thus arises (McPheeters and Skirvin 1983). In all the preceding examples, therefore, it was not genetic mosaicism in itself that made for within-plant variability in stem, leaf, flower, or fruit features, but rather within-plant heterogeneity of the chimeral meristematic structure.

CONCLUSION: GENETIC MOSAICS AS RARE SOURCES OF SUBINDIVIDUAL VARI-
ATION   There seems to be little question that somatic mutations and ensuing genetic mosaics occur in both wild and cultivated plants, and that they
are fairly frequent in crop plants and ornamentals, particularly when the
plants are subjected to artificial conditions that contribute to disruption
of mechanisms of cellular control. Information on their occurrence in wild
plants, in contrast, is still very scanty despite the recent improvements in
detection tools and the increased research efforts aimed to detect it. It is
important to emphasize that investigations that have looked for genetic
mosaicism in wild plants have generally been conducted on species that are
long-lived, profusely branched, or propagating vegetatively to form large
colonies of hundreds or even thousands of ramets presumably originating
from a single genotype. In theory, these life-history traits should favor the
appearance and accumulation of somatic mutations, as well as the long-
term persistence of genetic mosaicism (Whitham and Slobodchikoff 1981;
Gill 1986; Klekowski 1988a; Gill et al. 1995), and some empirical evidence
supports these expectations (Klekowski 1988b; Klekowski and Godfrey
1989; Caetano-Anollés 1999). Klekowski and Godfrey (1989), for example, found that mutation rates in long-lived mangroves are considerably higher than comparable rates for plants with shorter life spans. In a
comparison of two fern species, Klekowski (1988b) likewise found that
the frequency of mutant ramets was considerably higher in the long-lived
clones of the species reproducing almost solely vegetatively via stolons
than in the species where sexual reproduction, and genets of sexual origin, were frequent. Very high genomewide mutation rates were reported
by Caetano-Anollés (1999) for heavily cloning, vegetatively propagated
Bermuda grass (*Cynodon*). The extreme scarcity of unambiguous cases
of genetic mosaics in wild plants in spite of biased research efforts favoring species where they are expected to be most frequent (i.e., long-lived,
clonal, profusely branched), therefore suggests that they are far less frequent among wild plants than among cultivated plants.

   Whitham and Slobodchikoff (1981) emphasized that the bulk of evidence for somatic mutations in plants came from the horticultural and
agronomy literature. They acknowledged the potential problems involved

in dealing with cultivated plants, although they did "remind the reader that Darwin's *Origin of Species* relied heavily on such literature" (288). While there are effectively no reasons to suspect that cultivated and wild plants differ in any essential way with regard to the biological details of the mechanisms taking place at the suborganismal level (e.g., arrangement of meristemal layers, cell-division patterns), it is not unlikely that the two groups differ widely in the *frequency* with which certain biological phenomena are expected to occur. It seems likely, for example, that proneness to somatic mutations and stable genetic mosaics represented an advantageous trait for the success of the domestication process through facilitating the vegetative propagation of desirable qualities, which is consistent with the observation that about 57% of the 8,800 plant varieties cultivated in Europe in 1899 probably were originally derived from bud sports propagated vegetatively (Whitham and Slobodchikoff 1981). In view of this, it is not unreasonable to suggest that currently cultivated plants probably represent a heavily biased subsample of the plant world, characterized by a particularly high incidence of genetic mosaicism. Furthermore, continuous vegetative propagation of many cultivated plants for centuries or even millennia, uninterrupted by meioses, may have accentuated the frequency and persistence of stable mosaics, and extended the natural lifetime of chimeral genotypes, as suggested by Franks et al. (2002) for some grapevine cultivars (see also Riaz et al. 2002). As noted by Tilney-Bassett (1963, 281), "the large number of chimeras in cultivation today is a tribute to the vigilance of gardeners by whom they have been found and propagated, partly because of their interest as rare sports, and partly because of their ornamental value."

The likely rarity of genetic mosaics among wild plants, along with their presumably infrequent expression as phenotypic mosaics, suggest that genetic mosaicism is, at best, a very minor source of within-plant variation in phenotypic traits of reiterated structures in wild plants. Some quantitative evidence presented in chapter 3 is consistent with this interpretation. Given that, as noted earlier, somatic mutations are expected to accumulate in large, long-lived plants where the cumulative number of cell division cycles is very large (Klekowski 1988a), then if genetic mosaicism were a major source of within-plant variability in reiterated structures one would predict such variability to increase from small, short-lived plants to large, long-lived ones. This prediction may be tested by comparing the within-plant variability estimates for flower and fruit traits compiled in chapter 3. Woody and herbaceous taxa are similar ($P = 0.37$, Wilcoxon two-sample test) with regard to the magnitude of within-plant variability in floral traits, as measured with CV_{within} (table 3.2). Likewise, no differences exist ($P = 0.47$) between trees and shrubs in mean CV_{within} of fruit traits (table 3.3). In that sample of species, therefore, there is no suggestion of greater subindividual variability among comparatively larger and longer-lived plants.

Seed Crops

Seed crops represent a particularly widespread example of a naturally occurring genetic mosaic that, although unrelated in its origin to the propagation of somatic mutations, often leads to genetically based subindividual variation in the phenotypic attributes of reiterated structures. Studies using allozyme and, more recently, DNA-based genetic markers have revealed that multiple male parents are ordinarily involved in the fertilization of the ovules of an individual plant or even a single ovary (reviewed in Bernasconi 2004). Patterns of multiple paternity may result from pollen carryover or from multiple pollinator visits (Ellstrand 1984; Campbell 1998). In one study population of *Raphanus sativus*, for instance, multiple paternity occurred in all individuals and 85% of all fruits (Ellstrand 1984). In *Ipomopsis aggregata*, a minimum of 68% of fruits with two or more seeds were multiply sired (Campbell 1998), and 60% of fruits of *Glycine argyrea* were multiply sired (Brown et al. 1986). Whenever multiple paternity occurs, the developing seed crop borne by an individual plant represents a genuine, albeit ephemeral, genetic mosaic in respect to the genetic makeup of the seeds' embryos and endosperms. Since some seed traits are influenced by the paternally contributed genome of the embryo (Bernasconi 2004), multiple paternity of the seeds in a single crop could ultimately be responsible for some within-plant variance in seed traits. Multiple paternity may involve either a mixture of cross pollen donors or, in the case of self-compatible species, a mixture of self and cross pollen. Although genetic heterogeneity in seed traits is expected in both situations, the magnitude of its consequences in terms of within-plant variability in seed traits is expected to differ, as discussed below.

Multiple paternity exclusively involving a mixture of cross pollen donors, but not self pollen (e.g., in strictly self-incompatible species), most often generates a moderate to low amount of within-plant heterogeneity in seed characteristics. In *Lychnis flos-cuculi*, for example, 7% of the variation in seed mass within maternal families could be explained by paternal genotype (Biere 1991a). Experimental data obtained by Manasse and Stanton (1991) for the tropical perennial herb *Crinum erubescens* also support a role of variation in paternal genotypes in generating within-plant variation in seed mass. In their study, extreme seed-size variation in fruits occurred when mating pairs were inbred, either from selfing or biparental inbreeding, and relatively uniform seeds of intermediate size were obtained when pollen from several pollen donors was applied simultaneously to a flower. In this latter case, seed-mass variability in a fruit increased with increasing number of pollen donors involved in the fertilization of the ovules. The CV_{within} of seed mass increased regularly from the fruits pollinated using cross pollen from a single donor (CV = 51%) to fruits resulting from pollinations using pollen from four (CV = 59%) and ten (CV = 70%) donors. In *Rhamnus alpinus*, a temperate shrub, results of experimental pollinations involving single and multiple pollen donors also demonstrate that multiple paternity of the seeds of single fruits results in enhanced within-fruit variability in seed mass (Bañuelos and Obeso 2003; see also Marshall 1991). This study also revealed that the overall distribution of population-wide variance in seed mass was modified by experimentally manipulating the number of male parents siring the seed crop. In the single-donor treatment group, within-plant variation accounted for 42% of total seed-mass variance, and this magnitude rose to 66% in the multiple-donor treatment group.

The results from these experimental studies do not allow for generalizing about the *quantitative* importance of multiplicity of crossed paternal parents as a source of within-plant variation in seed traits under conditions of natural pollination. Its importance is expected to be relatively small, at least if one considers that controlled diallel crosses performed on a variety of species have generally shown that the quantitative effect of the paternal genotype on seed traits is quite modest in comparison to that of the maternal genotype and maternal conditions. For seed mass, one of the traits investigated most frequently in diallel crossing schemes, the effect of the paternal genotype has consistently been shown to be statistically nonsignificant or, if significant, quantitatively negligible. This has been shown, for example, in *Anthoxanthum odoratum* (Antonovics and Schmitt 1986), *Lychnis flos-cuculi* (Biere 1991a), *Nemophila menziesii* (Platenkamp and Shaw 1993), *Aquilegia coerulea* and *A. cazorlensis* (Montalvo and Shaw 1994; Castellanos et al. 2008), *Hydrophyllum appendiculatum* (Wolfe 1995), and *Asclepias incarnata* (Lipow and Wyatt 1999). For *Lupinus texensis*, however, Helenurm and Schaal (1996) reported a paternal component of seed-mass variance of 18%, thus showing that multiple paternity of the seed crop may occasionally become a major source of within-plant variance in seed mass.

Genetic heterogeneity of individual seed crops is expected to be a quantitatively more important source of within-plant variance in seed traits when multiple paternity involves a mixture of selfed and outcrossed pollinations. This is the ordinary situation in the numerous self-compatible species with mixed mating systems, whose seed crops are ordinarily made up of a variable mixture of selfed and outcrossed seeds (Barrett 1998). Even in species that regularly produce a significant fraction of selfed progeny, inbreeding depression is frequently expressed at the seed stage in the form, among other symptoms, of reduced size of selfed seeds in relation to outcrossed ones. Depending on the species, the ratio of selfed to outcrossed seed mass ranges between 0.37 and 0.98 in gymnosperms, and between 0.59 and 0.93 in angiosperms (Charlesworth and Charlesworth 1987). The coexistence of different-sized selfed and outcrossed seeds in the same seed crop that is typical of plants with mixed mating systems will therefore contribute to within-plant heterogeneity in seed size in these species. In addition, this component of within-plant heterogeneity in seed size may be spatially patterned if selfed and outcrossed progeny are not identically distributed within the parent plant, as may occur if outcrossing rates (i.e., the proportion of seeds originating from cross fertilization) differ among different plant sectors or even at the reduced spatial scale of individual fruits. Few studies have explicitly examined the possibility of within-plant variation in outcrossing rates, and these provide contrasting results. O'Connell et al. (2004) found no heterogeneity in outcrossing rate within the relatively large crowns of western red cedar (*Thuja plicata*). In contrast, Carromero and Hamrick (2005) found outcrossing rate to vary predictably among seeds located at different positions along the linear inflorescence of the biennial herb *Verbascum thapsus*. Working at a very small scale of within-plant variation, Horovitz et al. (1976) found that seeds at different locations within single pods of *Lupinus nanus* differed predictably in their average outcrossing rates. In the legumes of this species, mean outcrossing rate increases from 30% of seeds located at the most basal positions of the pod up to 65% in the most distal positions. In the heterocarpic *Crepis sancta*, the outcrossing rate of the larger, non-dispersing achenes produced at the periphery of the capitulum (91%) is greater than that of the smaller, dispersing achenes produced in the central parts (79%; Cheptou et al. 2001). The importance of the component of intraplant variance in seed size due to the mixture of selfed and crossed fertilizations in the same crop is expected to vary greatly among species and populations within species, depending on the extent of inbreeding depression in seed mass and the relative proportions of selfed and out-crossed seeds (i.e., the mating system).

Pseudomosaics

One or more genotypes originating from distinct zygotes can partly fuse or grow so intimately associated that they come to look superficially as if they were one and the same unitary plant. These close aggregates of different individuals exhibiting the appearance of unitary plants are not genetic mosaics in a strict sense, because their genetically distinct parts actually originate from different zygotes. For this reason, I will designate them here as pseudomosaics. Their prevalence in nature is still poorly known except for the few well-studied cases noted below, and their peculiar origin clearly sets them apart from most cases of intraindividual variability considered in this book. Nevertheless, they are probably a more frequent source of intra-"plant" variability than hitherto recognized, which justifies giving them some brief consideration. It is also useful to consider these pseudomosaics and some of their most obvious ecological correlates because, as noted earlier, some putative instances of sensu stricto genetic mosaics attributed to somatic mutations might actually reflect the coalescence of separate genetic individuals early in their ontogeny, which, in the long run, may eventually make them appear to be a unitary plant or clone.

Several mechanisms can give rise to seemingly unitary plants that are actually made up of different individuals. One of these is zygotic polyembryony of seeds, whereby multiple embryos are formed within a single ovule and enclosed within a single seed coat (Klekowski 1988a). This condition, which occurs more frequently in gymnosperms than in angiosperms (Chamberlain 1935; Klekowski 1988a), might in theory produce "plants" that are made up of more than one genetic individual, since all seedlings from the same polyembryonic seed are frequently viable. Nevertheless, they are not expected to be frequent in nature, since supernumerary seedlings generally grow poorly and have very high mortality (Martínez-Gómez and Gradziel 2003).

One would expect pseudomosaics to originate most frequently from spatial proximity of independent seeds during their germination and early seedling establishment. If several seeds end up on the same spot after dispersal and germinate very close to each other, and eventually all or most of them reach adulthood, the resulting adult plants may have extensively fused trunks that appear to be a single individual. In addition, this spatial proximity enhances the opportunities for natural root grafts to form, which allows for a certain level of physiological interconnectedness among the genotypes involved, in the form of water and nutrient sharing (Graham and Bormann 1966). Schuster and Mitton (1991) provide a photographic illustration of extensive xylem and phloem grafting between the main stems of two genetically distinct members of a cluster of *Pinus flexilis*. Plants whose seeds are ordinarily dispersed in multi-seeded packages or those that, although dispersed singly, end up in dense postdispersal aggregations, are therefore the most likely candidates to originate pseudomosaics. These include fleshy-fruited plants whose seeds are dispersed by frugivorous birds and mammals that produce extremely dense local aggregations of seeds in their feces, and those dispersed by scatter-hoarding birds and mammals that make multiseeded caches. Well-documented examples of pseudomosaics are consistent with this prediction. In several species of vertebrate-dispersed Neotropical strangler figs (*Ficus* spp.), seedlings and/or juveniles frequently coalesce and eventually produce what looks like a single "tree" whose parts are genetically dissimilar (Thomson et al. 1991; see also Thomson et al. 1997 for a downward revision of earlier figures on the prevalence of pseudomosaics in these species). In the California chaparral, seeds of *Prunus ilicifolia* are commonly dispersed by coyotes (*Canis latrans*), whose feces may contain more than 60 seeds of that species. Seedling survival is rather high in these dense aggregations, and spontaneous root grafts between emerging seedlings and juveniles are quite frequent, which eventually gives rise to a high frequency of physiognomic shrubs that are actually made up of several genetically distinct individuals (Bullock 1981). As a rule, nonflying mammals that feed on fleshy fruits (berries, drupes) disperse seeds in the form of extremely dense local aggregations (Herrera 1989a), so patterns similar to those reported by Bullock for *Prunus ilicifolia* are probably much more common in nature than ordinarily acknowledged.

The most detailed empirical information to date on pseudomosaic plants comes from a series of investigations by Tomback and associates on species of North American and European pines that have seeds dispersed by nutcrackers (*Nucifraga* spp.). These birds harvest ripe seeds and bury them in caches for later use. Unretrieved seeds may germinate and produce either single- or multitrunk trees (Tomback 1982; Linhart and Tomback 1985). Multitrunk trees may be of two growth forms, namely single-genotype multistemmed trees, or clusters of genetically distinct trees resulting from the fusion of two or more independent plants in the same nutcracker cache. This latter form is the most frequently reported in the literature. Frequencies of multiple genotypes in multitrunk pine trees are 18–81% in *Pinus flexilis* (Linhart and Tomback 1985; Schuster and Mitton 1991; Carsey and Tomback 1994), 66–83% in *P. albicaulis* (Linhart and Tomback 1985; Furnier et al. 1987), and 70% in *P. cembra* (Tomback et al. 1993). Multiple genotypes also sometimes occur with nonnegligible frequency in multitrunk plants of wind-dispersed pines (*P. ponderosa*, *P. aristata*, and *P. contorta*), but in these cases they possibly originate because nutcrackers and seed-storing rodents make multiseed caches of wind-dispersed seeds (Torick et al. 1996).

Some evidence suggests that, as expected, within-"plant" variation exhibited by genetic pseudomosaics tends to be slightly larger than that of single-genotype conspecifics. Thomson et al. (1997) studied the flowering phenology at the level of individual branches of strangler figs to determine if there was greater flowering asynchrony among branches of pseudomosaic trees formed by spontaneous grafting of genetically different individuals. They found that some pseudomosaic trees were internally more variable with regard to flowering phenology than genetically homogeneous ones. Multiple-genotype tree clusters are also internally more variable than single-genotype tree clusters in bird-dispersed pines (Diana Tomback, personal communication). Each genotype in a cluster may show differences in traits related to radial growth, canopy development, pollen-cone production, and seed-cone production and presentation (Feldman et al. 1999; Diana Tomback, personal communication). Although part of this variation is probably due to genetic differences, environmental factors and competition between genotypes are also likely to play a role in enhancing intra-"plant" variability in the case of multiple-genotype clusters. It must be noted that the average genetic relatedness among the different genotypes involved in pine tree clusters is significantly greater than that of randomly chosen plant pairs in the population (Schuster and Mitton 1991; Carsey and Tomback 1994). This most likely stems from the combination of coordinated dispersal by nutcrackers of half-sib progenies from the same mother plant, and the increased likelihood of root and stem grafting in genetically related individuals (Graham and Bormann 1966; Schuster and Mitton 1991). Under these circumstances, genetic effects on the within-"plant" phenotypic variability of these pseudomosaics, even if they occur, are expected to be relatively modest.

Organ-Level Developmental Plasticity

Genotypes are not necessarily strictly mapped into phenotypes. Phenotypic constancy is not a cue for the absence of genetic variation, nor does phenotypic variation inevitably require genetic differences. In canalization, different sets of genes may consistently produce a single phenotype even if exposed to variable environmental conditions or genetic backgrounds (Flatt 2005). Likewise, in phenotypic plasticity, phenotypic variation does not necessarily reflect the existence of underlying genetic variation (Pigliucci 2001). Canalization and phenotypic plasticity can be seen as two aspects of the same phenomenon (Flatt 2005), and may be intuitively envisaged as representing opposite extremes on a hypothetical continuum in the nature of the correspondence between genotypic and phenotypic spaces (or "genotype-phenotype map"; Wagner and Altenberg 1996). Because I am concerned in this chapter with the causes of phenotypic variation, I focus on plasticity alone, leaving aside canalization for the moment (I return to it in chapter 10). This section briefly introduces organ-level phenotypic plasticity, the main class of remote causes of within-plant variation in phenotypic characteristics of reiterated organ traits. A number of excellent reviews covering the voluminous literature on phenotypic plasticity have appeared over the last two decades (Schlichting 1986; Schlichting and Pigliucci 1998; Wells and Pigliucci 2000; Agrawal 2001; Pigliucci 2001; West-Eberhard 2003; DeWitt and Scheiner 2004a). I provide here only a succinct treatment of the subject, focusing on the facet of plasticity that involves reiterated structures and takes place at the subindividual level. This section sets the stage for chapter 6, where organismal mechanisms of within-plant variation are considered in detail. As discussed there, much within-plant variation in organ traits is the outcome of ontogenetic contingency acting in combination with organ-level phenotypic plasticity and associated developmental reaction norms.

Definitions and Background

The long-standing consensus on the core meaning of the phenotypic plasticity concept is shown by the following sample of definitions skimmed off some authoritative sources and presented in chronological order. Pheno-typic plasticity has been defined as "shown by a genotype when its expression is able to be altered by environmental influences" (Bradshaw 1965); "variation in phenotypic expression of a genotype that occurs in response to particular environmental conditions and which enhances the capacity of the individual to survive and reproduce under those conditions" (Sultan 1987); "the ability of a single genotype to produce more than one alternative form of morphology, physiological state, and/or behavior in response to environmental conditions" (West-Eberhard 1989); "a measure of how different the phenotypes produced [by a given genotype] in distinct environments are from each other" (Pigliucci 1996); the fact whereby "the same set of genes can yield different phenotypic outcomes when exposed to distinct environmental conditions" (Pigliucci 1998); the fundamental property of an organism whereby "a single genotype can produce different phenotypes in different environments" (Sultan 2000); "the ability of an organism to express different phenotypes depending on the biotic or abiotic environment" (Agrawal 2001); "the property of a given phenotype to produce different phenotypes in response to distinct environmental conditions" (Pigliucci 2001); and "the environmentally sensitive production of alternative phenotypes by given genotypes" (DeWitt and Scheiner 2004b). Leaving aside some subtle differences in ancillary elements of the concept that have shifted with the years (e.g., the requirement of adaptive value imposed by Sultan [1987], later to be relaxed by Sultan [2000]), the consistency of the definitions of phenotypic plasticity over four decades denotes a persistent agreement on the three core elements of its meaning: phenotypic plasticity refers to the capacity of (1) a genotype to produce a range of (2) phenotypic variants in response to (3) environmental variation.

I strictly adhere to this mainstream definition. As a shortcut to overcome the practical difficulties of obtaining sets of genetically identical individuals (Valladares 2003, 449), some authors have relaxed or dropped altogether the requirement of genotypic uniformity in their practical applications of the "phenotypic plasticity" concept, and have estimated phenotypic plasticity by common-garden measurements of the total phenotypic variation exhibited by conspecific individuals from the same population of unknown genetic relatedness (e.g., Valladares et al. 2000, 2002; Balaguer et al. 2001). This procedure conflates phenotypic variation genuinely due to genotype-specific phenotypic plasticity with variation due to genotypic heterogeneity and genotype × environment interactions. Interpretations based on "phenotypic plasticity" estimates obtained using such an approach should thus be taken with considerable caution.

It has long been acknowledged that, in plants, plasticity may be expressed in whole-plant traits, such as growth habit, size, and fecundity, as well as in the traits of reiterated structures such as leaves, flowers, and fruits (Goodspeed and Clausen 1915; Salisbury 1942; Schmalhausen 1949; Stebbins 1950; Bradshaw 1965; Sultan 1987). Provided that all modules of an individual plant are genotypically identical, within-plant variation in phenotypic traits of reiterated structures will exemplify genuine phenotypic plasticity that can most easily be observed without recourse to experimentation. In fact, some of the classic examples of phenotypic plasticity involve subindividual variation in organ traits, such as the discontinuous variation in leaf characteristics (heterophylly) shown by certain aquatic plants (Bradshaw 1965; Sultan 1987). These classic examples notwithstanding, phenotypic plasticity studies have mainly focused on the effects of environmental variation on whole-plant traits related to the number of modules (e.g., size, fecundity), rather than on the phenotypic consequences of environmental variation for the traits of individual modules. Among the latter, only leaf traits have been considered with some frequency in plasticity studies (Sultan 2000; Pigliucci 2001). Investigations of phenotypic plasticity in flowers, fruits, and seeds are remarkably scarce. These biases are possibly the outcome of the obvious, extensive variability often exhibited by leaves, but also of the widespread prejudice that phenotypic traits of reproductive organs (flowers, fruits, seeds) are very robust to alterations in the environment (Stebbins 1950; Trewavas 1986, table 1). This view was synthesized by Stebbins (1950, 73): "the relative plasticity of certain vegetative characteristics and constancy of reproductive ones has long been realized by plant systematists; upon it is based the greater emphasis in classification on reproductive characteristics as compared to vegetative characteristics."

Table 5.1 presents a selection of examples where organ-level phenotypic plasticity in response to variation in environmental conditions has been demonstrated either experimentally or observationally by means of phenotype-environment correlations under controlled conditions. The abundant information on leaves faithfully reflects the bias toward these organs in phenotypic plasticity studies. A broad variety of morphological, structural, and functional leaf traits respond to variation in aspects of the abiotic environment including light level, light quality, photoperiod, and temperature. Examples involving flower and fruit traits, although scarcer, clearly counter the view that these organs' phenotypes are particularly robust to environmental variation. Variations in temperature, light and nutrient levels, and soil moisture induce measurable organ-level phenotypic responses that affect, among other things, the number of floral parts; the size of the corolla; the production and composition of nectar; the size, color, and form of fruits; and the size of seeds. The examples gathered in table 5.1 represent instances of "developmental phenotypic plasticity," in which plasticity stems from "the fact that there are some windows of time during ontogeny when the organism is prone to alter its developmental trajectory in response to the external environment" (Pigliucci 1998). In the particular case of organ-level phenotypic traits, it is the individual organs or modules, rather than the whole organism as an aggregate of modules, that are "prone to alter" their development in an autonomous way according to some programmed developmental response to environmental change. I thus refer to these instances as "developmental organ-level phenotypic plasticity," and to the functions linking organ-level phenotypic traits to environmental factors as "organ-level developmental reaction norms," discussed in the next section.

Organ-Level Developmental Reaction Norms

Every genotype is characterized by its own specific phenotypic modifications in response to different environments, or "norm of reaction" (Schmalhausen 1949; see Sarkar 2004 for a historical review of the concept). Experimental studies of plant phenotypic plasticity usually use sets of genetically identical individuals (e.g., clonal copies obtained by propagating root or stem cuts) or families of controlled parentage (e.g., full-sib progenies obtained by controlled pollinations). Such sets of replicate individuals sharing a similar genetic background are split into groups, each of which is exposed to different environmental conditions, and the phenotypic traits of individuals in each experimental group are measured after some time of exposure to experimental environments. Results of these studies are typically depicted as reaction-norm plots, in which the average phenotypic characteristics of the different treatment groups are plotted against the environmental axis subject to experimental variation (DeWitt and Scheiner 2004b). This empirical description of a reaction norm can be modeled as a function, with the value of the environmental variable as

TABLE 5.1 **Selected examples of organ-level phenotypic plasticity, in which variation in some external environmental factor induces predictable phenotypic responses at the level of individual organs.**

Organ—type of phenotypic response	Species	Environmental factor	Organ trait responding to environmental variation	Reference
Leaves—discontinuous	Callitriche heterophylla	Submergence in water	Form, ovate vs. linear	Deschamp and Cooke 1984
	Cyamopsis tetragonoloba	Day length and temperature	Form, simple vs. trifoliate	Sparks and Postlethwait 1967
	Paradrymonia ciliosa	Light level	Anisophylly, equality vs. disparity of leaves in a node	Dengler and Sánchez-Burgos 1988
Leaves—continuous	Arabidopsis thaliana	Light and nutrient levels	Weight	Pigliucci et al. 1995
	Betula pendula, Corylus avellana, Lonicera xylosteum	Light level	Specific leaf area	Kull and Niinemets 1993
	Chenopodium album	Light quality (red to far-red ratio)	Specific leaf area	Causin and Wulff 2003
	Eichhornia crassipes	Light quality (red to far-red ratio)	Petiole length and degree of inflation	Richards and Lee 1986
	Fraxinus excelsior, Populus tremula, Corylus avellana, Tilia cordata, Fagus sylvatica Ilex aquifolium, Rhododendron ponticum	Light level	Specific leaf area	Niinemets and Kull 1998
		Light level	Thickness, specific leaf area, chlorophyll content	Niinemets et al. 2003
	Littorella uniflora	Submergence in water	Length and width	Robe and Griffiths 1998
	Nothofagus fusca	Light level	Size	Hollinger 1996
	Phaseolus vulgaris	Light level	Nitrogen content	Pons and Bergkotte 1996
	Persicaria maculosa, P. hydropiper, Polygonum caespitosum	Light level	Leaf area	Sultan 2003
	Proserpinaca intermedia	Temperature, photoperiod	Form, degree of division	Kane and Albert 1982
	Proserpinaca palustris	Submergence in water	Form, degree of division	Schmidt and Millington 1968
	Ranunculus nanus	Light level, temperature	Form and size	Menadue and Crowden 1990
	Veronica anagallis-aquatica	Water velocity	Area, specific leaf area, total chlorophyll	Torres Boeger and Poulson 2003

	Species	Environmental inducer	Responsive trait	Reference
Flowers—continuous	Campanula rapunculoides	Temperature	Corolla size, nectar quality	Vogler et al. 1999
	Cultivated Lycopersicon esculentum	Temperature	Number of petals, stamens, carpels	Sawhney 1983
	Ipomopsis longiflora	Temperature, water stress	Corolla size, nectar production	Villarreal and Freeman 1990
	Ipomoea trichocarpa	Temperature	Corolla length	Murcia 1990
	Lythrum salicaria	Soil moisture	Calyx length, stigma-anther separation, pistil length	Mal and Lovett-Doust 2005
	Nicotiana tabacum	Soil quality	Corolla length and spread	Goodspeed and Clausen 1915
	Sandersonia aurantiaca	Temperature	Corolla size and shape, pedicel length	Catley et al. 2002
	Silene coeli-rosa	Temperature	Number of sepals, petals, stamens	Lyndon 1979
	Spergularia marina	Nutrient level	Petal size and number	Delesalle and Mazer 1996
	Trifolium repens	Temperature	Nectar secretion rate, sugar concentration, sugar composition	Jakobsen and Kristjánsson 1994
Fruits—continuous	Cultivated Malus domestica	Light level	Size, color	Jackson et al. 1971; Tustin et al. 1988
	Cultivated Capsicum annuum	Temperature	Form	Aloni et al. 1999
	Cultivated Prunus salicina	Light level	Size	Murray et al. 2005
Seeds—continuous	Polygonum persicaria	Light level	Size	Sultan 1996

Note: Examples were chosen to illustrate the broad variety of both environmental inducers and responsive traits.

argument and the phenotypic value of the trait as function value (de Jong 1990). In this view, reaction norms describe the shape of the functional relationship that, for a given genotype, links phenotypic trait values with environmental variables, and provide a synthetic assessment of the nature and scope of that genotype's programmed developmental responses to environmental variation.

As noted above, plant traits considered in phenotypic plasticity studies may refer to either the plant as a whole (e.g., height, biomass, fecundity) or individual organs (e.g., leaf or flower size). Idealized representations of reaction norms for continuously and discontinuously varying organ traits are shown in figure 5.3. Due to practical limitations, the vast majority of reaction norms reported in the literature are based on phenotypic data obtained from only two or three distinct environments, and the shape of the phenotype-environment relationship over the whole range of conditions faced by a typical individual of a species is accurately known in few cases. These studies suggest, however, that plant phenotypic traits are often linked to variation in environmental factors by fairly smooth curves more or less akin to those shown in figure 5.3 (Kane and Albert 1982; Pigliucci et al. 1995; Catley et al. 2002; Mal and Lovett-Doust 2005).

Corresponding to the relative scarcity of studies focusing on the phenotypic plasticity of organ traits noted above, reaction norms have been established less frequently for organ-level traits than for plant-level traits. Despite this, however, nearly all studies listed in table 5.1 demonstrate functional relationships linking leaf, flower, fruit, or seed traits to variation in environmental factors. As an example, figure 5.4 shows the variation of average corolla length in *Ipomopsis longiflora* in response to variation in ambient temperature, for three different levels of water stress. This graph also illustrates that the phenotype-environment relationship can be highly context-dependent, depending on the particular values taken by other environmental factors. In the example in figure 5.4, the shape of the corolla length–temperature relationships depends on water-stress levels. Interaction (i.e., nonadditive) effects of environmental factors on reaction norms are probably widespread, and have also been shown, among others, by Kane and Albert (1982) for temperature and photoperiod on leaf shape in *Proserpinaca intermedia*, Menadue and Crowden (1990) for temperature and light level on leaf form in *Ranunculus nanus*, and Catley et al. (2002) for temperature and irradiance on corolla size and shape in *Sandersonia aurantiaca*.

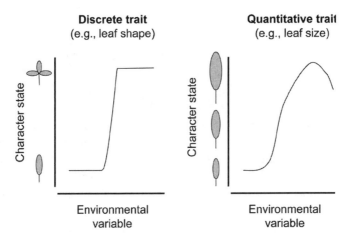

FIG. 5.3 Idealized organ-level reaction norms, depicting the shape of developmentally programmed phenotypic responses of organ traits to variation in environmental variables. Both discontinuous (e.g., heterophylly) and continuous (e.g., organ size) organ-level variation can be framed in terms of organ- and trait-specific reaction norms. In practice, the distinction between continuous and discontinuous responses is sometimes rather tenuous, as exemplified by the continuum of leaf shapes exhibited by certain heterophyllous plants (Wells and Pigliucci 2000).

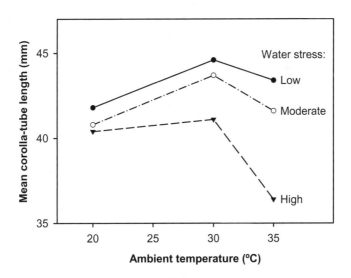

FIG. 5.4 Relationship between the mean corolla-tube length of *Ipomopsis longiflora* and ambient temperature under controlled conditions, evaluated at three levels of water stress. Drawn from data in Villarreal and Freeman 1990.

Phenotypic Plasticity and the Plant's Internal Milieu

At a macroscopic level, organ-level phenotypic plasticity implicates, by definition, the developmental responses of organs to variation in factors of the plant's *external* environment such as light, temperature, photoperiod, day length, nutrients, and moisture (table 5.1). These external factors, however, merely act as signals whose spatial or temporal variation induce more or less localized alterations in the plant's *internal* milieu. Such internal physiological alterations are what eventually trigger the observable plastic developmental responses in individual organs. There is considerable evidence indicating that the link between environmental signals and organ-level plastic responses is effected by plant hormones, and that endogenous hormone levels play a key role in the regulation of environmentally induced developmental plasticity (reviews in Crane 1964; Trewavas 1986; Voesenek and Blom 1996; Wells and Pigliucci 2000; Minorsky 2003). In addition to hormones, other substances may play similar regulatory roles in phenotypic plasticity, but the information available is considerably scarcer. Soluble sugars, for example, can act as regulatory molecules helping to control the development of leaves, fruits, and seeds (Gibson 2005; Rolland et al. 2006).

The most detailed and conclusive evidence of hormonal regulation of phenotypic plasticity at the organ level comes from numerous experimental studies of heterophylly in aquatic plants (reviews in Deschamp and Cooke 1984; Wells and Pigliucci 2000; Minorsky 2003). In all species studied so far the application of abscisic acid (ABA) initiates the production of aerial-type leaves by submerged meristems. Gibberellic acid (GA) affects leaf morphology in a way opposite to ABA, as its application induces submerged-type leaves in shoots grown aerially. Interactions between different hormones may also occur. In *Ludwigia arcuata*, treatment with ABA induces the formation of aerial-type leaves on submerged shoots, while treatment with ethylene results in the formation of submerged-type leaves on shoots grown aerially (Kuwabara et al. 2003). Endogenous ethylene concentration was higher in submerged shoots than in terrestrial ones, while endogenous ABA concentration exhibited the opposite pattern. When the two hormones are simultaneously applied to terrestrial shoots, the effect of ethylene is suppressed by ABA, and only terrestrial-type leaves form. In submerged leaves, the endogenous level of ethylene was unaffected by the addition of ABA, while in aerial leaves the endogenous level of ABA was considerably reduced by experimental addition of ethylene, resulting in the production of submerged-type leaves.

In contrast to the abundant information on the hormonal regulation of leaf type in heterophyllous plants, the role played by hormones as endogenous regulators of the phenotypic plasticity of other organs remains essentially unexplored in wild plants. The following examples from cultivated plants show, however, that endogenous levels of different hormones are also expected to play decisive roles in the regulation of the developmental plasticity of leaves of nonheterophyllous plants, flowers, and fruits. Variable cytokinin influx into leaves mediated by variable transpiration rate was implicated in the plastic responses of *Phaseolus vulgaris* leaves to variable shading (Pons and Bergkotte 1996). This mechanism has been confirmed for *Nicotiana tabacum* and *Arabidopsis thaliana*, where within-plant gradients in light intensity and leaf photosynthesis are associated with gradients in cytokinin levels, and cytokinins appear to act as an endogenous signal involved in the regulation of whole-plant photosynthetic acclimation to light gradients in canopies through their effects on leaf functional properties (Boonman et al. 2007). Tomato plants (*Lycopersicon esculentum*) grown at a low temperature regime produced flowers with a greater number of petals, stamens, carpels, and ovary locules than those grown at a higher temperature regime (Sawhney 1983). Experimental application of GA induced an increase in the number of petals, stamens, carpels, and ovary locules, but the effect was much greater on plants grown at high temperature than on plants grown at lower temperature, implying that at least part of the effect of lower temperatures on flowers was effected through increased levels of endogenous gibberellins. Garrod and Harris (1974) arrived at a similar conclusion regarding the hormonal mechanism underlying the increase in petal number of cultivated *Dianthus* flowers following exposure to low temperature. Auxins, GA, cytokinins, ABA, and ethylene have been implicated at various stages of fruit growth in many species (Crane 1964; Yonemori et al. 1995; Pérez and Gómez 2000; Ozga and Reinecke 2003), and spraying of developing fruit crops with GA is becoming an increasingly popular agronomic practice for increasing the size at harvest of the fruits of different species (Pérez and Gómez 2000; Chang and Lin 2006; Kappel and MacDonald 2007).

Summary

Widespread occurrence of organ-level phenotypic plasticity shows that there is no need of invoking subindividual genetic mosaicism to account for within-plant variation in the phenotypic characteristics of all kinds of reiterated organs, including those like flowers whose development has often been considered particularly robust to environmental changes. Within-plant variation can be parsimoniously explained in terms of predictable (i.e., programmed) plastic developmental responses expressed by individual organs in response to variations in both the external environment (e.g., light, temperature) and the internal environment (e.g., levels of endogenous hormones or other substances acting as growth factors), or in other words, the external signals and the internal inducers, respectively. As a general phenomenon, organ-level phenotypic plasticity provides the conceptual basis, and the necessary developmental context, for arguing in the next chapter that a number of the organismal mechanisms that account for within-plant variation stem from the joint effects of location in the plant, previous developmental history, and localized environmental features on the expression of plant organ phenotypes, a phenomenon known as "ontogenetic contingency" (Diggle 1994).

Organismal Mechanisms of Subindividual Variability

Ontogenetic contingency, the interplay between inherent architecture and environmental milieu, and developmental stochasticity are mechanisms responsible for within-plant variability of reiterated structures.

Chapter 5 was concerned with inducers of subindividual variability, that is, those cases where reiterated parts vary because the organism's genes vary or the parts experience varied environments. I contend that, in nature, the importance of these two inducers are vastly different. While genetic mosaicism amounts to little more than a biological oddity, microenvironmental variation acting in combination with organ-level developmental plasticity is responsible for most of the observed variation among organs of an individual. Also of very great importance is the existence of programmed seriation, as when fruits at the base of an inflorescence are largest and size diminishes distally. But such programmed seriation is often the result of spatial and temporal internal gradients, much like the response of the plant to external (microenvironmental) heterogeneities. In this way, organ-level developmental plasticity, in combination with the spatial and temporal variation in the internal and external microenvironments to which the organs borne by the same plant are exposed during initiation and development, will eventually crystallize into the *realized* within-plant variation in organ features that I reviewed in chapters 2 and 3.

This chapter takes up the organization of the response of the organism to its changing environment as well as the organization that it generates

in and of itself. The organ-level microenvironment is shaped by variables
that can be both external and internal to the plant. On one hand, variation
in the external microenvironment is mainly related to spatial and temporal differences in the physical environment as it is perceived by each
individual aerial organ of a plant. This is the concept of "phylloclimate"
as defined by Chelle (2005; despite its name, the concept applies to any
sort of aerial organ, not just leaves). The phylloclimate is described by
physical variables such as spectral irradiance, temperature, and features
of around-organ air (e.g., wind speed, temperature, relative humidity).
On the other hand, different organs in the same plant may also perceive
different internal microenvironments reflecting differences between distinct parts of the plant in, for example, the expression of some genes, photosynthetic yield, hormonal concentrations, chemical signals, secondary
compounds, water availability, or concentration of nutrients in the xylem.
Obviously, certain aspects of the internal plant microenvironment to
which an organ is exposed will be influenced by variation in one or more
parameters of the external microenvironment. For example, as described
later in this chapter, the amount of photosynthates locally available to
a developing fruit or seed is closely related to the light environment to
which it and the surrounding photosynthesizing leaves are exposed, and
on the impact of herbivores on these same leaves.

   In addition, there is an essential temporal component imposed on
spatial within-plant variation in the internal and external microenvironment faced by individual organs. The buildup of plant bodies proceeds by
the ordered, sequential addition of parts at the apex of growing vegetative or reproductive shoots. This means that different structures on the
same plant axis initiate and develop at different times and thus, possibly,
under contrasting external and internal microenvironments. This is nicely
exemplified by some deciduous trees from temperate and boreal regions
(e.g., *Betula*, *Populus*, *Liquidambar*) that have twigs with two morphologically distinct kinds of leaves. The first, more-basal leaves form in early
spring from overwintering buds formed late in the preceding growing season, while the more-distal leaves are later produced from the apical meristem without overwintering as leaf primordia (Critchfield 1960; Clausen
and Kozlowski 1965; Smith 1967). In this example, the different times at
which primordia formed ("preformed" vs. "neoformed," in Ray's terminology [1987]) are ultimately responsible for morphological dimorphism.
But growth by sequential addition of parts along an elongating axis also
has the unavoidable consequence that older, basally located structures
precede younger, distally located ones. The former thus have the capac-

ity to affect the growth and developmental trajectories of the latter via modifications of their internal microenvironments or, in other words, to make the internal microenvironment faced by late-developing parts in the same axis different from that experienced by earlier-developing parts, as discussed in detail later in this chapter. This happens, for example, when basally located, early-developing fruits deplete or monopolize available resources for growth (Diggle 2003), or when basal, mature leaves in a shoot emit chemical signals informed by the external microenvironment that modify the characteristics of later-developing, more distal leaves (Lake et al. 2001, 2002; Yano and Terashima 2001).

Spatial and temporal variation in the internal and external microenvironment faced by initiating and developing organs, in combination with the capacity of the individual plant's genotype to modulate/modify the development and morphogenesis of individual organs in response to variation in features of its microenvironment (i.e., organ-level phenotypic plasticity, chapter 5), is the predominant source of within-plant variation in characteristics of reiterated structures. Most of this chapter is devoted to describing the many ways that organs, in combination with variation in influential variables of external and internal microenvironments, eventually give rise to realized within-plant variation. I group these multifarious mechanisms under the common heading of "ontogenetic contingency," a concept coined by Diggle (1994) to refer to the joint effects of position in the plant, previous developmental history, and environment on the expression of plant phenotypes.

A second major mechanism contributes to within-plant variation in features of reiterated structures and is considered in this chapter. Organ-level developmental instability is defined here as the inability of an organ's developmental program to generate final copies whose features conform exactly to those predicted by the organ-level reaction norm in response to variation in the internal and external microenvironment. In other words, organ-level developmental instability accounts for the fraction of phenotypic variance in organ features that remains unexplained after accounting for the organ's specific reaction norm.

Ontogenetic Contingency

In many, possibly most, instances the phenotype of a reiterated structure ultimately depends on the complex interplay of when and where it is produced within the three-dimensional construction of the plant bearing it, and on the environmental milieu in which it develops. That environmental milieu, in turn, consists of three main elements: (1) the external biotic and abiotic environment, for example, CO_2 concentration, incident radiation, herbivores, and pollinators; (2) the internal environment, for example, nutrient, hormone, and chemical signals in the phloem; and (3) the preceding developmental events experienced by the structure or its immediate surroundings, for example, fruit set, inflorescence elongation, and herbivory. The joint effects of location, previous developmental history, and environment on the expression of plant organ phenotypes has been termed "ontogenetic contingency" (Diggle 1994, 1997), the phenomenon whereby "the developmental fate of a primordium depends upon when and where it is produced and what events have preceded it during the ontogeny of the organism" (Diggle 1994, 1364). Although this concept originally arose from Diggle's detailed investigations of variable sex expression (i.e., the relative proportions of male and hermaphroditic flowers produced) in the sexually labile plant *Solanum hirtum* (1994), its applicability extends well beyond this rather restricted context (Watson et al. 1995; Pigliucci 1998).

I review in this section the main effects and mechanisms generating within-plant variation in characteristics of reiterated structures by adopting the framework of Diggle's concept of ontogenetic contingency. Although the interplay of the three factors involved—position, developmental history, and external and internal environment—is explicitly considered whenever applicable, spatial position is the main organizing theme. Whenever within-plant variation in the characteristics of a reiterated structure may be directly or indirectly linked via some mechanism to its position in relation to the plant as a whole, or to other reiterated structures of any kind, it is appropriate to refer to the existence of some architectural effect. My usage of this term is thus more encompassing than that of Diggle (1995, 542), who circumscribed "architectural effects" to describe "morphological variation that can be ascribed to the position of a flower (and its organs) within an inflorescence, rather than to effects of resource supply." Furthermore, there is no particular reason to restrict architectural effects to flowers or fruits, since within-plant variation in leaf characteristics can also be explained in terms of architectural effects, as shown below.

Organization of This Section

I consider below the effects and mechanisms that, subsumed into the concept of ontogenetic contingency, contribute to within-plant variation in characteristics of reiterated structures. To help the reader navigate this very long section, table 6.1 outlines its organization. The environment's influence on organ traits can be abiotic-external (phylloclimate-driven effects), biotic-internal (architectural effects), or biotic-external (interactions with animals). Architectural effects driven by the internal environment are subdivided into those operating directly and indirectly on the organs' characteristics. In direct architectural effects, position alone is responsible for variation in the characteristics of reiterated structures, regardless of any possible effects on, for example, spatial variation in resource supply. This meaning is close to Diggle's definition (1995) mentioned above, as well as its more-recent version (Diggle 2003, 64), where she defined architectural effects as variation "due to positional variation inherent in the architecture of plant axes." I also include among direct architectural effects other types of mechanisms that are not necessarily related to positional variation along plant axes, such as place-dependent growth suppression and physical constraints on development (table 6.1).

Under indirect architectural effects I include those effects whereby position in the plant is responsible for variation in the characteristics of reiterated structures only through some intermediate effects on the supply of nutrients, hormones, chemical precursors, or signals, or any other substance relevant to the initiation, development, or maturation of the structure. Indirect architectural effects are subdivided into those operating at the scale of the whole plant and those operating at the scale of organs (e.g., fruits) or groups of organs (e.g., inflorescences, infructescences). Effects in the former group are mainly associated with plant sectoriality.

TABLE 6.1 **Taxonomy of the main effects and mechanisms that, subsumed into the "ontogenetic contingency" concept, are responsible for generating within-plant variation in features of reiterated organs.**

Influential environment	Class of effects	Proximate mechanisms involved
Abiotic, external	Phylloclimate-driven	Plastic organ-level responses to microenvironmental variation
Biotic, internal	Indirect architectural	Sectoriality at the scale of the whole plant
		Resource-mediated interference among organs and organ groups
	Direct architectural	Place-dependent suppression of growth
		Physical constraints on growth/development
		Purely positional
Biotic, external	Interactions with animals	Systemic induction of defenses
		Alteration of the spatiotemporal distribution of resources

Those in the second group are mostly related to anatomical and geometric peculiarities of individual organs or groups of organs (e.g., inflorescences, infructescences). The distinction between these two groups of indirect architectural effects is partly a matter of convenience, rather than reflecting a real dichotomy. On one hand, organ-level effects are unavoidably subordinated to processes operating at the whole-plant level. On the other, the hierarchically nested distribution of plant parts may conceivably give rise to a seamless continuum of architectural and resource allocation effects at every possible within-plant spatial scale, running from the organ up to the whole-plant level (Obeso 2004a).

Phylloclimate-Driven Effects

Chelle (2005) defined phylloclimate as the physical environment perceived by the individual aerial organs of plants. Although *phyllo* is the Greek noun for "leaf," the term coined by Chelle refers to the microclimate perceived by all classes of aerial organs, not just leaves. The set of physical variables describing phylloclimate are related to radiation, surrounding air, and organ temperature, all of which are known to have some effect on a variety of plant functions, as summarized in table 6.2. Phylloclimate depends on the interplay of three different agents: the atmosphere, as a determining factor of the canopy microclimate; the soil, through its effects as a reflector of energy and via thermal inertia; and the plants themselves via their three-dimensional structure, which implies a complex penetration of fluxes as a result of the spatial distribution of sources and sinks of mass and energy (Oke 1987; Chelle 2005). Due to this later component, phylloclimate-driven variation can also be seen as a particular instance of architectural effect. Within-plant variation in one or several of the physical variables that define phylloclimate is expected to generate some concomitant variation in the traits of individual organs, as a consequence of the latter's capacity to produce distinct phenotypic variants in response to changes in influential parameters of the external microenvironment such as radiation or temperature. One would expect larger plants (e.g., shrubs, vines, or trees) to perceive a broader phylloclimate space, and therefore it might be predicted that the relative importance of phylloclimate-driven variation as a mechanism causing within-plant heterogeneity should be greater among these growth forms. This does not mean, however, that the phenomenon is restricted to them. Phylloclimates can also be significantly heterogeneous even at the relatively reduced scale of small herbaceous

TABLE 6.2 **Physical variables describing phylloclimate, or the external physical environment perceived by individual aerial organs of plants, and their respective effects on plant functions.**

	Variables	Plant functions potentially affected
Radiation	Spectral irradiance	Photosynthesis
	Ultraviolet	Photomorphogenesis
	Photosynthetically active radiation	Stomatal opening
	Near infrared	Energy budget
Surrounding air	Wind speed	Photosynthesis
	Temperature	Thigmomorphogenesis
	Humidity	Stomatal opening
	CO_2 content	Energy budget
Organ temperature	Surface temperature	Growth and development
	Internal temperature	Photosynthesis

Source: Modified from Chelle 2005.

plants, as documented by Chelle (2005 and references therein) for young maize plants.

Some of the relatively coarse-grained, extrinsic gradients of within-plant variation mentioned in chapter 4 can be most parsimoniously explained in terms of organ-level developmental and morphogenetic responses to systematic within-plant variation in one or more components of phylloclimate, typically irradiance and temperature. This applies particularly to the case of variation in continuous leaf traits. A sizeable literature shows that many leaf traits vary depending on the light environment to which they are exposed during initiation and/or subsequent development (Lynch and González 1993; Klein et al. 1991; Niinemets and Kull 1998; Le Roux et al. 1999), and it is reasonably well established that systematic within-plant gradients in leaf size, mass per unit area, and foliar nitrogen concentration are largely the consequence of leaf-level developmental responses to the variable light environment (Valladares and Pearcy 1999; Balaguer et al. 2001; Niinemets et al. 2003; Valladares 2003). The horizontal within-plant gradients in leaf traits associated with compass directions, mentioned in chapter 4, can likewise be explained as a consequence of individual-leaf responses to compass-dependent variation in phylloclimate variables. For example, in the California chaparral shrub *Heteromeles arbutifolia*, leaves located in different orientations on individual shoots differ significantly in their anatomical and functional properties, and the variation is related to consistent differences in the light environment (Valladares and Pearcy 1999).

Within-plant variation due to organ-level responses to variation in the external physical microenvironment has been most thoroughly studied for leaves, but it is expected to occur in other organs as well. Janzen (1982b, 1982c) suggested that, if a seed's size is at least in part determined by the amount of sunlight received by its fruit, then within-plant variation in seed size might be partly the outcome of heterogeneity in the light environment. Field observations and growth-chamber experiments have shown that the sugar composition of the floral nectar produced by individual plants is a plastic character that varies significantly with ambient temperature (Freeman and Head 1990; Villarreal and Freeman 1990). Within-plant heterogeneity in average temperature of the air surrounding individual flowers can thus generate some within-plant variation in nectar composition. Variation in ambient temperature can also induce changes in flower or inflorescence size, as frequently documented for ornamentals grown under controlled greenhouse conditions (e.g., *Dendranthema*, Karlsson et al. 1989; *Viola*, Pearson et al. 1995; *Campanula*, Niu et al. 2001; *Chrysanthemum*, Nothnagl et al. 2004) and, although much less often, for wild species. In *Ipomopsis longiflora*, mean corolla-tube length of plants grown under controlled growth-chamber conditions was greater for plants at 30°C than for those at either 20°C or 35°C (fig. 5.4). Differences in mean flower size between the north- and south-facing sides of individual trees (Perfectti and Camacho 1999) might be the immediate consequence of flower-level developmental responses to within-plant variation in air temperature (Alonso 1997a).

Indirect Architectural Effects: Whole-Plant Level

As a consequence of plant growth taking place by means of the reiterated production of morphological subunits by meristems, individual plants are decentralized organisms that can be viewed as a collection of subunits hierarchically assembled at a range of scales (chapter 1). Although the study of plant modularity was traditionally the realm of morphologists (reviews in White 1979, 1984), the range of consequences of modularity considered by ecologists has broadened considerably in recent times to include aspects related to the demography, life history, resource allocation, evolution, and physiology of plants (White 1984; Watkinson and White 1985; Watson 1986; Tuomi and Vuorisalo 1989; Vuorisalo and Mutikainen 1999). Among these, physiological aspects are particularly germane to the issue of within-plant variation in the characteristics of reiterated structures.

Empirical evidence gathered in the last few decades has shown that plants consist not only of morphological subunits, as traditionally acknowledged (White 1984), but also of physiological subunits that may behave semiautonomously (Watson and Casper 1984; Watson 1986; Sprugel et al. 1991). In this view, the plant can be seen as an assemblage of semiautonomous "integrated physiological units" (IPUs; Watson and Casper 1984) made up of identifiable arrays of morphological subunits that function as relatively autonomous structures with respect to assimilation, distribution, and utilization of carbon. This subdivision into semiautonomous IPUs was originally envisaged as a consequence of restrictions in the possible pathways of photosynthate movement between sources and sinks, caused by one or more of the following: branches receive assimilate only from localized regions of the main body of the plant; lateral assimilate movement among structures declines with increasing plant size, structural complexity, and degree of maturity; and vascular architecture poses a rigid constraint on patterns of assimilate translocation (Watson and Casper 1984). The original carbon-centered IPU model was subsequently expanded to incorporate within-plant compartmentalization in other plant constituents and resources (Watson 1986; Sprugel et al. 1991; Orians et al. 2000; Orians and Jones 2001; Viswanathan and Thaler 2004). These studies have led to the development of a more-encompassing theory of plant sectoriality where the IPU model is enriched with the explicit recognition that vascular architecture rather rigidly dictates the patterns of movement of water, minerals, photosynthates, chemical signals, hormones, and induced chemical defenses within a plant, restricting the movement of these substances along certain pathways but not along others (Orians and Jones 2001; Orians et al. 2002, 2004). Plant sectoriality thus means that physiological activities are integrated within some parts of plant structures (e.g., within the same branch), but the same activities display a strong degree of independence at other structural scales (e.g., among distinct branches; Sprugel et al. 1991; Vuorisalo and Hutchings 1996; Brooks et al. 2003).

The transport of assimilates among leaves and shoots, from leaves to flowers or fruits, and to and from roots and shoots, is characterized by its marked longitudinal confinement in the vascular bundles associated with the leaf traces, and, for this reason, it is ultimately controlled by vascular architecture (Wardlaw 1968; Watson and Casper 1984; Watson 1986; Sprugel et al. 1991; Marshall 1996). The capacity of different parts of a physically coherent plant structure to be either integrated or independent is largely determined by vascular structure, with parts belonging to the same sector being directly interconnected by vascular strands (Stieber and Beringer 1984; Watson 1986; Orians et al. 2004). It is thus the underlying vascular architecture of the stem that determines the precise physiological link between a source and a given sink, rather than any particular metabolic feature of the individual sink (Marshall 1996). For example, within a given shoot, mature source leaves will export photosynthates to specific leaves or parts of leaves based on orthostichy—the phyllotactic arrangement of leaves into longitudinal ranks along the stem, which is correlated with connectivity of vascular traces. Export will be greatest between orthostichous leaves (i.e., along the same rank) because they share vascular traces, intermediate between leaves in adjacent orthostichies, and minimal or even absent between leaves in opposite orthostichies, for they lack vascular connectivity (Watson 1986; Preston 1998). Vascular restrictions do not apply exclusively to the movement of assimilates but also to the translocation of other substances (Watson 1986; Sprugel et al. 1991; Davis et al. 1991; Schittko and Baldwin 2003).

Orians and Jones (2001) suggested that vascular architecture is a key determinant of intraplant resource heterogeneity. They proposed a functional model whereby the combination of intrinsic (plant sectoriality) and extrinsic factors (local environmental variation) might be used to predict patterns of within-plant heterogeneity in food quality from the viewpoint of animal consumers. I focus here on the issue of within-plant variation in the characteristics of reiterated structures without reference to its possible influence on resource use by animals, which is treated in chapter 8. The restricted or spatially very limited lateral movement of water, nutrients, and photosynthates, in combination with differential acquisition capacity of different IPUs, is expected to be a major source of within-plant variation for those reiterated structures that are strong metabolic sinks, like flowers, fruits, and seeds. To the extent that certain features of reiterated structures (e.g., size, nutrient concentration) will vary depending on the supply of carbon, water, and minerals available to each individual structure during its initiation and development, within-plant variation in these features may be partly explained in terms of the conceptual model shown in figure 6.1. This model applies particularly to structures that behave consistently as metabolic sinks, such as flowers, fruits, and seeds, and much less so to those that are relatively autonomous (at least with regard to carbon), such as leaves. Furthermore, the model is valid only for species or situations where the initiation and development of reiterated structures rely on resources originating from current photosynthesis and nutrient

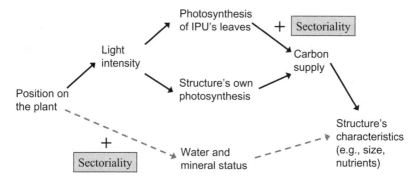

FIG. 6.1 Simple model depicting the chain of causal mechanisms that, acting in combination with plant sectoriality, lead to within-plant variation in characteristics of reiterated structures that may partly contribute to their own carbon requirements, but generally behave as net photosynthate sinks (e.g., flowers, fruits, and seeds). *Black continuous arrows* depict mechanisms whereby light-dependent position effects can generate within-plant variance in carbon-related characteristics of reiterated structures. *Gray dashed arrows* depict effects designated as "bottom-up" sectoriality in the text.

uptake, rather than on previous seasons' stored resources. Use of stored resources may totally abolish physiological autonomy of plant parts, as demonstrated by Lacointe et al. (2004) for deciduous walnut trees (*Juglans regia*). In their study, paired branches in the same tree were exposed to two contrasting light regimes over the whole growing season, and then separately labeled with $^{14}CO_2$ and $^{13}CO_2$ in September, so that the photosynthates from each branch could be traced independently at the same time. Before leaf fall, branch autonomy was nearly total, with virtually no carbon movement between branches. In the subsequent spring, in contrast, within-plant carbon transfers totally abolished branch autonomy.

As depicted by black arrows in figure 6.1, two concurrent mechanisms related to within-plant variation in light availability may act synergistically to enhance within-plant patchiness in carbon-dependent characteristics of reiterated structures (e.g., size). Both are motivated by the relationship between within-plant variation in light availability (due to self-shading, compass orientation, height above the ground, depth within the crown, etc.) and variation in leaf photosynthetic capacity (Hollinger 1989, 1996; Ellsworth and Reich 1993; Bassow and Bazzaz 1997; Le Roux et al. 1999; Casella and Ceulemans 2002; Meir et al. 2002). Within-plant variability in carbon-fixation capacity of leaves, acting in concert with restricted or spatially very limited lateral movement of photosynthate, will cause within-plant patchiness in the carbon supply available to initiating or developing structures. One of the basic "rules" of assimilate distribution within-plants is that reproductive sinks are principally supplied by their local source leaves (Marshall 1996). Photosynthates necessary for fruit and seed production on a branch are mainly or entirely provided by the branch itself, and sectoriality of nutrient transport may be as extreme as to lead to individual fruits receiving resources almost exclusively from the nearest leaves or those on the same orthostichies (Watson and Casper 1984; Sprugel et al. 1991). Plant sectoriality changes along with the normal shifts in carbohydrate demand that accompany fruit and seed production, tending to increase during the growing period of these metabolically very demanding sinks (Lacey and Marshall 1992; Preston 1998). This effect accentuates the influence of within-plant variation in local resource availability on the characteristics of these reiterated structures. Although within-plant patchiness in local photosynthate availability is expected to affect mainly traits of strong metabolic sinks such as fruits and seeds, this does not mean that other subindividually variable, less demanding organs cannot be similarly affected. In *Aesculus hippocastanum*, variation in carbohydrate availability translates into concomitant variation in the volume and sugar concentration of nectar produced by flowers (Wykes 1952); hence within-plant patchiness in photosynthate availability can also induce subindividual variation in nectar features.

Restrictions in the movement of photosynthate from source leaves to reproductive sinks is not the only way that within-plant heterogeneity in the light environment eventually causes heterogeneity in carbon supply to these reiterated structures (fig. 6.1). Flowers, fruits, and seeds are also photosynthesizing structures themselves (Blanke and Lenz 1989; Aschan and Pfanz 2003). In flowers, for example, the photosynthetic capacity of green sepals at the flowering or early postflowering stages can be comparable or even exceed that of leaves (Williams et al. 1985; Vemmos and Goldwin 1994; Smillie et al. 1999; Salopek-Sondi et al. 2000; Aschan et al. 2005), and photosynthesis by developing reproductive structures can contribute between 2 and 65% to their own carbon budget (review in Obeso 2002). It is therefore expected that within-plant variation in light intensity will also affect the photosynthetic capacity of these structures, and hence the magnitude of their carbon contributions to their own growth and maintenance costs will also be subject to light-dependent position effects.

The importance of light-dependent position effects and sectoriality as determinants of within-plant variance in carbon-related characteristics of reiterated structures (black arrows in fig. 6.1) should be greatest for struc-

tures that have high carbon requirements, have relatively long developmental periods, or are produced by plants with marked within-plant heterogeneity in light availability. Fleshy fruits produced by trees and shrubs
possess all these characteristics combined, and one would therefore
expect them to be particularly prone to within-plant variation induced
by sectoriality. Results from cultivated fruit trees are consistent with this
expectation and support the model in figure 6.1. For peach (*Prunus persica*), Génard and Baret (1994) found a very broad within-tree gradient
of diffuse and direct light transmitted to shoots. Fruits from the upper
parts of the peach tree were more purple, less orange red, and less firm
and had a higher sucrose content, a lower citric acid content, and a higher
pH, than fruits from the lower parts, and this variation was correlated with
their light regimes (Génard and Bruchou 1992). Fruits exposed to light
mainly in the afternoon were more purple, less yellow and orange, and
firmer, and had a higher citric acid content and lower sucrose and malic
acid contents, than fruits exposed to light in the morning. Also for peach,
Génard (1992) and Souty et al. (1999) found that variation in leaf area
and number of leaves around fruits gave rise to within-plant variability in
the growth rate, final size, and concentration of sugars and organic acids
of fruits, presumably because of the close relationship existing between
shoot leaf area and carbon assimilation. In apple trees, intraplant variation in fruit size, weight, soluble solids, and starch content are closely
correlated with variation within the crown in photosynthetic photon flux
(Barritt et al. 1987; Tustin et al. 1988). Fruits located in canopy positions
receiving greater amounts of photosynthetically active radiation tended to
be significantly larger, and to have a higher concentration of soluble solids in the pulp, than fruits located in shadier locations. Furthermore, fruit
size, weight, and starch and soluble-solid contents were directly correlated
with average specific leaf weight at the same position in the tree crown.
Kappel and Neilsen (1994) used hemispherical photography to measure
within-plant variation in the light microclimate of pear (*Pyrus communis*) fruit clusters, and found that the percentage of sky visible above each
cluster was related to fruit growth patterns and fruit size and quality at
harvest. When whole branches of pear trees were artificially shaded for
the whole fruit development period, fruits borne by these branches had
slower growth and were eventually smaller than those borne by control
branches exposed to full sunlight (Garriz et al. 1994). Although none of
these studies allow for separating the direct (via effects on fruit photosynthesis) and indirect (via effects on leaf photosynthesis in combination

with sectoriality) influences of the light environment on fruit traits, they do clearly support a prominent role of heterogeneity in the within-plant light environment in causing within-plant variation in fruit characteristics.

Sectoriality also occurs in the transport of resources from roots to shoots (Orians et al. 2002; Orians et al. 2004), and architectural characteristics of plants may also limit the horizontal movement of water and nutrients, not just that of photosynthates. This effect, which might be termed "bottom-up" sectoriality, acts in conjunction with carbon-supply heterogeneity to enhance within-plant variation in the characteristics of reiterated structures, and is depicted by dashed gray arrows in figure 6.1. There is extensive evidence that soil nutrient availability varies spatially at a variety of scales that are relevant to individual plants (Beckett and Webster 1971; Arp and Krause 1984; Robertson et al. 1988; Boerner and Koslowsky 1989; Grigal et al. 1991; Pelletier et al. 1999). Geostatistical analysis has demonstrated that the rooting zone of an individual plant can have much variation in nutrient availability, and nitrate and ammonium concentrations in woodland soil may vary two- to fivefold at scales of only 20 cm (Hodge 2004). Plants exploit this patchiness in soil nutrients through a combination of developmental (root proliferation, consisting of the initiation of new lateral roots) and physiological (increased nutrient uptake capacity of individual roots) responses of those roots that are exposed to nutrient- rich patches (Jackson et al. 1990; Robinson 1994; Hodge 2004). In addition, the transport of nutrients within the plant may be sectorial, with preferential transport from roots to leaves and branches with the most direct vascular connections, as demonstrated by experiments tracing the upward movement of markers applied to roots (e.g., dyes in water solution, and isotopically labeled nitrogen or phosphorus sources; Rinne and Langston 1960; Stryker et al. 1974; Hay and Sackville Hamilton 1996; Orians et al. 2002; Orians et al. 2004). For example, Orians et al. (2004) demonstrated experimentally that in saplings of *Populus* and *Acer*, nutrients are preferentially transported from specific roots to specific branches. Rinne and Langston (1960) used a split-root technique and ^{32}P as a label to follow the upward movement of nutrients when plants of peppermint (*Mentha piperita*) were fed the labeled solution through specific portions of their root system. When ^{32}P was fed to either half (two of the four main vascular bundles) or a quarter (just one of the four main vascular bundles) of the root system of a peppermint plant, phosphorus showed very little or no lateral movement to those aerial parts of the plant that were not directly linked to the particular vascular bundles that had received the nutrient solution. In sugar beet, each vascular ring in the root is in direct connection with a certain number of leaves, and leaves are preferentially connected with regions of the root belonging to the same orthostichy (Stieber and Beringer 1984).

The combination of horizontal patchiness in soil nutrient availability and "bottom-up" sectoriality will create within-plant heterogeneity in the characteristics of reiterated structures through differential supply of the water and nutrients required for photosynthesis and growth. This possibility was first explicitly advanced by Orians and Jones (2001), and subsequently demonstrated experimentally by Orians et al. (2002) for tomato (*Lycopersicon esculentum*). These authors manipulated nutrient levels to two different root zones of split-root tomatoes. Fertilizer was applied either to lateral roots from one sector of the root system or to the remaining root system. They hypothesized that leaves and leaflets within the fertilized sector would be larger and contain lower concentrations of phenolics, since fertilization typically causes an increase in growth and a decrease in investment in carbon-based defenses. Results confirmed predictions, as leaves vertically aligned above the fertilized lateral roots, which had direct vascular connections to those roots (as corroborated by dye transport patterns), were larger and had lower concentrations of phenolics than did leaves without direct vascular connections. In addition, the production and growth of lateral shoots was greater in the "fertilized" sectors. As shown by this experiment, restricted transport of nutrients coupled with spatial variation in below-ground nutrient availability can lead to differential rate of resource supply to specific leaves and branches, leading to differential growth and chemistry within a plant.

Plants differ in the magnitude of sectoriality. Interspecific differences may be related to phylogeny, growth habit, or the environment to which the species is adapted, and intraspecific variation may be due to individual differences in genotype, age, or developmental stage (Watson and Casper 1984; Sprugel et al. 1991; Lacey and Marshall 1992; Lötscher and Hay 1996; Marshall 1996; Preston 1998). For individual plants, even transient environmental conditions can temporally modify their degree of sectoriality (Zwieniecki et al. 2003). The existence of all these sources of variation at the inter- and intraspecific levels implies that the potential importance of sectoriality as a source of within-plant variation in characteristics of reiterated structures will vary among species, among individuals of the same species, and even temporally for the same individual. The model in figure 6.1, therefore, will apply to variable degrees to different species, conspecific individuals, or even the same individual plant at different times or in different environmental conditions. Sectoriality is facilitated by open vascular systems, in which the sympodia (axial bundles with associated leaf and branch traces) are entirely or essentially discrete, with only occasional interconnections consisting of minor or accessory bundles. Since open vascular systems are characteristic of gymnosperms and dicotyledons, sectoriality is expected to occur most frequently in plants belonging to these groups, and only rarely in monocots (Vuorisalo and Hutchings 1996). One would therefore expect that sectoriality should be a more important source of within-plant variation in gymnosperms and dicots than in monocots, and hence that the model in figure 6.1 should apply more specifically to the former two groups.

Within eudicots, the extent of sectoriality varies also depending on life history, habitat type, and taxonomic affiliation. Clonal species, for example, appear less sectorial than nonclonal ones (Watson and Casper 1984; Price et al. 1996), and interspecific differences in sectoriality have been related to variation in drought adaptations, growth speed, shade tolerance, and colonizing ability (Watson 1986; Orians et al. 2004). Simulated herbivory experiments have revealed that, under the stress imposed by artificial defoliation, resource translocation among branches may range from negligible to extensive depending on the species (Janzen 1976; Stephenson 1980; Shea and Watson 1989; Obeso 1998a; Mitchell et al. 2004), which probably reflects species-specific differences in the individual plant's capacity to laterally translocate resources when strong sinks (defoliated branches bearing developing fruits) put heavy demands on them. Apart from some indirect suggestions, however, there are remarkably few empirical data on the extent of interspecific differences in sectoriality and their ecological and phylogenetic correlates. Orians et al.'s comparative study of five species of *Betula*, *Populus*, and *Acer* (2004) represents an outstanding exception (see also Wheeler et al. 2005; Ellmore et al. 2006). These authors traced the movement of dye from isolated roots to transpiring branches, and performed split-root experiments with application of ^{15}N-labeled ammonium nitrate to either a single, isolated lateral root or the bulk of the root system. Because transport of nutrients between sectors requires flow through intervessel pit pairs of adjacent xylem vessel elements, they also quantified the area of intervessel pits, the number of pits per unit vessel wall area, and the percentage of vessel wall area that pits occupy. Their results revealed noticeable interspecific differences in sectoriality. The two *Betula* species were not sectorial, with tracers applied to isolated roots being likely to accumulate in all branches. In contrast, *Populus* and *Acer* species were sectorial, with tracers tending to accumulate preferentially in particular branches. In accordance with these findings, *Betula* species had the largest number of intervessel pits per unit wall area and the largest percentage of vessel wall area that pits occupy. Orians et al.'s results (2004) are important in that they reveal a decisive role of wood anatomy in explaining interspecific differences in sectoriality (see also Ellmore et al. 2006; Zanne et al. 2006).

Indirect Architectural Effects: Organ and Organ-Group Levels

In addition to the architectural effects operating at the whole plant level, other classes of architectural effects operating at considerably finer scales within individuals (e.g., within inflorescences, within infructescences, within fruits) also help generate within-plant variation in reiterated organ traits. Most often, flowers, fruits, and seeds are not produced singly, but grouped into inflorescences, infructescences, and multiseeded fruits. Leaves may also be produced as close aggregations of two or more units, as in species with whorled or opposite leaves. As shown in chapter 4, a considerable fraction of all within-plant variance for flower, fruit, and seed traits takes place quite locally, occurring in the same inflorescence or infructescence, or even in the restricted confines of individual multiseeded fruits. Variation in organ traits at this small spatial scale is generally due to some combination of indirect (resource-mediated) and direct architectural effects. These two mechanisms are considered here and in the next section.

Within acropetally developing inflorescences (e.g., racemes), the size of flowers, flower parts, fruits, and seeds, as well as the probability of fruit and/or seed maturation per flower, generally decline from proximal to distal positions (Lee 1988; Wolfe 1992; Diggle 1995, 1997). Dating back at least to Goodspeed and Clausen (1915) and Salisbury (1942), this well-known declining trend has frequently been attributed to a concomitant reduction in the amount of resources available for flower and fruit development, due to competition for resources between basally and distally developing structures. Basal flowers within an inflorescence typically are the first to reach anthesis, and may be pollinated and initiate fruit development long before distal flowers are mature. Because of this temporal precedence, basal fruits may become stronger sinks and more able to obtain resources from the parent plant in comparison to later-developing distal flowers and fruits. Furthermore, basal flowers and fruits are closer to the source of those resources and may be provisioned preferentially for that reason. Position-dependent differences between flowers, fruits, or seeds of the same inflorescence in their access to the resources necessary for growth and maturation are therefore a frequent mechanism underlying fine-scale within-plant variation in the characteristics of these reiterated structures. A direct demonstration of this effect was provided by Binnie and Clifford (1999). Using radioactively labeled carbon, they demonstrated that in inflorescences of *Phaseolus vulgaris*, reproductive organs at proximal sites on the raceme always received greater proportions of ^{14}C-photosynthate, and were shed significantly less often, than did reproductive organs at distal sites. Considerable indirect evidence comes from experiments showing that, after increasing the amount of resources available to distal flowers and fruits through removal of basally located structures, the normal morphological or developmental difference between basal and distal organs either vanishes or is much reduced, with distal structures becoming indistinguishable from or much more similar to those located at basal positions (Stephenson 1981; Wyatt 1982; Lee 1988; Diggle 1995, 1997; Ashman and Hitchens 2000; Medrano et al. 2000). These experiments support the view that within-inflorescence variation in organ traits will often be due to differential access to resources by the different organs as a consequence of their different positions. These indirect, resource-mediated architectural effects on within-inflorescence variation of organ traits have been demonstrated experimentally for corolla length and spread in *Nicotiana tabacum* (Goodspeed and Clausen 1915); corolla size and ovary length in *Solanum hirtum* (Diggle 1995); sepal and petal length in *Sidalcea oregana* (Ashman 1992); petal, stigma, and stamen length in *Arabidopsis thaliana* (Diggle 1997); fruit seediness in *Pancratium maritimum* (Medrano et al. 2000) and *Aquilegia canadensis* (Kliber and Eckert 2004); fruit size in tomato (*Lycopersicon esculentum*; Bangerth and Ho 1984); fruit-pulp mass in *Phytolacca rivinoides* (Byrne and Mazer 1990); and seed size in *Hydrophyllum appendiculatum* (Wolfe 1992) and *Lolium perenne* (Warringa, de Visser, and Kreuzer 1998; Warringa, Struik, et al. 1998). In all these instances, an extra provision of resources to distally located structures eliminated or mitigated the position-dependent gradient in organ characteristics and therefore reduced the magnitude of within-inflorescence variation.

Resource-mediated position effects are also expected to operate within the spatially restricted scale of individual fruits, particularly when these originate from ovaries where ovules are arranged linearly (e.g., Fabaceae, Brassicaceae) and thus differ in their distance from the resource supply source. In legume pods and crucifer siliques, for example, ovules located at different positions with respect to the base of the fruits often differ in their probability of producing a seed or in the size of the seeds eventually produced, and within-fruit variation in seed size may become one of the major sources of within-plant variation in these species (table 4.1). These patterns have frequently been attributed to competition for resources among developing seeds (Schaal 1980; Hossaert and Valéro 1988; Nakamura 1988; Rocha and Stephenson 1990). Nevertheless, few experiments demonstrate convincingly that competition for resources actually is the sole or even the main mechanism originating position-dependent variation in seed size within fruits with linearly arranged seeds (i.e., demonstrating that, when the resources available to individual fruits increase and all other potentially influential factors are kept constant, within-fruit variability in seed size is reduced), and other factors may be equally or more important in accounting for observed patterns. These factors include number of pollen donors siring seeds, order of ovule fertilization, intensity of microgametophyte competition, and genetic relatedness of zygotes (Stanton 1984; Mazer et al. 1986; Rocha and Stephenson 1990, 1991; Marshall 1991; Mohana et al. 2001; Mena-Alí and Rocha 2005), all of which may combine in intricate ways to modify the extent and outcome of mother-offspring conflict and sibling rivalry (Uma Shaanker et al. 1988; Ravishankar et al. 1995) and, as a consequence, alter patterns of within-fruit variation in seed size (Bañuelos and Obeso 2003; Obeso 2004b). The maternal parent and its offspring have conflicting interests in the extent of maternal resources allocated to developing seeds. As maternal parents are equally related to all their offspring, they should nourish equally all the seeds. In contrast, offspring are selected to be more selfish, compete for resources, and demand as much resources as they can from the parent, because they are less closely related to their sibs in the same fruit than to themselves (Trivers 1974). Individual developing fruits provide a perfect arena for the expression of parent-offspring conflict and sibling rivalry (Uma Shaanker et al. 1988; Ravishankar et al. 1995; Obeso 2004b). Some supporting evidence for the role of these factors in generating within-fruit variability in seed size comes from species where indirect architectural effects and unequal access to resources can be safely ruled out due to the radial disposition in the ovary of only a few ovules, as in *Rhamnus alpinus* (Bañuelos and Obeso 2003) and *Ilex aquifolium* (Obeso 2004b).

Direct Architectural Effects: Suppression of Growth and Space Constraints

Direct architectural effects are expected to be important determinants of within-plant variation in organ size (Wolfe 1992; Diggle 1995, 1997, 2003). Organ size reflects both cell number and cell size, and cell expansion and proliferation play central roles in organ-size determination (Mizukami 2001); hence any proximate mechanism that generates within-plant heterogeneity in growth rate will eventually translate into within-plant variation in organ size. These mechanisms include place-dependent suppression of growth and space constraints.

Place-dependent suppression of growth plays an important role in plant morphogenesis through modulation of the size and shape of organs (Basile and Basile 1993), and can also contribute to generate within-plant variation in organ traits. Spatial constraints due to morphological characteristics may give rise to place-dependent suppression of growth. This is the case, for example, of space constraints operating within fruiting heads, pods, and cones of many species. Woody or leathery fruits may physically constrain seeds in different positions of the fruit to different extents, thus generating within-fruit variation in seed size for purely mechanical reasons, as suggested for *Pisum sativum*, *Raphanus raphanistrum*, *Theobroma cacao*, and other species (Glendinning 1963; Stanton 1984; Gutiérrez et al. 1996; Fukuta et al. 2006). This effect has been clearly illustrated by McGinley et al. (1990) for lodgepole pine (*Pinus contorta*), where space constraints within cones were a major proximate cause of within-tree variation in seed mass. In this species, individual seed mass declines significantly from the base to the tip of the cone. Because of the shape of the cone, scale size declines from the base to the tip of the cone, and within a cone there is a positive correlation between scale size and individual seed mass. Small scales at the distal end of the cone are so compressed that they are unable to hold any seeds at all. In this species, therefore, within-cone seed mass variation appears to be a direct consequence of variation in the size of cone scales at different points of the cone. Similar architectural effects are expected to operate in other cone-producing gymnosperms and in angiosperms that produce conelike fruiting structures (e.g., Betulaceae, Proteaceae, Casuarinaceae), where the space available for growth of individual seeds can vary greatly within fruits depending on their position in relation to the extremes of the cone axis. In *Banksia spinulosa*, the decline in seed size from basal to distal positions of the conelike infructescence runs parallel to a decline in follicle size, and the smaller follicles in the apical third of the infructescence may constrain the size of enclosed seeds (Vaughton and Ramsey 1997).

Differential physical constraints on the development of seeds located on different positions within fruits or inflorescences can also operate in species without woody fruits, as nicely shown by some elegant experiments done on the grains of several cereal species. In rice, barley, oats, and wheat, the final size of individual seeds is significantly reduced when the available physical space for seed expansion is artificially reduced by insertion of small foreign objects (pebbles, Styrofoam pellets) between the glumes of flowers just after pollination (Grafius 1978). This indicates that the hydrostatic pressure of developing grass seeds is rather weak and that, for purely mechanical reasons, within-inflorescence variation in the size of the lemma and the palea (the two structures immediately surrounding seeds in the inflorescence of grasses) will set variable upper limits to the size of individual seeds in the spikelets of grasses, being thus ultimately responsible for a significant fraction of within-inflorescence variation in seed size (Grafius 1978; Tibelius and Klinck 1987). This has been confirmed experimentally for *Triticum aestivum* and *T. durum* (Millet 1986). Within-spike variation in the size and shape of individual grains closely mirrors the variation in the size of the floret cavity from which they developed, as estimated by injecting floret cavities with liquid silicone.

Seed-size gradients commonly occur within the fruiting heads of Asteraceae. In species of this family, regardless of whether they have heteromorphic seeds or not, seeds located in central whorls are typically smaller than those in peripheral whorls (Fick and Zimmerman 1973; McGinley 1989; Ruiz de Clavijo 1995; Imbert et al. 1996; Munshi et al. 2003), which can also be attributed to physical constraints on growth operating differentially at different locations in the infructescence. In cultivated sunflower (*Helianthus annuus*), for example, Fick and Zimmerman (1973) reported a 30% decline in average seed weight from the peripheral whorls of fruiting heads to the central ones. Similar centripetal decreases in seed size seem commonplace in sunflower capitulae (Munshi et al. 2003). These radial gradients are also related to architectural constraints mediated by place-dependent suppression of growth due to lack of space in the central positions of the receptacle. Florets in the heads of Asteraceae are arranged spirally on the receptacle, with the distance between successive florets, and hence the space available for the development of individual achenes, increasing as the spiral moves out from the center. This results in a higher number of seeds per unit area at the center of the receptacle than at the edge. Within-head gradients in seed size are thus expected to arise just because seeds in the central whorls experience growth suppression derived from intense spatial constraints (McGinley 1989).

Space constraints can also be responsible for the very small-scale variations in leaf size and shape of anisophyllous plants, as documented in detail for *Aucuba japonica* (Ali and Kikuzawa 2005). In this evergreen understory shrub, the two leaves in an anisophyllous pair have different sizes not because their respective precursor primordia differed in size (in fact, the smaller leaf of the pair arises from the initially larger primordium), but rather because space limitations constrain the expansion of the primordium closer to the inflorescence axis but not the opposite one. That anisophylly frequently stems from a simple physical constraint is further supported by the observation that, in some species, anisophyllous leaf production may switch over to isophyllous development once the constraint that causes the inequality of the primordia or of the growing conditions is lifted, as discussed by Dengler (1999) for *Acer pseudoplatanus* and *Populus deltoides*. In *Acer pseudoplatanus*, for example, the first pair of primordia are borne in the transverse plane, and they are of equal size. The second primordium pair are initiated in the median plane, where the dorsal leaf primordium is physically constrained between the lateral shoot axis and the parent shoot. The leaves eventually arising from dorsal leaf primordia are smaller than those arising from ventral leaf primordia. Anisophylly disappears during the second year of growth, when the new leaf primordia are formed without this constraint.

Other Direct Architectural Effects

As noted earlier, position-dependent variation in features of flowers, fruits, or seeds sometimes originate from differential access to resources rather than from differential position itself. In other instances, however, the position-dependent patterning persists regardless of resource status, which clearly suggests that differences are an unavoidable developmental consequence of position alone (Wolfe 1992; Diggle 1995, 1997). This aspect has been much less thoroughly studied than position-related variation in the probability of setting fruits or seeds, but a number of examples show that variation in flower and fruit traits in acropetal inflorescences often takes place regardless of resource availability (reviewed by Diggle 2003). In *Fragaria virginiana*, variation of floral traits with position in the inflorescence is essentially unaltered by increased resource availability (Ashman and Hitchens 2000). Similar results were reported by Bawa and Webb (1983) for stamen number in *Muntingia calabura*, Wolfe (1992) for flower size in *Hydrophyllum appendiculatum*, Diggle (1995) for anther length in *Solanum hirtum*, Diggle (1997) for stigma length in *Arabidopsis thaliana*, and Wolfe and Denton (2001) for fruit size and seed number per fruit in *Linaria canadensis*. In these cases, within-inflorescence variation in flower or fruit features should be exclusively attributed to direct architectural effects (Diggle 1995, 1997, 2003). In other cases, although within-inflorescence variation persisted after artificial increases of the resources available to flowers and fruits, the magnitude of the variation was lessened. This has been found, for example, for ovary length in *Solanum hirtum* (Diggle 1995); stamen length in *Arabidopsis thaliana* (Diggle 1997); calyx, corolla, anther, ovary, and mature fruit size in *Mimulus guttatus* (Diggle 1995); fruit size in tomato (*Lycopersicon esculentum*; Bertin et al. 1998); and ovule number in *Saxifraga longifolia* (García 2003). Both direct and indirect architectural effects are at work in these species in determining within-inflorescence variation, and organ variation responds to an "architectural effect" plus a "resource treatment effect" sensu Diggle (1995, 1997).

Direct architectural effects are not restricted to acropetal inflorescences, and the direction of positional trends do not always involve reductions from basal to distal positions. The inflorescences of oats (*Avena sativa*) and rice (*Oryza sativa*) differ from that of other small-grain cereals in being a panicle rather than a spike, and their development proceeds from the uppermost terminal spikelet downward to the base of the panicle. In these species, the uppermost, primary seeds are significantly heavier than the more basal, secondary ones (Jeng, Wang, et al. 2003; Rajala and Peltonen-Sainio 2004), and in the case of rice, total protein content declines from distal to basal seeds (Liu et al. 2005). In oat panicles, the size disadvantage of basal seeds still persists after artificially removing the distal florets shortly after pollination (Tibelius and Klinck 1987; Rajala and Peltonen-Sainio 2004). In soybean (*Glycine max*), the regular decline in total protein content of seeds occurring from apical to basal nodes in the plant (fig. 4.3) remains unchanged after application of nitrogen fertilizer at different growth stages of the plant (Bennett et al. 2003), which demonstrates that it is position in itself, rather than competition among seeds in basal and apical nodes for limiting nitrogen, that is responsible for within-plant variation in seed composition in this plant.

There is still another direct architectural mechanism that applies to within-inflorescence variation in fruit and seed features. Fruits and seeds are linked to flowers and, more specifically, to the gynoecium, by ontogenetic continuity, and some floral traits such as ovary size, ovule number, and ovule size are expected to be correlated with the consequent fruit traits, namely fruit size, fruit seediness, and seed size, respectively. In the absence of differential fertilization or ovule abortion, ovary size and ovule number should therefore covary, respectively, with fruit size and seed number. This is supported by the strong correlations frequently reported between ontogenetically linked characters of flowers and fruits (Primack 1987; Kang and Primack 1991). A flower-color dimorphism in the annual *Linaria canadensis* is linked to variation in flower size, with flowers of the light blue morph being about 40% smaller than flowers of the dark purple morph. As a consequence, fruits originating from light flowers are heavier than fruits from dark flowers (Wolfe and Sellers 1997).

Simply because of the close correlations expected between ontogenetic earlier and later characters, one would predict that some of the within-infructescence positional effects exhibited by fruit or seed traits may eventually prove to be subsidiary, unavoidable consequences of architectural effects previously exhibited by flowers within inflorescences. For example, the size of the ovary and of individual ovules declines from basal to distal positions in the inflorescences of many species (Bawa and Webb 1983; Macnair and Cumbes 1990; Warringa, Struik, et al. 1998; Ishii and Sakai 2002). This floral variation may subsequently translate into concomitant declines in fruit or seed size even in the absence of competition for resources or any direct architectural effect specifically pertaining to seeds or fruits. Different lines of evidence support this expectation. In Diggle's review (2003), for example, all studies that looked for within-inflorescence variation in both ovary and fruit characters reported similar directions of change, that is, similar architectural effects, from proximal to distal positions for the two types of characters. In tomato (*Lycopersicon esculentum*) infructescences the decrease in the size of ripe fruits from basal to distal positions, although arising in part from competition for resources among developing fruits (Bertin et al. 1998), runs parallel to a decline in the number of cells in the ovary of flowers at anthesis (Bohner and Bangerth 1988). In *Lolium perenne*, the steep decline in seed size from basal to distal positions within spikelets mainly arises from positional differences in growth rate, which, in turn, are ultimately due to differences in ovule dry weight at anthesis (Warringa, Struik, et al. 1998).

Within-infructescence variation in fruit seediness may also be in some instances an epiphenomenon of previous within-inflorescence variation in ovule number. Species with racemose inflorescences exhibit a predictable acropetal decline in ovule number per ovary (Diggle 1995; Mazer and Dawson 2001; Ishii and Sakai 2002; Guitián et al. 2004). Steady reductions in ovule number per flower also occur from early (primary) to later (secondary and tertiary) positions in dichasial inflorescences, as shown by Buide (2004) for *Silene acutifolia*. This sort of variation implies that, even without differences among fruits in the probability of ovule abortion due to insufficient resources or pollination, positional variation in ovule number per ovary is sufficient in itself to give rise to positional variation in fruit seediness (Diggle 1995). In *Clarkia unguiculata*, for example, the position-dependent pattern of within-infructescence variation in number of seeds per capsule nearly perfectly matches that of within-inflorescence variation in number of ovules per ovary (Mazer and Dawson 2001), thus suggesting that variation in fruit seediness was just an unavoidable sequel to variation in ovule number, rather than a fruit-related architectural effect in itself.

Proximate Mechanisms of Direct Architectural Effects

The proximate mechanisms responsible for those direct architectural effects at the organ-group level that are unrelated to space constraints are largely unknown (Diggle 1995, 1997). Some plausible explanations implicate position-dependent variation in size of the vascular system, size and shape of meristems, and expression of genes involved in organ development. I briefly consider each of these explanations in turn.

The size of a plant organ is frequently correlated with the amount of vascular tissue supplying it (Carlquist 1969; Housley and Peterson 1982; Cui et al. 2003), and under conditions of nonlimiting resource availability and high photosynthetic rates, the number and size of seeds produced by individual plants or single infructescences may ultimately be limited by the capacity for translocating photosynthates of the conducting tissues that connect source and sink organs. An example of this is the close linear dependence of kernel dry weight per ear on phloem cross-sectional area in the peduncle across 26 varieties of winter wheat (*Triticum aes-*

tivum; Nátrová and Nátr 1993). Similar relationships between vascular tissue volume and seed size or number have been documented for oats (*Avena sativa*; Housley and Peterson 1982) and rice (*Oryza sativa*; Cui et al. 2003).

Since the diameter of inflorescence axes and the cross-sectional area of vasculature typically decline distally, it has been suggested that a progressive reduction in meristem size or in the number, size, or characteristics of vascular bundles as the inflorescence develops may be at the root of positional effects on size-related characteristics of late-produced flowers and fruits (Byrne and Mazer 1990; Wolfe 1992). In contrast to reproductive structures in basal positions, those produced distally are borne on stems of smaller diameter that contain less vascular tissue, which may limit delivery of the water and nutrients needed to make larger fruits (Diggle 1995). This mechanism might explain, for example, the decline in fruit size and seed number per fruit from the bottom to the top of the infructescence in *Linaria canadensis*, since that decline runs parallel to a decline in the thickness of the stem and there is a close correlation between fruit size and stem diameter at the fruit's node (Wolfe and Denton 2001). It would seem that explanations of this sort should not apply to those architectural effects that are characterized by larger distal structures in relation to the basal ones (*Impatiens capensis*, Waller 1982; *Myrosmodes cochleare*, Berry and Calvo 1991; *Aquilegia canadensis*, Kliber and Eckert 2004). Detailed anatomical studies of the vascular architecture of *Avena sativa*, however, reveal that this is not necessarily the case. In that species, increase in seed size from basal to distal positions in the inflorescence (Tibelius and Klinck 1987; Rajala and Peltonen-Sainio 2004) can also be related to an increase in the capacity of the vascular system to transport carbohydrates. The size of the vascular system does decline from the lower to upper internodes, but it does so less rapidly than the number of seeds produced by each node; hence each individual seed developing on the upper positions of the panicle is served by proportionally more vascular tissue than one in the lower positions (Housley and Peterson 1982).

Under most circumstances, the final size of any given organ is proportional to the size of the meristem out of which it has developed (Sinnott 1921; Whaley 1939; Abbe et al. 1941), a relationship dubbed Sinnott's law (Grafius 1978). Regular variation along plant axes in the initial size of primordia may therefore be at the origin of many direct architectural effects that involve variation in organ size and, in the case of multipart organs like flowers, variation in the number of parts, as shown recently by Doust

(2001) for the highly variable flowers of *Drimys winteri* (Winteraceae; see below). Sinnott (1921) showed that, in a single *Acer saccharum* tree, the sum of the leaf-blade volumes of the two leaves borne at a given node was directly and closely correlated with the cross-sectional area of the pith of the internode below, thus revealing a relationship between final leaf size and the size of the leaf primordia when the leaves differentiated. In corn (*Zea mays*) stems, leaf-blade width is directly correlated with the size (cell number) of the shoot apex from which the leaf primordia arose (Abbe et al. 1941). Most likely, a pervasive structural correlation between twig diameter and leaf primordia size also underlies the strong relationship existing across and within species between twig cross-sectional area and surface area of leaves borne by it (one of the so-called Corner's rules; White 1983; Brouat et al. 1998). Although the significance of variation in meristem size as a source of within-plant variation in organ size is expected to be more important for leaves than for flowers, fruits, and seeds (Sinnott 1921), this does not mean that it cannot sometimes apply to position-dependent variation in these structures (Stevens et al. 1972; Doust 2001). In *Linanthus androsaceus*, Stevens et al. (1972) showed that deviations from the normal corolla-lobe number (five) were related to variation in the size of shoot apices. Using two genetically selected lines for increased (SU, "selection up") and decreased (SD, "selection down") number of corolla lobes, these authors tested the hypothesis that apical size is a critical determinant of the number of organs a flower will form. As predicted by the hypothesis, plants of the SU line had significantly broader shoot apices than plants of the SD line. Furthermore, anomalies in corolla lobe number exhibited by SU plants consisted almost exclusively of flowers with more than five petals, whereas anomalies shown by SD plants consisted predominantly of flowers with fewer than five petals.

The role played by variation in primordium size in generating architectural effects is nicely illustrated by Cottrell and Dale's investigation (1984) on size and development of spikelets within the spike of barley (*Hordeum vulgare*). In this plant, final grain size declines from basal to distal positions in the spike, and seeds located in the lowermost positions can be twice as heavy as those at the tip. This pattern is not altered by removal of about half the developing grains in the spike, thus showing that within-spike variation in seed size is purely a direct architectural effect. This is confirmed by observations indicating that carpel size at anthesis varies with position along the spike, and that carpel size is correlated with final grain weight (Scott et al. 1983). Cottrell and Dale (1984) asked whether spikelets and florets along the spike are similar in size at given developmental stages up to carpel differentiation or whether carpel size differences are established very early in spikelet development, thus well before anthesis. By sequentially measuring spikelet primordia at different positions of the developing inflorescence, they demonstrated that for any given developmental stage, average size of spikelet primordia declined from basal to distal positions, as a consequence of development (i.e., spikelet differentiation) progressing relatively faster than growth (i.e., increase in primordium size) in the distal compared with the basal spikelets. They also found a close correlation between the final size of individual grains and the width of the corresponding spikelet primordia at the developmental stage that preceded the differentiation of the glume and lemma initials. Taken together, these results demonstrate that architectural effects on seed size in barley spikes are a consequence of the regular acropetal decline in spikelet primordium size that becomes established during the earliest stages of inflorescence differentiation as a consequence of position-dependent decoupling between developmental and growth rates. It is tempting to speculate that similar mechanisms operate in the spikes of other cereal species that, like barley, exhibit a decline in seed size from basal to distal positions in combination with a correlation between grain size and floret-cavity size. Differential spatial variation of growth and development along plant axes may provide a general proximate mechanism accounting for the many direct architectural effects characterized by regular variation of organs along axes. Depending on whether developmental and growth rates are decoupled along axes and, if they are, on their relative trends of variation along nodal positions, contrasting patterns of longitudinal variation in organ size are possible (fig. 6.2). The results of Cottrell and Dale (1984) on *Hordeum vulgare* would exemplify an scenario similar to the one depicted in figure 6.2b.

Regular position-dependent variation in the shape of meristems, acting in combination with variation in meristem size, can also give rise to direct architectural effects involving variation in size and form of structures. This has been elegantly shown in Doust's exemplary study of the development of *Drimys winteri* (Winteraceae) flowers (2001). In this species, as in other basal angiosperms, there is substantial within-plant variation in the number of floral organs (perianth parts, stamens, and carpels). In *D. winteri* much of this floral variation is position-dependent, and is accounted for by differences between terminal and lateral flowers in the same inflorescence. Terminal floral meristems are generally larger and more circu-

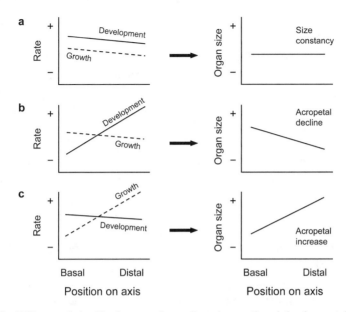

FIG. 6.2 Different relationships between the gradients in growth and developmental rates of organ primordia along plant axes (*left column*) may give rise to contrasting longitudinal patterns of organ size variation (*right column*) and, therefore, to direct architectural effects of different signs. Size constancy along axes would be expected to occur whenever growth and development vary in unison along axis positions (*a*). In contrast, decoupled within-axis gradients of growth and developmental rates may give rise to either acropetal declines (*b*) or increases (*c*) in organ size, depending on the relative trends of variation of growth and development. Only simple linear relationships are shown, but this model could easily be extended to accommodate other types of functions linking growth and developmental rates (e.g., nonlinear), which would likewise result in different trends of within-axis organ variation.

lar than lateral floral meristems, which are smaller and elliptical in shape because of space constraints and the pressures exerted by floral bracts. They are initiated in the axil between its subtending bract and the inflorescence axis. Differences in meristem size translate into organ-number differences, with terminal flowers initiating on average seven more organs than lateral flowers. Differences in meristem shape, in turn, determine variation in floral symmetry and organ arrangement. During the early development of terminal flowers with more or less circular floral meristems, organ primordia are initiated sequentially at a regular divergence angle. In contrast, in lateral flowers with elliptical meristems, organs are preferentially initiated toward the poles. As a result, flowers at different positions in the inflorescence differ rather predictably in aspect ratio and phyllotactic patterns of their floral organs.

Position-dependent gene expression can also eventually give rise to direct architectural effects, as when the effects of some mutations vary with flower position (Diggle 1995, 1997). In *Arabidopsis thaliana*, for example, *leafy* and *apetala*-1 cause the transformation of proximal flowers to inflorescences, whereas distal flowers are only partially affected. In contrast, the effects of some organ identity mutations (e.g., alleles of *ap2*) become more severe in distal flowers within inflorescences. In *Avena sativa*, architectural effects on seed size are largely mediated by predictable within-inflorescence gradients in cell division rate and, eventually, cell number per seed (Rajala and Peltonen-Sainio 2004), which may be related to concomitant position-dependent variations in the expression of one or more genes in a set involved in the control of organ size (Mizukami 2001).

There is growing recognition that seed development in general, and particularly the growth phase characterized by highest cell division rate and accumulation of storage products, is subject to regulation by a highly complex network involving a variety of genes, transcription factors, signaling molecules, and signal-responsive hormones, each of which may interact with others, act independently of others, or even have opposite effects (Brocard-Gifford et al. 2003; Borisjuk et al. 2004; Hills 2004; Gibson 2005). Although the nature of the possible cues driving differential gene expression remains largely unknown, it seems reasonable to suggest that position-dependent variation in any external or internal cue with the capacity to modulate one or more crucial steps of the regulatory network (e.g., signaling molecules such as sugars; Gibson 2005) might ultimately explain some direct architectural effects on seed characteristics that are fully or largely independent of variations in resource supply. This explanation was advanced by Bennett et al. (2003) to account for the gradients in protein content and fatty acid composition of seeds occurring within soybean (*Glycine max*) plants (fig. 4.3). These gradients were resistant to increases in nitrogen availability, as the application of fertilizer did not enhance the amount of protein precursors in the lower nodes. These authors suggested that genes coding for these proteins may be under the influence of some environmental factor(s) that vary within-plants in a position-dependent fashion, and hypothesized that temperature and spectral quality of the light might be playing this role of positional cues influencing gene expression.

Further evidence supporting a role of differential gene expression in determining direct architectural effects comes from detailed studies conducted on rice (*Oryza sativa*) seeds. In this species, grains located on distal branches of the panicle are larger and contain more amylose than those on basal branches (Umemoto et al. 1994; Jeng, Wang, et al. 2003; Zhang et al. 2003; Liu et al. 2005). Within-inflorescence differences in size and amylose content of rice seeds are mainly due to position-specific differences in the activity of enzymes involved in starch synthesis in the endosperm during seed development, particularly the granule-bound starch synthase encoded by the *waxy* gene, which is expressed in the seed endosperm (Umemoto et al. 1994; Umemoto and Terashima 2002; Jeng, Wang, et al. 2003; Hirano and Sano 1998, 2000). Variation in the amylose content of rice endosperm is closely related to the posttranscriptional regulation of the *waxy* gene, and *waxy* expression is sensitive to variation in environmental factors like temperature (Wang et al. 1995; Hirano and Sano 1998; Lin et al. 2005). Within-panicle variation in the size and amylose content of rice seeds might therefore be a consequence of position-dependent variation in the expression of *waxy* associated with microenvironmental variation. Additional support for the hypothesis of genetic control of within-panicle variation in the characteristics of rice seed is provided by a comparison of wild type cultivar Tainung 67 and its sodium azide–induced mutant SA419 (Jeng, Wang, et al. 2003; Jeng, Tseng, et al. 2003). In marked contrast with the wild type, in the mutant the activity of enzymes involved in starch synthesis varies little between the grains located on distal and basal positions of the panicle, and as a consequence, the effect of position in the panicle on both the size and the amylose content of grains is negligible.

Interactions with Animals: Systemic Induction of Defenses

Animals form part of the plant's external environment, one of the three interacting forces inherent to ontogenetic contingency. Any interaction with animals that has the capacity to modify the source-sink relationship within-plants (e.g., herbivores, pollinators, seed predators), or to induce chemical changes that are not homogeneously distributed over the plant, can ultimately modify the magnitude and the spatial pattern of within-plant variation in the characteristics of reiterated structures at both the whole-plant and organ-group levels. Depending on the type of interaction and on specific details, interactions with animals can either enhance or reduce within-plant variation.

In addition to the two main causal pathways depicted in figure 6.1, there is another sectoriality-related mechanism whereby physiological compartmentalization of plants, acting in concert with localized herbivory, promotes within-plant variance in chemical characteristics of reiterated structures, particularly leaves. Secondary metabolites found in leaves can be constitutive (permanent) or induced in response to localized damage by herbivores or pathogens. Induced responses can be systemic or local, when they occur in the same organ that suffered damage (Karban and Baldwin 1997). In the case of local responses, variation between different parts of a plant in their biographies of herbivory generates patchiness in chemical properties of organs, whose duration will depend on the persistence of the induced defense, and whose degree of heterogeneity will depend on the spatial patterning of the damage and the extent of lateral transport of the defensive compounds. Systemically induced defenses can also generate within-plant heterogeneity in chemical characteristics. In this case, the damaged plant tissue produces a signal that is transmitted systemically throughout undamaged parts of the plant, causing the induction of a defensive response elsewhere (León et al. 2001). The molecular signals that elicit the induction of secondary metabolites at remote sites are spread through the plant vascular system; hence the spatial pattern of within-plant distribution of induced systemic responses to localized damage depends closely on vascular architecture (Orians 2005). This has been demonstrated for species of *Populus*, *Nicotiana*, and *Solanum* (Davis et al. 1991; Jones et al. 1993; Shulaev et al. 1995; Orians et al. 2000; Schittko and Baldwin 2003; Viswanathan and Thaler 2004). In these species, systemic chemical induction tended to be greatest in orthostichous leaves, lower in leaves in adjacent orthostichies, and minimal in leaves in opposite orthostichies. In other words, plant vascular architecture caused systemic induction to be sectorial, rather than uniformly distributed throughout the plant, thus enhancing within-plant variability in the concentration of defensive secondary metabolites. In these and other similar cases (Shelton 2005), damage by herbivores may significantly enhance within-plant variation.

Although induction of chemical defenses by herbivore damage was first demonstrated for proteinase inhibitors (Green and Ryan 1972), three decades of active research in this field have shown that almost every secondary metabolite known to play a defensive function in plants is also subject to induction (Karban and Baldwin 1997, table 3.2). This means that, in principle, sectoriality-mediated heterogeneity in systemic induction can affect most types of secondary compounds and, therefore, that their within-plant heterogeneity in concentration is partly attributable to that phenomenon. The degree of within-plant variability in secondary metabolite concentrations attributable to this effect, however, is expected to be influenced by the specific distribution and magnitude of the damage. Orians and Jones (2001) predicted that as damage extent increases, plants will first become internally more heterogeneous and then less heterogeneous. If damage is localized within one vascularly connected sector, heterogeneity would be maximal, but as damage spreads across most or all sectors, heterogeneity should decrease. Some empirical evidence is consistent with this intuition, as several studies have shown that severe wounding may effectively reduce the constraining effects of vascular architecture on the movement of the induced compounds (Jones et al. 1993; Rhodes et al. 1999; Viswanathan and Thaler 2004).

Within-plant variation in the concentration of secondary metabolites may also take place at the reduced within-leaf level (chapter 2). Sectoriality-mediated heterogeneity of systemic induction might contribute to generating within-plant variation in concentration of secondary metabolites at this restricted spatial scale as well, although this aspect remains essentially uncharted territory. Within leaves, systemic induction may affect only some particular leaflets or leaf halves, and patterns of within-leaf variation can be satisfactorily explained by consideration of detailed aspects of vascular architecture, specifically the number of vascular bundles provisioning each leaf and how they are shared among contiguous orthostichies (Rhodes et al. 1999; Orians et al. 2000; Viswanathan and Thaler 2004). Nevertheless, within-leaf heterogeneity in concentration of secondary metabolites may in some instances be unrelated to the spatial pattern of vascular bundles in the leaf. In *Raphanus sativus*, Shelton (2005) found that the extreme within-leaf patchiness in the concentration of glucosinolates (accounting for 57% of total variance) was spatially random and did not exhibit spatial autocorrelation at the within-leaf scale.

Interactions with Animals: Spatiotemporal Distribution of Resources

The influence of herbivory on within-plant variation can extend beyond its effects on the systemic induction of defenses. By altering the overall amount of resources available for the development of reiterated structures, or modifying the spatiotemporal distribution of these resources within plants, herbivores may contribute to modifying the extent and spatial distribution of within-plant variation. For example, by feeding on the subtending leaves that supply resources to inflorescences, herbivores may produce steeper within-inflorescence acropetal gradients in resource availability, to the detriment of late-season, distally located flowers and fruits. Although a number of studies have demonstrated effects of foliar herbivory on *mean* flower, fruit, and seed traits (Frazee and Marquis 1994; Aizen and Raffaele 1996; Strauss et al. 1996; Strauss et al. 2001), I am not aware of any investigation specifically relating leaf herbivory to levels of within-inflorescence variation. The relationship suggested here, however, is supported by results of some experiments showing that the magnitude of within-inflorescence variation increases after manipulations mimicking foliar herbivory. In *Aquilegia canadensis*, several flower and fruit traits decline with flower sequence in the inflorescence. To ascertain the relative importance of architectural effects and resource competition as causes of this pattern, Kliber and Eckert (2004) reduced resource availability to experimental plants by removing basal rosette leaves and cauline leaves subtending flowers. Although these authors did not test for the influence of simulated herbivory on the magnitude of within-inflorescence variability, a simple reanalysis of some of their data clearly supports the predicted relationship. For each trait and experimental group (control vs. defoliated), I computed the ratio between the largest and smallest means of the four flower-sequence positions considered. These ratios can be used as rough estimates of within-inflorescence variability. The ratio of within-inflorescence extreme values was larger for defoliated inflorescences than for controls for all traits (fig. 6.3), thus supporting the predicted direct relationship between foliar herbivory and within-inflorescence variation through reduction in resource availability.

Seed predation on immature seeds can also modify within-inflorescence variation in fruit or seed traits, but its effects generally run opposite to that of herbivory on basal leaves. By removing competing metabolic sinks, seed predators may indirectly favor fruits and seeds located more distally and initiated later, which may eventually translate into decreased within-inflorescence variation in size-related fruit and seed traits. Some results of Marshall et al. (1985) for *Sesbania* are consistent with this scenario. In the perennial *Sesbania drummondii*, the incidence of seed predators varied greatly between two study seasons. In one of the years the incidence of seed predators was negligible, and there was a significant decline in seed size toward the distal extreme of the inflorescence. In the other year, seed destruction by seed predators was extensive, and the size of seeds did not vary significantly with position in the inflorescence, which is consistent with an "homogenizing" effect of seed predation.

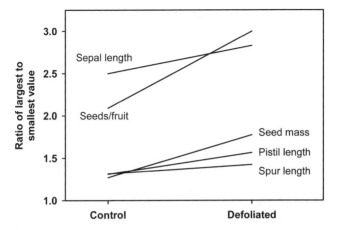

FIG. 6.3 By reducing the amount of resources available for flower and fruit development, foliar herbivory may increase the magnitude of within-inflorescence variability in those flower, fruit, and seed traits that decline with position in the inflorescence as a consequence of resource competition. For five traits that decline with flower sequence position in the inflorescences of the perennial herb *Aquilegia canadensis*, the graph illustrates the difference between control and defoliated (basal rosette leaves removed) plants in the ratio between the largest and smallest means of the four flower-sequence positions. Based on Kliber and Eckert 2004. Numerical data were obtained from graphs in their figs. 2 and F1.

In cases where within-inflorescence variation in flower and fruit traits is at least partly due to depletion of resources by early-developing structures, pollinators can also contribute to modify the magnitude of within-plant variability, and geographical or seasonal variation in pollinator abundance can result in variation in the extent of within-plant variability. In species where fruit set is related to pollinator visitation, the degree of resource depletion within inflorescences will be directly related to pollinator visitation and pollination success, and one would predict that within-inflorescence variation will increase with increasing pollinator visitation. This prediction is supported for *Hydrophyllum appendiculatum*, where increased pollination intensity results in enhanced within-plant variability in seed weight and inflorescence size (Wolfe 1992). Differences between flowers of the same plant in the identity of pollinating agents may also give rise to within-plant variation in fruit and seed features whenever pollinators differ in the quality or quantity of pollen delivered to stigmas, a possibility that does not seem to have been explored to date. In species where such effects occur, individual plants whose flowers are pollinated by more diverse pollinator arrays should be expected to exhibit larger variation in those fruit and seed features that are contingent on the quality and quantity of pollen received.

Organ-Level Developmental Instability

Developmental stability, or "developmental homeostasis," is the ability of organisms to produce consistent phenotypes despite fluctuations in their internal and external environments (Mather 1953; Lerner 1954). Markow (1995, 105) offered a second definition of developmental stability as "the situation achieved when an organism has adequately buffered itself against epigenetic perturbations, displaying its developmentally programmed phenotype." In the first definition, the expression "consistent phenotypes" should not be taken to mean "identical phenotypes," but rather phenotypes conforming consistently with those predicted as a function of the prevailing environmental conditions and reaction norms for the organs involved (i.e., the "developmentally programmed phenotype" in Markow's definition). Homeostasis and its opposite, developmental instability, can be used in relation to either whole organisms or individual organs. I am mainly concerned here with the latter meaning. Failure to produce organ phenotypes that are closely consistent with those expected from a predetermined developmental and morphogenetic plan may contribute to within-plant variation in the characteristics of reiterated structures.

Organ-level developmental instability may be seen as the inability of the developmental and morphogenetic program inherent to an organ to eventually generate copies whose features conform exactly to those predicted by the organ-level specific reaction-norm function for the current internal and external environment. As defined in chapter 5, organ-level reaction norms reflect developmental plasticity and represent *average* functional relationships linking organ phenotypes with the external and internal microenvironments. They are better seen as probabilistic relationships rather than errorless, perfectly deterministic functional relationships. In this view, trait value departures from the *average* organ-level reaction norm (i.e., noise around the functions depicted in figure 5.3) denotes developmental instability. In other words, organ-level developmental instability accounts for the fraction of within-plant variance in phenotypic organ features that remains unexplained after the organ's specific reaction norm to changing internal and external microenvironment has been comprehensively accounted for.

For purely stochastic reasons, the phenotypes of two genetically identical organs exposed to identical internal and external microenvironments during their initiation, growth, and maturation, may still differ to some extent due to "developmental noise" or "developmental error." Such variability seems inevitable in biological systems, at least because of a random component in chemical reactions within cells (McAdams and Arkin 1999; Elowitz et al. 2002; Blake et al. 2003). Temperature-dependent stochasticity in chemical reactions within cells ("thermal noise"; Soulé 1982), and particularly stochasticity in the set of reactions that control the abundance of gene products, will generate heterogeneity in the response of a population of cells to inducing stimuli and limit the accuracy of cellular and developmental processes. Living cells possess low copy numbers of many components, including DNA and important regulatory molecules; thus stochastic effects or "noise" in gene expression may account for differences between cells that are otherwise genetically identical (Raser and O'Shea 2005; Kærn et al. 2005). In clonal populations of genetically identical unicellular organisms, stochasticity in gene expression leads to substantial phenotypic variation (Elowitz et al. 2002; Sumner and Avery 2002; Raser and O'Shea 2004), and similar mechanisms might also account for the instability of the phenotypes in multicellular organs. Even though cells of multicellular organisms possess distinctive molecular devices that are dedicated to filtering noise in gene expression (Martinez Arias and Hayward 2006) and buffering morphogenetic responses from the destabilizing effects of stochastic processes (Queitsch et al. 2002), these molecular mechanisms themselves may also be susceptible to a certain error rate. This would allow some unfiltered stochasticity of gene expression to cascade into the macroscopic phenotypic scale and translate into developmental noise and phenotypic instability. For example, in *Arabidopsis thaliana* organ size is regulated through coordination of cell division and expansion involving a broad range of transcription factors and other signaling molecules, and is controlled by the action of an organ-size checkpoint gene (Mizukami 2001). Stochasticity in the expression of any gene involved in the regulatory network, and particularly of the organ-size checkpoint gene, if not fully cancelled by some molecular filtering mechanism, may eventually generate developmental noise and random deviations in organ size.

There is yet another reason for unavoidable random scatter of organ trait values around the average predicted by organ-level reaction norms. Organ development takes some time, and the internal and external environments do not remain constant during the whole period of initiation and development. With regard to the external environment, consider for example that individual developing leaves, flowers, or fruits are not permanently in exactly the same spot (supporting structures grow, and are irregularly shaken by wind), which means that the developmental and morphogenetic program underlying the organ-level reaction norm will "perceive" a succession of slightly different phylloclimates rather than a static, invariant one. This inherently noisy external microenvironment will likely contribute to the appearance of noisy phenotypic responses around the mean values predicted from consideration of average phylloclimate.

Measuring Instability: Systematic and Random Components of Within-Plant Variation

Studies of developmental instability in plants are comparatively scarce in relation to the remarkable number of investigations conducted on the subject in animals in the last few decades (Parsons 1990; Markow 1995; Møller 1997; Swaddle 2003). Furthermore, almost all the few plant-oriented studies of developmental instability have focused on fluctuating asymmetry (Freeman et al. 1993). Fluctuating asymmetry is a particular symptom of developmental instability that involves slight departures from identical expression of a trait in (1) paired organs located at the two sides of some obvious axis of bilateral symmetry, such as opposite leaves in the same node or petals at both sides of zygomorphic flowers; or (2) the two halves of bilaterally symmetrical structures like leaves (Møller and Eriksson 1994; Evans and Marshall 1996; Cowart and Graham 1999; Perfectti and Camacho 1999; Llorens et al. 2002; Díaz et al. 2003). The appealing simplicity of the a priori null hypothesis of developmental stability tested by fluctuating asymmetry studies (i.e., perfect bilateral symmetry) probably explains much of the extraordinary popularity enjoyed by these investigations (Palmer 1996).

Fluctuating asymmetry has been traditionally used as a proxy for (Mather 1953), or interchangeably with (Markow 1995; Møller and Shykoff 1999) developmental instability, although it actually represents only a particular subclass of a broader, more-encompassing phenomenon. Recognizing this distinction is particularly pertinent in the case of plants. While fluctuating asymmetry is often the only instability measure available in animals, in plants a multiplicity of developmentally equivalent structures, peculiarities of growth form, and diverse intrinsic gradients of phenotypic variation (e.g., regular variation along stem or inflorescence axes; chapter 4), combine to open a considerable range of possibilities for measuring developmental instability beyond fluctuating asymmetry (Freeman et al. 1993; Freeman et al. 1996; Sherry and Lord 1996b; Alados et al. 1999; Alados et al. 2001). However, fluctuating asymmetry is for the most part not directly relevant to the issue of differences among reiterated organs of the same plant; hence I do not give separate consideration to it here (for treatments of fluctuating asymmetry in paired or bilaterally symmetric plant organs see, e.g., Møller and Eriksson 1994; Jennions 1996; Møller and Shykoff 1999).

In comparison to the abundant literature on methods for measuring and analyzing developmental instability *within individual organs* (Palmer and Strobeck 1986; Merilä and Björklund 1995; van Dongen et al. 1999), the more general issue of measuring developmental instability *among organs* has received relatively little attention (Freeman et al. 1993). Whenever some predictable within-plant gradient in the characteristics of an organ exists (e.g., in relation to position along a stem or inflorescence axis), the within-plant variance of the trait under consideration can be decomposed into two distinct components, namely the fraction due to developmentally predictable, patterned variation attributable to the aggregate effect of ontogenetic contingencies impinging upon organ development, and the fraction due to random deviations around this pattern. This distinction was already acknowledged by Pearson in 1901, when he distinguished between variation due to organ differentiation related to function, position on the individual, or season of production, and variation due to chance events alone. In a footnote aimed as a rebuttal to some of Bateson's criticisms (1901) of his publication on homotyposis, Pearson wrote (287): "A diversity due to differentiation and a variability due to chance are quite distinct things. The one is the result of dominating factors which can be isolated and described; the other of a great number of small factors, varying from organ to organ, and incapable of being defined or specified. Indeed, upon each dominating factor of differentiation is superposed such a chance variability." Half a century later, Paxman (1956) called these the "differentiation" and "instability" components of within-plant variation, respectively. Freeman et al. (1993), referring to these same components of within-plant variation, distinguished between a "developmental invariant" providing an a priori idealized state of variation, and "developmental instability" around such invariant, represented by deviations away from the ideal. For simplicity, I refer to these here as the "systematic" and "random" components of intraplant variation. When it exists, the systematic component may be interpreted as the realization of some putative organ-level reaction norm in response to variation in some external (e.g., incident light) or internal (e.g., hormone or nutrient levels) factors that, by varying predictably along the gradient considered, induce a predictable variation of organs (e.g., regular decline in leaf size with nodal position along the stem, or predictable changes in specific leaf weight with depth within the canopy).

Consider the simplest case whereby a quantitative character of a re-iterated structure, for example, leaf size, is linearly related to its position along some intrinsic or extrinsic plant gradient, for example, nodal position along a stem or depth into the canopy (chapter 4). By designating leaf area as y and position on the gradient as x, and using the standard notation for linear models, the area of leaf i may be modeled as

$$y_i = a + b \cdot x_i + \varepsilon_i, \quad 6.1$$

where $b \cdot x_i$ represents the systematic component and ε_i stands for the random component of variation (fig. 6.4a). In the particular case when $b = 0$, there would be no organ-level plastic response to position-dependent changes in internal or external influential factors (i.e., a flat organ-level reaction norm), and all observed variation in the trait would consist of purely random deviations around a constant mean value a, regardless of position along the gradient (fig. 6.4b).

Equation 6.1 may be generalized to accommodate any functional relationship linking organ trait value and its coordinate on the intrinsic or extrinsic plant gradient under consideration,

$$y_i = f(x_i) + \varepsilon_i. \quad 6.2$$

This generalized equation might describe, for example, the organ-level norm of reaction of a continuous character (fig. 5.3), around which some developmental stochasticity is also to be expected. Assuming that sampling and measuring error are both negligible (or separately accounted for), the phenotypic variance remaining after statistically accounting for any existing functional relationship of the type $f(x)$ is attributable to developmental instability. This approach to the study of developmental stability was first formulated and applied in Paxman's pioneering study (1956) on differentiation along stem nodal positions and developmental stability in

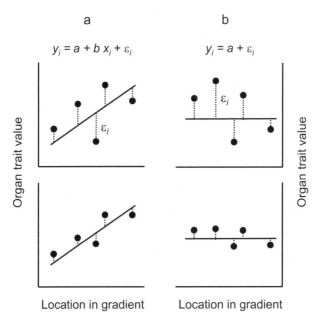

FIG. 6.4 Schematic representation of the systematic and random (developmental instability) components of within-plant variation in trait value of a reiterated structure in relation to some extrinsic (e.g., light intensity) or intrinsic (e.g., nodal position along a branch or linear inflorescence) plant gradient. In *a*, within-plant trait variance is due to the combination of a systematic trend of variation (represented by a linear relationship) and a random component attributable to developmental instability (random departures from the regression line, ε_i). In *b*, no systematic component of within-plant variation exists ($b = 0$), and all variance is due to random deviations around a constant trait value. Upper graphs denote situations of greater developmental instability than lower ones.

the flowers and leaves of *Nicotiana rustica*, and has been adopted by later investigators (Freeman et al. 1993; Sherry and Lord 1996a). In the particular case of $f(x)$ = constant, that is, when no discernible functional relationship exists between the phenotype of an organ and its position along the plant gradient considered, all variance would therefore be attributable to developmental instability. Using Paxman's terminology (1956), figure 6.4a depicts a significant differentiation component defined by the regression line, plus a developmental instability component denoted by the scatter of points around the regression line. Figure 6.4b, in contrast, reflects only instability, since the differentiation component would be null (regression slope equals zero).

It is important to emphasize that function $f(x)$ in equation 6.2 provides only a subtraction criterion for the dissection and analysis of within-plant phenotypic variability. Under that explicit formulation, the random component of variation will closely reflect "developmental instability" if and only if all other possible causes related to systematic variation have been either ruled out or statistically accounted for (Paxman 1956; Freeman et al. 1993). This exclusion of alternative causes may come about through either (1) testing and rejecting one or more hypothesized functional relationships, leading to acceptance of the condition that $f(x)$ is constant, as in Sherry and Lord's study of floral instability (1996a) referred to below; or (2) if some hypothesized functional relationship is supported (i.e., $f(x)$ is not constant), statistically removing the effect of such relationship. Developmental instability will thus account for, *at most*, the fraction of variation that remains (statistically) unexplained after verification or falsification of one or more hypothesized trends of variation implicating other plausible internal or external causes. In other words, the fraction of within-plant variation that cannot be explained in terms of genetic mosaicism, architecture, plastic developmental patterns, or some complex combination of these, can be attributed to developmental instability. Because of its residual nature, estimates of the relative importance of developmental instability in causing within-plant variation will therefore be as good and reliable as our estimates of the shape of $f(x)$ for the particular set of conditions under consideration. If the hypotheses of functional relationships tested are insufficient, poorly chosen, or biologically unrealistic, the importance of developmental instability will be overestimated. For this reason, estimates of developmental instability based on subtracting the variability imputed to predictable patterns must be interpreted as upper limits rather than actual estimates, since there is always the possibility that some architectural or developmental effect contributing to systematic variation went unrecognized.

Developmental Instability and Variability of Reiterated Structures

Although developmental instability has often been considered an important cause of within-plant variation of reiterated structures in plants (Heslop-Harrison 1959; Roy 1963; Sakai and Shimamoto 1965; Barrett and Harder 1992), I am aware of few studies that have directly addressed its relative importance in comparison to other causes of within-plant variation.

The general framework depicted in figure 6.4, which partitions observed within-plant variance into its systematic and random compo-

nents, is also useful in examining the significance of developmental instability in generating within-plant variation in discrete, meristic traits. Two
illustrative examples are provided by studies of intraplant variation in
organ numbers in flowers of *Actaea rubra* (Ranunculaceae; Lehmann
and Sattler 1994) and *Drimys winteri* (Winteraceae; Doust 2001). In the
first species, the number and position of petals and stamens vary considerably among flowers of the same plant as a consequence of homeotic
transformations of floral primordia. This variation is not predictably associated with the intrinsic gradient represented by position of the flower in
the inflorescence (upper, middle, and basal positions). Consequently, the
respective $f(x)$ for these floral traits in relation to inflorescence position
would be constant, and developmental instability would account for all
within-plant variation in floral organ number in this species. In contrast,
for plants of *Drimys winteri* the number of floral organs varies predictably
depending on the position of the flower in the inflorescence (terminal
vs. lateral); hence $f(x)$ is not constant, and partitioning within-plant variance into its systematic and random components is justified. Doust (2001,
table 7, and unpublished data) partitioned the total within-plant variance
in organ numbers into the systematic component due to position in the
inflorescence and the random component attributable to developmental instability. His results indicate that, in *Drimys winteri*, developmental
instability may account for up to 73–98% of within-plant variance in petal
number and 74–100% of variance in carpel number, but only 18–38% of
within-plant variance in stamen number.

As noted in chapter 2, within-plant variation in floral organ number is
not restricted to species that, like *Actaea* or *Drimys*, belong to families in
or near the magnoliid clade, which are well-known for their variable floral
merosity. Variability in floral merosity also occurs in species belonging to
families or genera where floral part number is a constant, taxonomically
diagnostic character (family-specific "floral formulas"). In the few cases
where this floral inconstancy has been examined in some detail, developmental instability seems a major cause of within-plant variation in meristic floral traits (*Nyctanthes, Linanthus, Ipomopsis, Spergularia*; Roy 1963;
Huether 1968, 1969; Ellstrand 1983; Ellstrand and Mitchell 1988; Mazer
and Delesalle 1996).

Continuous traits are more amenable to the dissection of within-plant
variance into systematic and random components, but there have been few
attempts at studying intraplant variation from that perspective. In *Clarkia
tembloriensis*, Sherry and Lord (1996a) found no significant relationship

between nodal position and either overall flower size or length of floral organs. In this case, therefore, $f(x)$ is constant, and within-plant variation in these floral features could properly be interpreted as random fluctuations around mean values mainly reflecting developmental instability. All floral traits exhibited considerable within-plant variation, as judged by the relatively large mean CV_{within} values for style length (CV_{within} = 10.7%), short filament length (8.2%), long filament length (7.5%), petal length (9.0%), petal width (10.0%), sepal length (6.6%), hypanthium length (16.3%), and ovary length (12.0%; Sherry and Lord 1996a, table 6). These figures fall within the range of variation of CV_{within} for floral traits for a large number of species summarized in table 3.2 and figure 3.2, which may be an indication that these latter predominantly reflect variation stemming from developmental instability.

Investigations of intrinsic and extrinsic gradients of within-plant variation in leaf traits provide useful information to assess the relative importance of developmental instability in determining within-plant variation. I estimated the systematic and random components of variation in leaf structural and compositional traits by dissecting total variance in the trait into components explained (R^2) and unexplained ($1 - R^2$) by fitted functional relationships $f(x)$ linking the trait with intrinsic (nodal position) or extrinsic (e.g., irradiance) within-plant gradients (table 6.3). When the systematic component of leaf trait variation is evaluated in relation to intrinsic gradients, estimates of developmental instability are relatively high, ranging between 0.29 and 0.65 (mean = 0.44). When variation in relation to extrinsic, light-based gradients is considered, the corresponding figures are much lower, in the range 0.03–0.37 (mean = 0.17). These figures are similar to Cowart and Graham's estimates (1999) for the proportion of within-plant variation in *Ficus carica* leaf width and lobe length attributable to random developmental variation (0.08–0.21), obtained using a method based on fluctuating asymmetry measurements. Estimates of developmental instability shown in table 6.3 obtained after accounting for systematic variation related to extrinsic and intrinsic gradients refer to different species, and thus are not strictly comparable. Nevertheless, it is tempting to interpret the difference between the two groups of estimates as indicative of more precise systematic variation and, consequently, smaller developmental instability, of leaf trait in relation to extrinsic gradients (light microenvironment) rather than to intrinsic gradients (nodal position).

Despite the presumed importance of developmental instability as a source of within-plant variation, direct quantitative evidence of its pro-

TABLE 6.3 **Estimates of systematic and random variation in leaf structural and compositional traits.**

Type of within-plant gradient	Species	Leaf trait (y)	Within-plant gradient (x)	Components of within-plant variance (proportion of total)		Reference[c]
				Systematic[a]	Random[b]	
Intrinsic	Cecropia obtusifolia	Nitrogen concentration	Nodal position on stem	0.61	0.39	3
	Clarkia tembloriensis	Leaf length	Nodal position on stem	0.48–0.64	0.36–0.52	11
	Ficus insipida	Nitrogen concentration	Nodal position on stem	0.60	0.40	3
	Heliocarpus appendiculatus	Nitrogen concentration	Nodal position on stem	0.71	0.29	3
	Sassafras albidum	Chlorophyll content	Nodal position on stem	0.51–0.53	0.47–0.49	6
	Urera caracasana	Nitrogen concentration	Nodal position on stem	0.35	0.65	3
Extrinsic	Betula pendula	Specific weight	Relative irradiance	0.63	0.37	2
	Corylus avellana	Specific weight	Relative irradiance	0.80–0.91	0.09–0.20	2, 5
	Crataegus monogyna	Specific weight	Relative irradiance	0.97	0.03	8
	Fagus orientalis	Specific weight	Integrated quantum flux density	0.91	0.09	12
	Fagus sylvatica	Specific weight	Integrated quantum flux density	0.87–0.92	0.08–0.13	12
	Fagus sylvatica	Specific weight	Relative irradiance	0.85–0.91	0.09–0.15	5, 8
	Fragaria virginiana	Specific weight	Integrated light energy	0.96	0.04	9
	Fraxinus excelsior	Specific weight	Relative irradiance	0.88	0.12	5
	Ilex aquifolium	Specific weight	Relative irradiance	0.85	0.15	8
	Ilex aquifolium	Specific weight	Integrated quantum flux density	0.70–0.84	0.16–0.30	10
	Juglans regia	Specific weight	Daily photon flux	0.70–0.77	0.23–0.30	1, 7
	Lonicera xylosteum	Specific weight	Relative irradiance	0.79	0.21	2
	Nothofagus fusca	Specific weight	Photon flux density	0.71	0.29	4
	Populus tremula	Specific weight	Relative irradiance	0.83	0.17	5
	Prunus avium	Specific weight	Relative irradiance	0.91	0.09	8

(continued)

TABLE 6.3 *(continued)*

Type of within-plant gradient	Species	Leaf trait (y)	Within-plant gradient (x)	Components of within-plant variance (proportion of total)		Reference[c]
				Systematic[a]	Random[b]	
	Quercus petraea	Specific weight	Relative irradiance	0.80	0.20	8
	Quercus pyrenaica	Specific weight	Relative irradiance	0.73	0.27	8
	Rhododendron ponticum	Specific weight	Integrated quantum flux density	0.81–0.83	0.17–0.19	10
	Sorbus aria	Specific weight	Integrated quantum flux density	0.86	0.14	8
	Sorbus aucuparia	Specific weight	Integrated quantum flux density	0.83	0.17	8
	Tilia cordata	Specific weight	Integrated quantum flux density	0.92	0.08	5

Note: Estimates determined by dissecting total variance of the trait into components explained (R^2) and unexplained ($1 - R^2$) by a fitted functional relationship $f(x)$ linking the trait with some intrinsic or extrinsic within-plant gradient (x) (fig. 6.4). The component of variance unexplained by $f(x)$ sets an upper limit to the variance due to developmental instability. All species listed are trees except *Clarkia tembloriensis* and *Fragaria virginiana* (herbs) and *Lonicera xylosteum* and *Rhododendron ponticum* (shrubs).

[a] R^2 for fitted $y = f(x)$.

[b] $(1 - R^2)$.

[c] 1, Klein et al. 1991; 2, Kull and Niinemets 1993; 3, Traw and Ackerly 1995; 4, Hollinger 1996; 5, Niinemets and Kull 1998; 6, de Soyza et al. 1990; 7, Le Roux et al. 1999; 8, Aranda et al. 2004; 9, Chabot et al. 1979; 10, Niinemets et al. 2003; 11, Sherry and Lord 1996b; 12, Fleck et al. 2003.

portional importance in relation to other causes is rather scarce, being mostly referred to variation in leaf traits, as shown by the preceding paragraphs. Nevertheless, two further lines of evidence give indirect support to the notion that developmental instability can often be a major cause of within-plant variation: (1) a direct relationship has frequently been demonstrated or inferred between factors known to increase developmental instability and the magnitude of within-plant variation; and (2) in all cases where spatial pattern of within-plant variation in a trait has been examined with a sufficient degree of resolution, most within-plant variation takes place at very small spatial scales and lacks spatial autocorrelation. I summarize the relevant evidence below.

If within-plant variation in the characteristics of reiterated structures is partly a consequence of developmental instability operating at the organ level, then the magnitude of the former will be affected by the same factors known to impinge on the latter. The level of developmental instability of an organism is influenced by both environment and genotype, as evidenced by different strains displaying different levels of stability under the same environmental conditions, and identical strains showing different levels of stability under different environmental conditions (Clarke 1993; Møller and Shykoff 1999). Increased homozygosity and disruption of coadapted gene complexes that arose through genomic coadaptation are the two main genetic factors thought to increase developmental instability (Mitton and Grant 1984; Palmer and Strobeck 1986; Clarke 1993; see Pertoldi et al. 2006 for a recent review of the relationship between developmental instability and "genetic stress"). In a meta-analysis of the effects of genetic factors hypothesized to increase developmental instability in plants (homozygosity, hybridization, mutation, and genetic differences among individuals), Møller and Shykoff (1999) showed variable but usually significant effects, although the number of studies included was generally small. Among environmental causes, a broad variety of factors have been reported to increase developmental instability in plants, including pollutants, light, temperature, moisture, soil nutrients, herbivory, parasitism, competition, and radioactive, electromagnetic, and ultraviolet radiation (Møller and Shykoff 1999). Considerable evidence supports a causal relationship between environmental stress involving these factors and increased developmental instability (Freeman et al. 1993; Parsons 1990, 1992; Markow 1995; Alados et al. 2002), so that it has become customary to use developmental instability as a biomonitoring tool for inferring the quality of the environment in which organisms are reared and for assessing environmental stress before fitness components become influenced (Freeman et al. 1996; Freeman et al. 1999; Freeman et al. 2005; Leary and Allendorf 1989; Palmer 1996; Kozlov and Niemelä 1999; Alados et al. 2001; for some objections and caveats see Anne et al. 1998; Heard et al. 1999; Duda et al. 2004; Murphy and Lovett-Doust 2004).

All else being equal, within-plant variability of reiterated structures would tend to be comparatively larger in hybrids (assuming hybridization generally disrupts coadapted gene complexes), in inbred populations, and in populations or individuals that are exposed to distinctly suboptimal values of one or more potentially influential environmental factors. Some studies support the predicted relationship between heterozygosity and within-plant variability. In *Clarkia tembloriensis*, plants from more homozygous populations tend to have greater within-plant variance in leaf size over developmentally comparable stem nodes than plants from more heterozygous populations (Sherry and Lord 1996b). Seyffert (1983) likewise found an inverse relationship between mean heterozygosity and within-genotype variability in the anthocyanin content of the flowers of *Matthiola incana*.

Experimental and observational evidence link environmental stress and within-plant variation. Investigations conducted under controlled greenhouse conditions have demonstrated that the magnitude of within-plant variation tends to increase after exposure of plants to some environmental stress. This effect was demonstrated for male plants of *Cannabis sativa* by Heslop-Harrison (1959), who showed that experimental exposure to stress in the form of cold night temperatures considerably increased within-plant variability in floral traits. Plants exposed to a temperature regime of 22°C during the light period and 10°C during the dark had considerably larger within-plant variance in tepal and stamen number than control plants remaining throughout at 22°C, while mean values for floral traits were similar for control and experimental plants. Similar effects have been reported for experimental plants of *Ipomopsis longiflora* grown under controlled conditions and exposed to different levels of water stress. In this case, the average coefficient of variation of corolla-tube length increases steadily from low through moderate to high water stress levels (Villarreal and Freeman 1990). In the tristylous species *Eichhornia paniculata*, plants of the midstyled morph exhibit variability in the position of short-level stamens, the extent of which varies depending on stress level (Barrett and Harder 1992). Plants grown in large pots, fertilized regularly, and getting sustained water levels (unstressed) had only 3% of flowers with abnormally elongated short stamens. In contrast, plants in small pots, unfertilized, and subjected to periodic drought stress had 31% of flowers with abnormal short stamens.

Correlative evidence from natural plant populations documenting relations between environmental stress and within-plant variability is more difficult to interpret unambiguously in terms of cause and effect than are results of greenhouse or growth-chamber studies. Such field data, however, are particularly valuable in that they demonstrate that variations in within-plant variability can take place in the range of environmental conditions actually faced by organisms in the wild. In *Lupinus perennis* and *Purshia tridentata*, within-plant variability in seed size increases under stressful conditions caused by intraspecific competition and herbivory (Krannitz 1997a; Halpern 2005). In three populations of *Linanthus androsaceus* studied by Huether (1969), the proportion of plants having one or more flowers deviating from the normal pentamerous corolla was far greater among plants that had been decapitated by ground squirrels at the seedling stage (30–33%) than among those that had escaped herbivory (3–8%). Huether did not interpret this difference as a consequence of herbivory per se; instead his explanation was based on the fact that chewed plants had to develop their new flower primordia later in the season, when both the length of daylight and the extreme temperatures are greater than would normally have been encountered by the plants. These two factors would act as an environmental stress on the plants, leading to an increase in the number of abnormal flowers produced. This interpretation was corroborated experimentally by Huether (1968). Seasonally increasing or decreasing levels of within-plant variability in floral traits has been found in other species and interpreted as a consequence of seasonally variable environmental stress (Roy 1959, 1963; Ellstrand and Mitchell 1988). In trees of *Nyctanthes arbor-tristis*, for example, Roy (1959) found that individual means for the number of petals per flower changed over the flowering season by only 2–4%, while the corresponding standard deviations changed by 20–40%. In the perennial herb *Helleborus foetidus*, the cymose inflorescences produce flowers continuously over the three-month-long winter flowering period. At a southeastern Spanish montane locality, within-plant variability in flower size was greater for flowers opening in February, when plants were exposed almost daily to freezing temperatures, than for those opening in March, when the ambient temperature was warmer and there were no frosts (fig. 6.5).

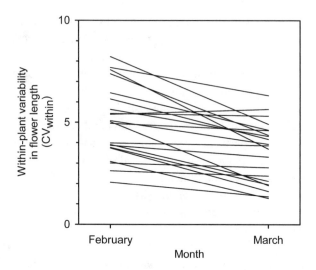

FIG. 6.5 Within-plant variability in flower size in the winter-flowering perennial herb *Helleborus foetidus* declines significantly ($F_{1,20}$ = 9.19, P = 0.007) from early- to late-opening flowers, a pattern that is consistent with the prediction of greater developmental instability under cold stress. Each line joins the early and late variability values of one plant (N = 21 plants). In February 39% of the days had frost, and the monthly mean of daily minimum temperature was 1.8°C. The thermal environment was considerably less stressful in March, when the corresponding figures were 0% and 5.8°C. C. M. Herrera, unpublished data.

The extent of developmental instability may vary among different parts of the same plant, and existing data on within-plant variation in the extent of variability are also consistent with an important role of developmental instability in generating within-plant variation. In *Ficus carica*, leaves located in the crown periphery are more asymmetrical than those inside the crown, and this effect was interpreted as a consequence of the fact that the outer crown is a more stressful environment for leaves than the inner crown, as the former are subjected to greater cold, heat, ultraviolet and visible light, and desiccation (Cowart and Graham 1999). Likewise, leaves located at the top of *Tectona grandis* trees are more developmentally unstable than those located at middle and lower crown positions (Bagchi et al. 1989). In *Annona cherimola* trees grown under Mediterranean-climate conditions, fluctuating asymmetry of south-facing leaves is greater than that of north-facing ones, probably a consequence of the more stressful conditions experienced by the former, sunnier locations (Perfectti and Camacho 1999). In addition to providing evidence supporting developmental instability as a source of within-plant variation, these examples also illustrate the point that, due to environmentally induced variation in the accuracy of developmental regulation, the amplitude of within-plant variation may be very labile and context-sensitive, varying not only among but also within plants. This implies that variability itself will often be spatially textured within plants, with more and less variable zones and, accordingly, with intraplant gradients in variability running between zones of differing variabilities.

Further suggestions that developmental instability may be an important source of within-plant variation come from studies showing that, when spatial patterns of within-plant variation in a trait are examined with a sufficient degree of resolution, a substantial fraction of within-plant variation takes place at very small spatial scales, is randomly distributed, and lacks spatial autocorrelation (chapter 4). One of these examples is provided by Luyssaert et al.'s study of intratree variation of cadmium concentration in leaves of *Salix fragilis* (2001), described in chapter 4. Randomness at even smaller spatial scales has been also reported. After herbivory by lepidopteran larvae, the distribution of induced glucosinolates in leaves of *Raphanus sativus* was spatially random and did not exhibit spatial autocorrelation at the within-leaf scale (Shelton 2005). Since it is reasonable to assume that over very small spatial scales organs or organ parts develop at about the same time and under similar internal and external microenvironmental conditions (i.e., position on the x-axis of fig. 6.4 held constant), then variation in organ traits at this restricted scale must predominantly reflect developmental instability derived from random effects occurring at the molecular and cellular levels.

Rate of development of organs has been suggested as a possible correlate of development instability, on the reasoning that a faster growth rate may result in more "developmental mistakes" and hence a decrease in developmental stability (Heslop-Harrison 1959). According to this hypothesis, one would expect within-plant variability to be greater for individuals or populations characterized by faster organ growth rates. Although the justification for a direct relationship between growth rate and developmental instability may be intuitively appealing, an inverse one would make equal sense, since faster growth implies shorter exposure to external disrupting agents and, therefore, decreased likelihood of failures in the accuracy of the developmental program. This possibility was suggested by Stebbins (1950), who pointed out that structures that are formed by short periods of meristematic activity, such as petals, are less subject to environmental influences and are likely to be less modifiable than characters that are formed over longer periods. According to this interpretation, therefore, one would expect within-plant variability to be smaller for individuals or populations characterized by faster organ growth rates. I am unaware of investigations providing suitable data to distinguish between these two opposite possibilities (but see Sherry and Lord 1996b for some evidence favoring a direct relationship between growth rate and instability).

Assessing the Importance of Different Causes: Some Prospects

The preceding sections have illustrated the broad variety of organismal mechanisms that may underlie within-plant variation in reiterated structures. The orderly dissection and classification presented should not lead us to lose sight of the fact that in any particular situation most of these mechanisms may be operating concurrently. Observed within-plant variation of a particular organ trait in a given species or population will invariably represent the aggregate outcome of the joint operation of all major factors, namely phylloclimate-driven developmental plasticity, direct and indirect architectural effects, and developmental instability. Differences in levels of within-plant variability—between organ types, traits of the same organ, populations, or species—will reflect in complex ways the confluence of distinct underlying factors, each acting with different intensity and at different times. To complicate things even further, interactions between influential factors (i.e., nonadditivity of the phenotypic effects of different causal agents) are also to be expected. This is illustrated, for example, by instances of plastic heteroblasty where patterns of leaf trait variation along shoots depend on light and temperature (Winn 1996b; Burns 2005; see below). Because of the presumable complexity of the causal network influencing within-plant variability, performing an accurate dissection of the relative importance of each factor is a formidable or even infeasible task in most instances, and it is not surprising that a comprehensive dissection and factoring of within-plant variation has not been attempted so far for any plant species, and that there is no published data set amenable to that sort of analysis. Nevertheless, less ambitious approaches aimed at examining the significance of only a few factors at a time are certainly feasible.

Manipulative experiments aimed at answering specific questions about the origin of within-plant variation exhibited by particular species provide a first step toward elucidating the relative importance of different mechanisms in generating subindividual variation. A good example of this approach is provided by Winn's experimental investigation (1996b) of the factors responsible for within-individual variation in leaf traits in the annual plant *Dicerandra linearifolia*. Leaves of this species exhibit within-plant variation in morphological, anatomical, and physiological traits, with consistent differences existing between the early leaves produced in winter at the more basal nodes and late ones produced in summer at the distal nodes. Because for this species temperature and leaf nodal position are confounded in nature (basal leaves are produced in winter, distal ones in summer), the effects of these factors on within-plant variation could be separated only by means of an experimental design conducted under controlled conditions. By growing plants in controlled chambers according to a factorial design involving high and low temperature levels, Winn was able to dissect the relative importance of two factors that potentially influence leaf variation, as well as their interaction: "programmed developmental change" and phenotypic plasticity at the level of individual organs in response to temperature. In the terminology used in this chapter, the former source of variation would correspond to an intrinsic gradient in leaf variation linked to a direct architectural effect. Results indicated that leaves produced by the same individual differed depending on *both* the environmental conditions during their development and the node at which they were produced, but the relative importance of these two factors differed considerably among traits. For five of the six traits examined (leaf thickness, leaf area, specific leaf weight, density of stomata, and chlorophyll content), the effects of node were much larger than the effects of temperature, and direct architectural effects appeared to contribute more to within-individual variation in these traits than did phenotypic plasticity. Furthermore, results of the experiment demonstrated that patterns of variation in leaf traits along nodal positions are not fixed but can be altered by temperature, as denoted by significant interaction effects between node and temperature on within-plant variation. A similar environment-dependent pattern of regular leaf trait variation along stems has been demonstrated experimentally by Burns (2005) for *Senecio lautus*. In this species, the shape of the relationships linking nodal position of the leaf with leaf length, area, degree of lobing, and mass per unit area depended significantly on light intensity. In plants in shaded environments the increase with nodal position of leaf area and leaf length was considerably steeper, and that of leaf mass per area and degree of lobing much shallower, than in control plants exposed to normal light conditions. By experimentally testing for interaction effects of nodal position and environmental variables on leaf traits, these studies were able to dissect the relative importance of plastic responses, direct architectural effects, and their interaction on organ traits.

For a given organ trait, plant species differ widely in levels of within-plant variability and in the relative contributions to this variability of differences among and within branches or stems (chapters 3 and 4). Comparative analyses testing hypotheses or predictions that relate interspecific differences to variation in influential factors may also shed light on the importance of the various sources of intraplant variation. This comparative approach may be useful, for example, to ascertain the importance of sectoriality as a cause of within-plant variation, and to examine whether its importance remains consistent across species or environments. Degree of physiological integration differs widely among species, thus denoting the existence of an integration-sectoriality continuum. This gradient seems related to intrinsic differences among species in vascular architecture and xylem anatomy, particularly the size, characteristics, and density of intervascular pits, the small openings in the walls of vascular elements that permit lateral water flow between adjacent vessels (Orians et al. 2004; Orians et al. 2005; Wheeler et al. 2005; Ellmore et al. 2006). On one hand, xylem anatomy has a strong phylogenetic component, and structural and functional xylem traits, including detailed morphological aspects of intervascular pits, are diagnostic features of entire plant families or even higher taxonomic categories (Cronquist 1981; Jansen et al. 2001; Boyce et al. 2004). On the other, the distribution of species and genera possessing different types of intervascular pits has clear ecological correlates. Drought-induced embolism (cavitation) has been related to the porosity of lateral intervessel connections, and plants with distinctly developed vestured pits (those having pit apertures with projections from the secondary cell wall) are mainly restricted to warm frost-free habitats such as lowland rain forests, deserts, savannas, and Mediterranean-climate areas (Jansen et al. 2003; Jansen et al. 2004). These ecological patterns may be interpreted as an indication that vestured pits are specialized structures acting to reduce the probability of water-stress-induced cavitation in environments where plants are likely to be subject to high transpiration rates and high negative xylem pressures (Jansen et al. 2004). From the preceding observations it may be predicted that, if sectoriality is actually a major factor influencing within-plant variation, then interspecific patterns of within-plant variation in woody plants, and particularly the component arising from differences among shoots or branches, should generally match the phylogenetic and ecological patterns of xylem traits that ultimately are the main determinants of sectoriality. For example, if structural restrictions to lateral water movement (e.g., reduced pit area per vessel; Wheeler et al. 2005) is a feature characteristic of the xylem of woody plants living in arid environments, one would then predict greater sectoriality in plants from these habitats, and hence greater overall within-plant variability in organ traits and/or a larger component of variability due to variation among stems and branches. Unfortunately, the comparative data on levels of within-plant variation summarized in chapters 3 and 4 are too scanty to test this prediction in a comparative context.

To this point, I have shown that subindividual variation in reiterated organ traits is widespread, extensive, and spatially organized; affects almost every conceivable trait; and is the visible outcome of a complex underlying web of remote causal agents and proximate mechanisms. Dealing with all these highly descriptive topics was necessary to lay the biological foundations for the analysis of the ecological and evolutionary implications of subindividual variation, which constitutes the ultimate goal of this book. The four chapters that follow show that subindividual variability often is a genetically based individual property; that within-plant variation in organ traits may have manifold consequences for foraging animals; and that such variation may influence the fitness of individual plants, thus creating distinct opportunities for animal-mediated selection on subindividual variability levels. Such variability being the composite outcome of a web of causes, any of these may eventually become the target(s) of selection on variability.

CHAPTER SEVEN

Subindividual Variability as an Individual Property

The Haldane-Roy conjecture is verified and extended: individual plants have not only their characteristic means, but also their characteristic standard deviations and characteristic spatial patterns of within-plant variation.

In 1959 Indian biometrician Subodh Roy published in *Nature* a one-page note whose succinct title read "Regulation of Morphogenesis in an Oleaceous Tree, *Nyctanthes arbor-tristis*." Despite its promising title, this contribution actually provided little in the way of information on morphogenesis, as it was exclusively concerned with summarizing the results of a detailed investigation of the variability of the number of petals in flowers of the species, based on the examination of an amazingly large sample consisting of 158,926 flowers (a full account of this and related work was presented in Roy 1963). Roy's original publication is remarkable not only because of its unbeatable sample size, but also because it has been nearly completely ignored by researchers during the five decades since its publication. It has received only four citations during the period 1960–2005 (according to ISI Web of Science database, accessed February 2006), and I suspect that very few publications appearing in *Nature* will ever equal that record. His obvious failure to arouse interest among peers notwithstanding, Roy was actually a pioneer in attributing seasonal changes in intraplant variability of numbers of floral part to temporally variable homeostasis. He wrote, "The variance of a metrical character may be as important a property of an organism as its mean, and should be measured on a number of species." Although his failed appeal to treat variability as another descriptive feature of organisms obviously was aimed at the reiterated, analogous structures produced by plants, it may also be considered a forerunner to the subsequently widespread use of the phenotypic variance of paired structures in (bilaterally symmetrical) animals as a measure of the developmental instability of individuals, a subject discussed in chapter 6.

Roy's studies of within-plant variation of flowers and leaves in *Nyctanthes*, along with those of Dronamraju (1961) on within-plant heterogeneity in style length in flowers of *Bauhinia acuminata*, and of Davis and Ramanujacharyulu (1971, and references therein) on within-plant variation in the handedness of floral estivation and leaf vernation, were all the direct outcome of one of the new lines of research initiated by J. B. S. Haldane shortly after he settled in India in 1957 (Clark 1968). According to one of his pupils, Haldane initiated this particular research "to understand the nature of organ regulation in living organisms" (Dronamraju 1987), and he clearly placed the results of Roy's research on within-plant variability of *Nyctanthes* flowers in the context of developmental instability. Referring to these results, and particularly to the fact that within-plant variability in petal number increased toward the end of the flowering period, Haldane (1959, 713) wrote: "If the size of pots made by a potter became more variable at the end of a day, we should say that he was getting tired. I do not know what we are to say about a plant." In addition, Haldane stressed elsewhere (1957, 312) that "individual plants not only have their characteristic means, but their characteristic standard deviations" and remarked that Roy had taken up "the problem of homotyposis where [Karl] Pearson left it in 1903" (see chapter 2 for the Pearsonian concept of homotyposis). Since this publication by Haldane antedated Roy (1959) by a couple of years, the former is probably to be credited as much as the latter with the paternity of the notion that intraplant variance should be considered as distinctive an individual trait as the mean. For this reason, and because the idea has gone essentially untested since its original formulation, I will refer to it here as the "Haldane-Roy conjecture." It is interesting to note that the attention paid by Haldane in his later life to the issue of within-plant (or within-genotype, for that matter) variability in organ characteristics denoted a significant shift of opinion about the importance of a phenomenon whose evolutionary significance he had previously categorically dismissed in one of his major works as being "irrelevant for the problem of evolution" (Haldane 1932; see chapter 1 for full quotation). Most likely this change of mind was not unrelated to the increasing significance conferred by some evolutionists during the 1940s and 1950s to the issues of developmental homeostasis and plasticity (Waddington 1941, 1959; Mather 1953; Lerner 1954; Berg 1959).

Haldane's and Roy's suggestion of considering within-plant variances in a given organ trait as another descriptor of the plants' phenotypes in addition to customary trait means (for similar views see Paxman 1956; Suomela and Ayres 1994) was initially motivated by the assumption that observed variability would largely reflect departures from some expected average value due to lack of developmental homeostasis. Nevertheless, characterizing individuals by their within-plant variances still holds considerable practical and theoretical interest even after acknowledging that within-plant variation will in most instances represent the aggregate outcome of a mixture of proximate mechanisms acting simultaneously, and not just the consequence of developmental instability alone, as shown in chapter 6. As noted there, some aspects of within-plant variation probably are an inevitable consequence dictated by direct architectural effects and space constraints (e.g., seed-size variation within cones and conelike fruits). In other cases, however, within-plant variability in organ traits may itself be a trait that has been shaped by special adaptation because it confers some fitness advantage to the plants that exhibit it in comparison to those that do not (Winn 1996a, 1999b; chapter 10). This would apply, for instance, to within-crown variation in leaf traits of trees, such as specific leaf area, nitrogen content, and photosynthetic features (Field 1983; Hollinger 1996; Kull 2002).

The possible adaptive significance of within-plant variability in the characteristics of leaves, flowers, fruits, and seeds is discussed at length in subsequent chapters. Nevertheless, before examining the fitness implications of within-plant variability in organ traits, it is essential to verify the generality of the Haldane-Roy conjecture. This implies testing whether, in most species and for most traits of reiterated structures, within-plant variability in a given organ feature is actually a distinctive trait of individual plants, or in other words, whether individual plants in the same population differ with regard to their respective intraplant variabilities. Regardless of their causes, individual differences in variability are the necessary raw material for (phenotypic) selection on variability to occur. The first section of this chapter is devoted to this crucial question. I also consider another facet of within-plant variability that, although it was not incorporated originally into the Haldane-Roy conjecture, does represent a logical extension of it: Conspecific individuals may differ with regard to the *shape* of their respective trait-value distributions. In addition, plants not only differ in the purely statistical properties of their within-plant trait-value distributions, but also in how such variation is spatially organized at small scales. The second section documents individual differences in the organization of within-plant variation in relation to intrinsic and extrinsic gradients, and in the apportionment of organ trait variance among and within axes of the same plant (e.g., branches, inflorescences). From an evolutionary perspective, individual differences in the statistical properties and spatial organization of within-plant variation would only make sense if they have some genetic component. In the last section of this chapter, I examine several lines of evidence showing that individual differences in the magnitude and characteristics of within-plant variation frequently have a genetic basis.

The Haldane-Roy Conjecture Verified

Since within-plant variability has been very rarely treated before as an individual attribute deserving consideration, there is almost no published evidence that can be used to test the Haldane-Roy conjecture and related notions. Variances or standard deviations for organ traits of individual plants are virtually absent from the ecological and botanical primary literature. For this reason, I rely largely in this section on reanalyses of data from my own studies or those of my associates. Table 7.1 summarizes information on the magnitude and statistical significance of individual differences in within-plant variability of leaf, flower, fruit, and seed traits for those species listed in tables 3.1 to 3.4 with raw data available and sufficiently large sample sizes per plant as to provide reliable estimates of CV_{within}. The magnitude of individual differences in variability is expressed in terms of both the observed range of individual plants' CV_{within} and the interquartile range, a robust measure of scale. In addition, whenever possible I tested the statistical significance of individual differences in variability by application of a version of Levene's test for comparing relative variabilities (Van Valen 1978).

The data in table 7.1 provide strong and unequivocal support for the Haldane-Roy conjecture for a variety of traits and a sample of species differing widely in growth form and taxonomic affiliation. Regardless of the species and the organ trait considered, and with relatively few exceptions

TABLE 7.1 **Variation among conspecific individuals in the magnitude of within-plant variability for leaf, flower, fruit and seed traits.**

Trait	Species	Within-plant variability (CV_{within})		Significance of individual differences[a]
		Range	Interquartile range	
Leaf traits				
Area	*Daphne gnidium*	12.7–26.9	7.3	****
	Daphne laureola	24.0–55.1	7.6	****
	Prunus mahaleb	37.9–77.6	12.2	****
Fresh mass	*Daphne laureola*	26.0–51.8	9.9	****
	Prunus mahaleb	37.8–95.7	10.6	****
Length	*Daphne gnidium*	7.3–17.4	4.0	**
Longevity[b]	*Thuja plicata*	12.1–37.3	10.2	ns
Number of teeth in margin[c]	*Nyctanthes arbor-tristis*	54.1–258.9	147.8	na
Specific weight	*Daphne laureola*	5.2–39.7	4.6	****
	Prunus mahaleb	6.6–30.1	3.4	****
Water content	*Daphne laureola*	1.0–9.5	1.5	****
Width	*Daphne gnidium*	7.4–17.7	5.7	****
Floral traits				
Petal number	*Nyctanthes arbor-tristis*	8.4–11.0	1.7	****
Petal length	*Silene acutifolia*	7.7–12.3	2.1	ns
Corolla length or diameter	*Daphne laureola*	6.5–23.7	4.1	*
	Helleborus foetidus	0.4–12.6	2.4	****
	Hormathophylla spinosa	5.6–12.9	3.0	*
	Ipomoea wolcottiana	3.1–31.9	7.3	****
	Lavandula latifolia	1.0–9.3	2.4	****
	Pancratium maritimum	1.6–19.0	6.0	**
Spur length	*Viola cazorlensis*	1.2–26.7	6.5	****
Nectar production rate[d]	*Epilobium canum*	6.5–116.7	na	na
Fruit traits				
Transversal diameter	*Arum italicum*	4.1–13.0	2.8	ns
	Berberis hispanica	7.2–19.6	4.4	ns
	Corema album	3.4–11.5	2.2	****
	Crataegus laciniata	6.2–9.3	1.5	ns
	Crataegus monogyna	8.4–15.3	4.2	****
	Daphne gnidium	2.9–12.9	2.9	****
	Daphne laureola	5.0–8.2	2.1	ns
	Gonzalagunia hirsuta	5.6–17.8	3.4	ns
	Guaiacum officinale	3.1–10.8	5.8	*
	Guazuma ulmifolia	4.9–15.8	4.4	**
	Hedera helix	5.7–12.1	2.6	*
	Juniperus communis	5.3–12.7	3.6	*
	Juniperus phoenicea	5.0–10.5	3.4	*
	Miconia prasina	5.7–11.9	4.8	**
	Olea europaea	3.7–7.6	1.0	ns
	Osyris lanceolata	3.4–10.1	2.1	****
	Phillyrea latifolia	3.6–10.5	3.2	****
	Rosa canina	4.8–35.4	3.2	*
	Viburnum lantana	7.3–12.1	3.2	ns

Fruit mass	*Hedera helix*	23.7–34.9	2.1	*
	Juniperus communis	4.0–31.9	9.5	*
	Juniperus phoenicea	16.0–34.8	8.2	ns
	Osyris lanceolata	10.4–57.0	12.9	****
	Rosa canina	12.5–33.2	7.7	*
Seed traits				
Elaiosome mass	*Helleborus foetidus*	32.4–84.6	14.4	****
Seed mass	*Asphodelus albus*	8.2–28.9	7.0	****
	Ateleia herbert-smithii	8.5–22.6	3.0	na
	Cassia grandis	7.0–13.0	2.0	na
	Crataegus monogyna	11.3–28.5	5.6	*
	Daphne gnidium	7.2–34.7	7.8	***
	Helleborus foetidus	6.0–51.1	9.1	****
	Ilex aquifolium	15.0–39.1	11.2	****
	Lavandula latifolia	12.6–33.2	10.8	****
	Narcissus longispathus	4.9–20.4	3.9	****
	Paeonia broteroi	5.0–61.0	15.5	****
	Pancratium maritimum	3.8–59.9	7.9	****
	Phillyrea latifolia	9.4–29.9	8.7	****
	Pinus sylvestris	14.1–20.1	3.6	**
	Prunella vulgaris	28.5–52.8	9.0	na
	Quercus coccifera	7.1–38.6	13.2	***
	Quercus ilex	5.5–28.0	5.7	***
	Smilax aspera	9.2–33.7	6.4	****
	Sorbus aucuparia[c]	11.1–24.2	3.9	****
	Tamus communis	8.7–54.2	13.1	****
	Taxus baccata	3.0–23.3	2.4	****

Note: Estimated with CV_{within}, as defined in chapter 3. Species and traits are a representative subsample of those listed in tables 3.1 to 3.4, and were selected among those with the largest average sample sizes per plant. Except where otherwise indicated, see those tables for data sources.

[a]Statistical significance of among-individual heterogeneity in the extent of within-plant variation was tested using Levene's test for relative variability (Van Valen 1978). *, $P < 0.05$; **, $P < 0.01$; ***, $P < 0.001$; ****, $P < 0.0001$; ns, not significant; na, original data not available for computations.

[b]Data from Harlow et al. 2005; B. Harlow, personal communication.

[c]Data from Roy 1963.

[d]From Boose 1997.

[e]B. Pías and M. Salvande unpublished data.

(10 statistically nonsignificant outcomes out of a total of 62 tests), the tests reveal that conspecific individuals differ significantly in their levels of within-plant variability. CV_{within} varied considerably among individuals of the same species, as denoted by broad ranges and large interquartile ranges. A within-plant variability continuum occurs in most species, with populations generally comprising phenotypically constant to highly variable plants. This is illustrated graphically in figure 7.1 for four selected examples taken from table 7.1. Within-plant variability (CV_{within}) ranged between 38 and 78% for leaf area in trees of *Prunus mahaleb*, between 1 and 9% for corolla length in *Lavandula latifolia* shrubs, between 5 and 20% for seed mass in the perennial herb *Narcissus longispathus*, and between 32 and 85% for elaiosome mass in the herb *Helleborus foeti-*

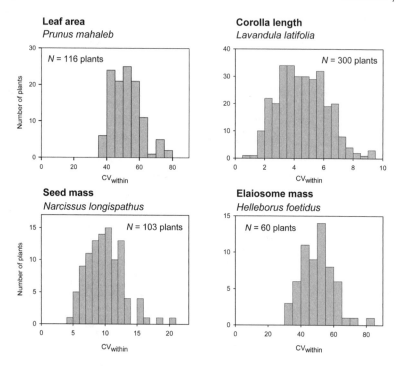

FIG. 7.1 Frequency distributions of within-plant variability estimates (CV$_{within}$, as defined in chapter 3) for representative leaf, flower, and seed traits in southeastern Spanish populations of *Prunus mahaleb* (a tree), *Lavandula latifolia* (a shrub), *Narcissus longispathus*, and *Helleborus foetidus* (perennial herbs). Note differences in scaling of horizontal axes. Data sources are shown in tables 3.1, 3.2, and 3.4 for these species and traits.

dus. The distributions of CV$_{within}$ presented in figure 7.1 are representative of the set of species listed in table 7.1. Regardless of the organ or trait under consideration, therefore, populations of most species are made up of individuals differing widely in the magnitude of within-plant variability in organ traits.

The preceding conclusion is further supported by the few published reports of individual differences in within-plant variability that I have been able to locate. For example, in the nickel-hyperaccumulating plant *Psychotria douarrei*, Boyd et al. (1999) found that individual shrubs did not differ significantly in mean nickel concentration in leaves, but differed widely in the extent of within-plant variation. In *Populus angustifolia*, the magnitude of differences in phenol content among leaves of the same shoot were shown by Zucker (1982) to vary strongly between trees.

For *Lychnis flos-cuculi*, Biere (1991a) reported that seed progenies from different maternal parents sown under controlled conditions differed significantly in their respective coefficients of variation of time to germination. In a population of *Phytolacca americana*, Armesto et al. (1983) found that CV_{within} for mean percent germination of seeds, as estimated from germinability of seeds from different racemes on the same plant, varied widely among plants, ranging between 25 and 99%. For *Ipomopsis aggregata*, Pleasants (1983) showed that individual plants differed significantly in within-plant variability of nectar production rate, as measured by their respective CV_{within}. In a test of the hypothesis that within-plant variability in nectar production rate increases with the number of open flowers per plant, Biernaskie and Cartar (2004, fig. 1) presented data revealing considerable differences among conspecifics in within-plant variability in nectar production rate. In conclusion, therefore, there is clear support for Haldane's and Roy's suggestion that individual plants not only have their characteristic means but also their characteristic variation around the mean, as well as their contention that subindividual variability should be treated as an individual property in itself. In addition, the fact that different plants in a population show different levels of within-plant variability ultimately implies that the component of population-wide phenotypic variance in an organ trait due to within-plant variation (Var_{within}, as defined in chapter 3) is not distributed equally among individuals.

The Conjecture Extended: Higher Moments of Within-Plant Distributions

The Haldane-Roy conjecture ultimately arises from the recognition that all the repetitions of a given organ produced by the same plant in a particular season, or over the course of its whole existence, are not identical. Consequently, a proper description of an individual plant's phenotype with regard to some trait of a reiterated structure will require information on the within-plant *statistical distribution* of organ trait values. This information should include not only the central tendency represented by the mean (first moment of the distribution), but also the scatter around the mean represented by the variance (the second moment). The incorporation of the within-plant variance to the description of individual plant phenotypes represents an improvement in relation to the usual way of describing them using the mean alone. Nevertheless, the variance describes only the amount of variation exhibited by an individual, yet it is insensitive to possible differences in the shape of within-plant distributions of trait values, an aspect that can also be useful in characterizing individual plants phenotypically. Acceptance of the fact that it is the *statistical distribution* of organ trait values that provides the best and most comprehensive phenotypic description of one individual, rather than any arbitrarily chosen moment of the distribution (traditionally, the mean), opens the way for one obvious extension of the Haldane-Roy conjecture: A complete phenotypic characterization of individual plants would require the inclusion of the third (skewness) and fourth (kurtosis) moments of the within-plant distributions of organ trait values in addition to the mean and the variance.

Skewness characterizes the degree of asymmetry of a distribution around its mean. A positive skewness implies a distribution with an asymmetric tail extending out toward more positive values, while a negative value signifies a distribution whose tail extends out toward more negative values. Kurtosis measures the relative peakedness or flatness of the distribution relative to a normal distribution. Skewness and kurtosis, therefore, are parameters that are informative about the shape of within-plant trait distributions that may provide additional information on individual-level characteristics of variability. Their incorporation represents a logical extension of the Haldane-Roy conjecture. In fact, Roy (1963) explored the possible use of measurements of skewness and kurtosis to characterize differences between plants in the frequency distribution of petal numbers. The question thus becomes: In addition to differing in the means and variances of organ trait values, do individual plants differ also in the shape of trait value distributions ?

In practice, incorporating the skewness and kurtosis coefficients (usually denoted by g_1 and g_2, respectively) of within-plant distributions of organ trait values into the description of individual phenotypes is less generally applicable than incorporating the mean or the variance. It will be statistically feasible and biologically meaningful only if the number of similar structures produced by a plant is large enough for the notion of "shape" of the within-plant distribution to make sense. This limitation will generally restrict the application of g_1 and g_2 to trees and large shrubs producing large numbers of reiterated structures. In these particular cases, however, there is evidence that individual plants also differ in the skewness and kurtosis of their within-plant trait distributions. In nine *Pinus contorta* trees studied by McGinley et al. (1990), two trees showed signifi-

cantly right-skewed seed-mass distributions ($g_1 > 0$), one tree was significantly left skewed ($g_1 < 0$), and the rest did not differ significantly from a
normal distribution. These authors did not present numerical data on the
kurtosis coefficients, but inspection of frequency distributions in their figure 1 suggests that individual trees also differed in this respect, since some
seed-mass distributions departed noticeably from normal-shaped curves
by being either too flat (platykurtic) or too narrow (leptokurtic). Comparing the frequency distributions of petal number per flower of *Jasminum multiflorum* plants, Roy (1963) found that some of them had positively skewed distributions, while others had distributions not departing
from normality. I was able to compute within-tree skewness and kurtosis
coefficients for leaf traits for 116 *Prunus mahaleb* trees with leaf variability data included in figure 7.1 and table 7.1. For leaf area, for example, 32
trees (27.6%) had significantly positively skewed distributions (g_1 range
= 0.7–2.0), and 13 trees (11.2%) had significantly leptokurtic distributions (g_2 range = 1.6–8.1), which clearly denotes the existence of considerable individual variation in the shape of distributions for this trait. I
obtained results similarly denoting individual variation in g_1 and g_2 for
other leaf characters in *P. mahaleb*, and for leaf, fruit, and seed traits of
other trees and shrubs included in table 7.1. This lends support to the suggested extension of the Haldane-Roy proposal: In plants producing large
numbers of reiterated structures, the skewness and kurtosis coefficients
of within-plant distributions could profitably be included as two further
descriptive statistics of individual phenotypes in addition to the mean
and the variance. In this way, a truly comprehensive statistical description
of individual phenotypes could be achieved by combining the first four
moments of the within-plant trait distributions.

Individual Differences in the Organization of Within-Plant Variation

Descriptors of within-plant variability based on the moments of trait value
distributions are of a purely statistical nature. They provide a numerical
description of the magnitude of variation and the shape of the trait distribution, but are insensitive to important biological details, such as individual differences in the distribution over time or space of organ trait values.
For a given organ trait, individual plants of the same population not only
differ in the amount of variability but also, as shown below, in the nature

of their respective intrinsic and extrinsic gradients of within-plant phenotypic variation. In addition, plants differ in the proportional contribution of variation among and within axes (e.g., stems, inflorescences) to overall within-plant variance in organ trait values, thus giving rise to individual differences in the spatial texture of within-plant variation. In short, individuals differ in how they deploy organ variants over time and space, which means that spatial and temporal patterns of deployment of subindividual phenotypic diversity should be considered as individual properties too.

Leaves

Several intrinsic and extrinsic gradients of within-plant variation in leaf traits were described in chapter 4. Although these gradients have rarely been examined on a per-plant basis, significant differences between conspecifics have invariably been found in the characteristics of such spatial relationships whenever they have been looked for.

De Soyza et al. (1990) studied the variation of leaf chlorophyll content within eight *Sassafras albidum* trees, and found remarkable individual heterogeneity in the within-plant pattern of variation of that leaf trait. They performed within-tree comparisons between leaves located in the outermost, sunlit shell of foliage ("sun" leaves) and those located near the primary trunk, deep into the crown ("shade" leaves). For all trees combined, sun leaves had higher average chlorophyll content per leaf surface area than shade ones, yet there was considerable individual scatter around this predominant trend: the chlorophyll content of sun leaves was significantly greater than that of shade leaves in five trees, significantly smaller in one tree, and did not differ significantly in two trees. De Boer (1999) studied the variation in pyrrolizidine alkaloid concentration in leaves of *Senecio jacobaea* at different positions along the stem, and reported the data separately for each of four study plants. Alkaloid concentration declined steadily from top to bottom leaves in three plants, but remained fairly constant in one plant. Harlow et al. (2005) demonstrated that, on average, leaf longevity increased significantly with depth in canopy in *Thuja plicata* trees. Individual trees, however, differed significantly in the slope of the leaf longevity–canopy depth regressions. Some trees exhibited very steep relationships while others had flat and nonsignificant relationships. Cowart and Graham's study of within-plant variation of leaf traits in *Ficus carica* trees (1999) revealed significant plant × crown position (inner vs. outer) and plant × height interaction effects on both leaf width and leaf lateral lobe length, which denoted heterogeneity among trees in their respective within-plant gradients in leaf morphology. In a similar vein, Perfectti and Camacho (1999) also reported significant tree × exposure effects on leaf size in cultivated *Annona cherimola*. In wild *Prunus mahaleb* trees, water content of individual leaves varies predictably within trees depending on both exposure and height above the ground (fig. 4.2). A detailed analysis of these data on a per-plant basis further revealed significant individual differences in both the exposure- and height-related gradients of intra-plant variation in that leaf trait.

Flowers

Few published reports have explicitly documented individual differences in spatial or temporal patterns of floral features, but these scanty data and some reanalyses of raw data from published studies unequivocally show that conspecific individuals generally differ in their organization of within-plant variation in floral traits. This is apparent, for example, in the significant time × maternal family interaction effect on number of ovules per flower found by Mazer and Delesalle (1996) in a greenhouse study of *Spergularia maritima*. In the perennial herb *Pancratium maritimum*, inflorescences produce four to nine large flowers that open sequentially. On average for the population, corolla length declines with blooming order in the inflorescence, each consecutive flower being on average about 2 mm shorter than the one that opened just prior to it (Medrano et al. 2000, and personal communication). I computed regressions of corolla length on order in the inflorescence separately for the different plants studied by Medrano et al. (2000), and found that regression slopes differed significantly among individuals, ranging between −4.6 mm/position (sequential decline) and +2.5 mm/position (sequential increase). This result denotes broad individual differences in both the sign and the magnitude of the within-plant trend of variation that links flower size with position in the opening sequence. Similar conclusions emerge from detailed investigations of intraspecific and intraplant floral variation in two species of Caryophyllaceae. In *Silene acutifolia*, Buide (2004) found that the number of ovules per flower declined from early (primary) position to later (tertiary) position in the inflorescence, and that this pattern of intrainflorescence variation remained consistent across populations and years of study. To look for possible individual differences, I reanalyzed a subset of Buide's raw data on a per-plant basis and found that plants differed significantly in the rate of decline in ovule number from primary through secondary to tertiary flowers in the inflorescence. The declining trend was not significant in some plants, and in those in which it was, the slope of the regression varied widely among individuals. Petal length and ovule number also decline regularly from primary to tertiary positions in the inflorescences of *Petrocoptis viscosa* (Navarro 1996, and personal communication). As in the case of *S. acutifolia*, reanalysis of the raw data for *P. viscosa* showed that individuals effectively differed in the slope and statistical significance of the relationship linking floral traits and inflorescence position. A last example of individual differences in the nature of intrinsic gradients of floral variation concerns the perennial herb *Polygonatum odoratum*. Flowers of this species are hermaphroditic, but the number of ovules per flower tends to decline from basal to distal positions along stems, thus exemplifying a clear intrinsic gradient in a floral trait along a plant axis (Guitián et al. 2004). On reanalysis of the original data, I found a significant plant × position interaction effect on the number of ovules per flower, the slope of the regression of ovule number on nodal position on stem ranging from −1.8 (acropetal decline) to +1.8 (acropetal increase). Although an acropetal decline in ovule number was the prevailing trend at the population level, a fraction of individuals departed from this general pattern by exhibiting either a reversed trend or no trend at all.

Fruits

I failed to locate published information allowing for an assessment of individual differences in intrinsic or extrinsic gradients of within-plant variation in fruit traits, either for wild or cultivated plants. Nevertheless, two large unpublished data sets for fleshy-fruited trees from the Iberian Peninsula do reveal that conspecific plants growing in the same population can be surprisingly heterogeneous with regard to the spatial scale at which within-plant variation in fruit traits takes place. *Sorbus aucuparia* is a slender deciduous tree whose fleshy fruits (berrylike pomes) are produced in infructescences located at the tip of branches. Bea Pías and Miguel Salvande (personal communication) conducted a detailed investigation of fruit variation in a large population of this species from the Sierra del Caurel, in northwestern Spain. For each of 44 study trees, they sampled ten ripe fruits from each of five infructescences on different branches (N = 2,200 fruit in total), then measured fruit length and width and counted the number of enclosed seeds. I did a similar study of fruit variation in the small tree *Crataegus monogyna* in the Sierra de Cazorla, southeastern Spain, in which I sampled and measured ten ripe drupes from each of four branches at main compass directions in 60 trees ($N = 2,400$ fruits in total). Variance partitions conducted separately for each tree demonstrate that conspecific trees differed widely in the relative importance of differences among and within branches of the same plant as sources of within-plant variance in fruit traits (table 7.2). In populations of the two species, trees where all within-plant variance in fruit traits occurred among fruits of the same branch coexisted with others where most within-plant variance was accounted for by differences among branches. These data demonstrate that, in these species at least, the "spatial texture" of phenotypic variation in fruit traits was also an individual trait, in the same manner as the magnitude of variation or the characteristics of intrinsic and extrinsic gradients.

Seeds

A number of studies have clearly shown that the spatial organization of within-plant variation in seed traits generally varies among conspecifics. In fourteen plants of *Asphodelus albus* studied by Obeso (1993), mean seed weight declined significantly from basal to distal positions in the inflorescence in ten plants and increased significantly in one plant, and there was no significant position-dependent gradient in seed size in three plants. In *Pastinaca sativa*, average seed weight declines from primary through

TABLE 7.2 **Conspecific trees of *Sorbus aucuparia* and of *Crataegus monogyna* differ widely in how the within-tree phenotypic variation in fruit traits maps onto the plant's architecture.**

	Percentage of within-tree variance due to differences among branches	
Species and trait	Range	Interquartile range
Sorbus aucuparia (N = 44 trees)		
Fruit length	0–67.9	27.7
Fruit width	0–60.1	25.9
Fruit seediness	0–50.7	20.3
Crataegus monogyna (N = 60 trees)		
Fruit length	0–33.8	11.8
Fruit width	0–78.4	16.1

Note: In both species, individual trees differed widely in the relative importance of differences between branches and among fruits on the same branch as sources of within-plant variance in fruit traits. Based on unpublished data from B. Pías and M. Salvande (*S. aucuparia*) and C. M. Herrera (*C. monogyna*).

secondary to tertiary umbels plants studied, although this pattern is not invariant in all plants (Hendrix 1984). Seeds produced by primary umbels were significantly heavier than those produced by secondary umbels in only nine of the ten plants studied by Hendrix, and seeds produced by secondary umbels were heavier than tertiary seeds in only eight plants. Furthermore, secondary and tertiary seeds' weight relative to primary seeds' weight both varied greatly among plants (ranges = 36–97% and 16–83%, respectively), thus denoting considerable individual heterogeneity in the steepness of the within-plant gradient in seed weight running from primary through secondary to tertiary umbels. For *Onopordum acanthium*, *Amaranthus retroflexus*, *Diplotaxis tenuifolia*, and *Tragopogon dubius*, Cavers and Steele (1984) and McGinley (1989) presented data showing considerable heterogeneity among individuals of the same species in seasonal patterns of variation in seed size. To the extent that seasonal variation in seed size in these species reflects architectural effects (chapter 6), these data illustrate individual differences in intrinsic patterns of seed-size variation within-plants. Individual differences may also involve variations taking place at much smaller scales. In *Raphanus raphanistrum*, the pattern of seed-weight variation by position within fruits was shown by Stanton (1984) to differ markedly among maternal families. In three- and four-seeded fruits, for example, seed size declined from basal to distal positions in the fruit in some families, while it did not vary in others.

Evidence from cultivated plants likewise shows that within-plant gradients of variation in the chemical composition of seeds may also differ among conspecifics. Brim et al. (1967) used nuclear magnetic resonance spectrometry to investigate within-plant variation in percent oil content in soybeans (*Glycine max*), and found a significant plant × position interaction effect on oil content. The trend of variation in oil content along nodal positions of the stem was not consistent among plants, or in other words, individuals differed with regard to their intrinsic, position-dependent gradients in this seed trait. Working also on soybeans, Marchetti et al. (1995) likewise found that individual plants differed in the characteristics of within-plant gradients in the amount of protease inhibitors in seeds. Similar inconsistencies among individual plants in patterns of within-plant variation in seed mass, oil content, and oil composition have been reported for sesame (*Sesamum indicum*; Mosjidis and Yermanos 1985) and sunflower (*Helianthus annuus*; Fick and Zimmerman 1973).

Individual differences in the spatial or temporal organization of within-plant seed variation may sometimes involve discrete characters. Plants of *Impatiens capensis* produce two types of seeds, originating from either cleistogamous (CL) or chasmogamous (CH) flowers. In one of his study populations, Waller (1982) found a highly significant plant × seed type interaction effect on seed mass, with CH seeds larger in some plants and CL seeds larger in others. This striking result reveals that not only the magnitude, but also the sign of within-plant differences in seed size among seed types may differ among individuals of the same population.

Genetic Basis of Differences in Within-Plant Trait Variability

This section summarizes evidence showing that differences among conspecifics in both the magnitude and the spatial organization of within-plant variation in organ traits often have a genetic basis. Individual differences in within-plant variation represent the realized outcome of the differential operation, importance, or characteristics of one or more of the mechanisms described in chapters 5 and 6, that is, organ-level reaction norms, ontogenetic contingency (phylloclimate-driven variation, direct and indirect architectural effects, responses to biotic factors), and developmental instability. If the factors underlying these mechanisms have a genetic basis, then observed individual differences will ultimately have a genetic basis too. It is not biologically unrealistic to suggest, for example, that individual plants of the same species may differ in the shape of their organ-level reaction norms, degree of physiological integration, three-dimensional arrangement of vascular bundles, density and characteristics of intervessel pits, geometry of fruit walls, homeostatic ability, or any other factor ultimately responsible for within-plant variation in organ traits. To the extent that these structural or functional differences have a genetic basis, then individual differences in the within-plant variation in organ traits caused by these factors will also be genetically based. I consider this indirect evidence in the first subsection below. The clearest evidence of a genetic basis of within-plant variation, however, is provided by a handful of studies considering organ variability itself as another phenotypic trait, and addressing its study by adopting classic quantitative genetics approaches. These investigations are considered in the second and third subsections below.

Indirect Support: Genetic Basis of Factors Causing Within-Plant Variation

I argued in the preceding chapters that the existence of organ-level developmental reaction norms is ultimately responsible for a significant portion of within-plant variation in organ traits. If individual plants differ genetically in the shape of their organ-level responses to variation in environmental variables, this provides a mechanism for genetically based variation among conspecifics in within-plant variation. One example of variable organ-level responses to the external environment is the differences between genetically distinct conspecific trees in their leaf-level response curves to variation in the light environment. Within-tree variation in leaf mass per unit leaf area and nitrogen content generally reflect leaf-level plastic responses to variation in the light environment (see references in chapter 4). In *Juglans regia*, genetically distinct lines differ in the functional relationship describing the response of leaf mass per unit area and nitrogen content to variable daily photon flux density (Klein et al. 1991). Intraspecific variation in the magnitude and spatial characteristics of leaf-shape variation along intrinsic plant gradients, such as those involved in heteroblastic species, may also have a genetic basis. In wild *Arabidopsis thaliana* plants, leaves change in size and shape from the juvenile through the adult stage according to a well-defined heteroblastic pattern, and a large number of genes have been identified that influence the developmental pattern that originates such sequential variation in leaf form (Tsukaya et al. 2000; Pérez-Pérez et al. 2002).

The extent and nature of plant sectoriality, another major factor contributing to within-plant variation in organ features, may also differ among conspecific plants, and these differences may have a genetic basis. Lötscher and Hay (1996, 1997) demonstrated experimentally that two genotypes of *Trifolium repens* differed in physiological integration, as revealed by their different capacities to translocate ^{32}P and ^{45}Ca from a single nodal root to shoot branches. These genotypic differences were most apparent when treatments were imposed that altered intraplant source-sink relationships (root severance and defoliation). In one genotype the imposed treatments had only minor effects on the translocation of nutrients from the nodal root to distant branches, thus denoting strong sectoriality. In the other genotype, in contrast, the treatments considerably enhanced lateral transport of nutrients to far-side branches, thus denoting weak sectoriality. These differences among genotypes in the extent of sectoriality were consistent with differences in the organization of the vasculature of their stolons. Genetically based differences in the extent of physiological integration (i.e., the inverse of sectoriality) have been also demonstrated for *Fragaria chiloensis* (Alpert 1999) and *Ranunculus reptans* (van Kleunen et al. 2000). These examples involve clonal herbs characterized by vigorous vegetative growth and the formation of large clones, and extrapolations to nonclonal plants such as trees or shrubs should be done with caution. Despite this, they are useful to illustrate that individual differences in sectoriality may have a genetic basis and, therefore, that genetic differences may also lie behind individual differences in the extent and spatial organization of within-plant variation.

Of all the factors accounting for within-plant variation in organ traits, developmental instability has been the only one frequently considered from the perspective of its possible genetic basis. By definition, random deviations from a systematic developmental trend (fig. 6.4) are not heritable, but this is not to say that none of the factors that influence the appearance of such deviations is inherited. The factors that influence the extent of noise at the molecular or cellular level, or those homeostatic mechanisms that correct for errors during development, may have a genetic basis (Palmer 1996). In fact, a few studies using classic population genetics crossing designs have demonstrated significant maternal and paternal influences on the extent of homeostasis in plants. These include studies by Paxman (1956) and Sakai and Shimamoto (1965) on within-plant instability in leaf and floral traits in *Nicotiana rustica* and *Nicotiana tabacum*, respectively, and Bagchi et al.'s study of leaf venation in isogenic lines of *Tectona grandis* (1989). These studies clearly indicate that the fraction of individual differences in within-plant variation in organ traits due to differences in developmental stability often has an heritable component.

Direct Support: Wild Plants

Traditionally, the level of within-plant variation in traits of reiterated structures has been not considered an individual property worthy of consideration; thus it is not surprising that there have been so few experimental investigations of wild plants that directly look for a possible genetic basis of levels of within-plant variation. The results of the few investigations that I have been able to locate are summarized below, in chronological order. All of them have invariably supported a genetic basis of variability in leaf, flower, and seed attributes.

Paxman (1956) conducted a pioneering investigation of the genetic basis of within-plant variation in leaf and flower traits by means of a set of diallel crosses among five varieties of *Nicotiana rustica*. He treated the within-plant variances of traits as ordinary characters, and then used analyses of variance to detect genetic effects in the usual ways. By this means, he was able to demonstrate significant heritabilities for within-plant variability in stamen length, pistil length, and leaf-shape index. Half a century later, Paxman's study remains exemplary not only for its insightful dissection of within-plant variation into systematic and random components, mentioned in chapter 6, but also for the elegance of his analytical treatment of results of diallel crosses to demonstrate the heritability of within-plant variation.

In another remarkable study, Seyffert (1983) investigated the genetic basis of within-plant variation in floral anthocyanin content of the annual *Matthiola incana*. The study was based on the analysis of 256 defined genotypes obtained from a full diallel cross of 16 pure lines representing all possible homozygous combinations of four biallelic loci involved in the synthetic pathway of floral anthocyanins. Results demonstrated the existence of a strong hereditary component to levels of within-plant variability in floral anthocyanin content, which were partly attributable to the direct effect on variability exerted by some specific loci and by their epistatic interactions.

Seburn et al. (1990) investigated floral variation in the tristylous aquatic plant *Eichhornia paniculata* by clonally propagating replicates of 14 different genotypes and growing them under common-garden conditions. Populations of this species exhibit considerable variability in floral traits, including the number and symmetry of tepals, and the degree of filament elongation in short-level stamens. In addition to the ordinary question of whether genotypes differed in mean floral traits, these authors also investigated whether they differed in the amount of floral variability. Overall within-genotype floral variability, as measured with the determinant of the trait covariance matrix, exhibited extreme variation among genotypes, thus demonstrating a genetic basis of intraplant variation in floral traits in this species.

Biere (1991a) used a full diallel crossing design to analyze the genetic basis of variation in time to germination among seeds of the perennial herb *Lychnis flos-cuculi*. Genotypes obtained from controlled crosses differed significantly in the variability of germination time exhibited by their seed progeny, the coefficients of variation of time to germination (CV_{within}, as used in this book) ranging between 31 and 72%. High or low CV_{within} values of maternal progeny mainly resulted from variability within each full-sib family and not just from variation among the full-sib families from a common female parent sired by different paternal parents. Differences between genotypes were statistically significant, as revealed by comparisons of the CV_{within} between pairs of progeny groups from different maternal genotypes, thus denoting the genetic basis of within-progeny variability in time to germination.

Winn (1996a) investigated the genetic basis of within-individual variability in leaf traits for the annual *Dicerandra linearifolia*. Plants from 24 paternal half-sib families were raised in growth chambers, and five leaf traits (area, thickness, chlorophyll concentration, chlorophyll *a:b* ratio, density of stomata) were measured for two leaves produced by each plant at different times in the seasonal cycle. There was significant within-individual variation in four of the five traits examined. Individuals differed in the magnitude of within-plant variation in leaf traits, and there was additive genetic variation for within-individual variability in leaf area, chlorophyll concentration, and chlorophyll *a:b* ratio.

Direct Support: Cultivated Plants

As already noted on several occasions in this book, intraplant variation in the characteristics of reiterated structures of economic value (e.g., fruits, seeds) has traditionally been a matter of concern to agronomists, fruit growers, and farmers in general, who have long endeavored to reduce this unwanted source of heterogeneity in crop products. This has prompted a considerable number of breeding efforts to develop cultivars or synthetic lines characterized by reduced within-plant organ variability. Published examples abound for cultivated plants grown under controlled uniform conditions demonstrating a genetic basis of differences in both the magnitude and the spatial organization of within-plant variation. I will not provide here a comprehensive review of the extensive literature available, but only a summary of representative examples.

Different clones or cultivars of the same species often differ characteristically in the magnitude of within-plant variation in organ traits, which clearly points to a genetic basis of such variation. Three poplar (*Populus*) clones studied by Casella and Ceulemans (2002), grown under uniform conditions, were similar in exhibiting vertical variation in leaf mass per unit area and leaf nitrogen concentration, but differed widely in their ranges of variation. For leaf fresh mass per unit area, for example, the within-clone ranges were 175–275, 150–425, and 175–400 g/m^2. In 17 poplar clones studied by Pellis et al. (2004), the degree of heterogeneity of leaf area along single shoots was a characteristic feature of each clone, and the coefficient of variation of leaf area within the same shoot ranged widely (37–86%). In sunflower (*Helianthus annuus*), the coefficient of variation of the oil content of single seeds within sunflower heads differed among varieties, and there was a significant variety × position effect on fatty acid composition of oil seeds, revealing that the spatial pattern of within-head variation in oil composition was not consistent across varieties (Zimmerman and Fick 1973). In oats (*Avena sativa*), seed-size variability and the shape of seed-size distribution depended on genotype (Doehlert et al. 2004). And in *Brassica napus*, the CV_{within} for the protein content of single seeds varied among plants depending on cultivar (Velasco and Möllers 2002).

Differences between cultivars, varieties, or isogenic lines of cultivated plants in the spatial organization of within-plant variation in organ traits are also indicative of a genetic basis. In *Nicotiana tabacum* the pattern of variation in leaf size along nodal positions in the stem varies considerably among varieties, with some varieties producing the largest leaves at intermediate positions and others at the most basal nodes (Sakai and Shimamoto 1965). The poplar clones studied by Casella and Ceulemans (2002) mentioned earlier not only differed in the magnitude of within-plant variation in leaf features, but also had characteristic, clone-specific vertical profiles of variation in leaf traits. Volatile terpenoid levels and composition in root and leaves of carrot (*Daucus carota*) are under genetic control (Simon 1982; Kainulainen et al. 1998). In a study of four genetic stocks of carrot, Senalik and Simon (1987) demonstrated that the content and composition of volatile terpenoids varied among roots and leaves of the same plant, and among different parts of leaves. Although these authors did not perform explicit analyses to test for genotype differences in the spatial pattern of terpenoids, their graphs clearly show that the within-leaf pattern of terpenoid emission was specific to each genotype. The amount and composition of the terpenoids emitted by the petiole, midrib, and leaf blade were similar in some genotypes but differed considerably in others. Habegger and Schnitzler (2000) further showed that the fine-scale pattern of intraleaf terpenoid distribution in carrot was cultivar-specific.

Studies of cultivated fruits and grain crops also reveal that differences in spatial patterns of within-plant variation in fruit and seed traits can be genetically based. The size of ripe tomato fruits (*Lycopersicon esculentum*) varies depending on position on the plant, and the relationship that links fruit size and nodal position varies among cultivars (Bertin et al. 1998). A similar contrast was documented by Rajala and Peltonen-Sainio (2004) for two oat (*Avena sativa*) cultivars, which exhibited different patterns of within-panicle variation in seed size. While in one cultivar seed weight declined only slightly from the primary to the secondary positions in the panicle, the intrapanicle gradient was quite steep in the other cultivar. In perennial ryegrass (*Lolium perenne*), different genotypes exhibit distinct intrinsic gradients of seed-size variation with position in the inflorescence (Warringa, de Visser, and Kreuzer 1998; Warringa, Struik, et al. 1998). Certain varieties of corn (*Zea mays*) exhibit position-dependent variation in the fatty acid composition of seed oil along the ear, while others do not (Jellum 1967). A significant variety × flower position effect has been reported for seed oil content in *Carthamus tinctorius* (Williams 1962), denoting that intrinsic within-plant gradients of seed variation were variety-specific. Calderini and Ortiz-Monasterio (2003) studied the effects of position in the spike on the macronutrient and micronutrient concentration of wheat (*Triticum aestivum*) grains. They compared patterns of seed chemical variation along spike positions in two cultivars and one synthetic hexaploid line, and found significant genotype × position interaction effects on Ca, K, P, and Mn concentration, demonstrating that genetically different lines exhibited contrasting intrinsic gradients of seed chemical composition along the spike. Similar results were obtained by Bramble et al. (2002) in a comprehensive study of the variance structure of single-grain protein content in four wheat cultivars in western Kansas. In addition to other sources of variability (e.g., field, plot), these authors evaluated the relative importance of variance among spikes within-plants, among spikelets within spikes, and among kernels within spikelets, as sources of variation in the protein content of seeds. Cultivars differed in the nature of within-spike gradients in seed protein content, and each of them had a characteristic spatial structure of within-plant variance in the trait under consideration.

Possibly the strongest evidence of a genetic control of differences in within-plant variation in seed traits has been provided by studies of variation in the size and characteristics of rice grains. Individual rice grains located at different positions in the panicle differ predictably in size and in their amylose and starch content (chapter 6), and within-inflorescence patterns of seed variation are cultivar-specific (Zhang et al. 2003; Liu et al. 2005), which clearly points to an underlying genetic basis. Direct supporting evidence was provided by Jeng, Wang, et al.'s study of wild type cultivar Tainung 67 and its artificially induced mutant SA419 (2003). The two cultivars showed different starch and amylose accumulation patterns in relation to grain positions on the panicle. In the wild type cultivar there was a decline in amylose and starch accumulation in the grains located on proximal secondary branches in comparison with grains located on distal primary branches. In the mutant cultivar SA419, in contrast, grain position had little effect on the contents of starch and amylose of grains located on different branches, which clearly demonstrated that differences between cultivars in patterns of within-plant variation in seed characteristics are genetically based.

Summary and Closing Remarks

The evidence presented above documents two aspects that are fundamental to the analysis and interpretation of the ecological and evolutionary significance of within-plant variation in organ traits to be undertaken in the following chapters. On one hand, conspecific individuals ordinarily differ in the extent and spatial organization of intraplant variation in phenotypic traits of reiterated structures. On the other, there is compelling evidence that these individual differences in magnitude and organization of variation are genetically controlled. Verification of the Haldane-Roy conjecture, and its reinforcement by indications of a genetic basis of within-plant variation, should impel us to change the ways in which we characterize individual plant phenotypes. Almost without exception, this has been traditionally accomplished by using exclusively the mean of a sample of organ trait values (e.g., in phenotypic selection studies or investigations of geographical variation). If individual plants not only have characteristic means but also characteristic standard deviations, then some measure of intraplant variation (e.g., variance, standard deviation) or relative variability (CV_{within}) should routinely be used in addition to the mean to properly characterize individual plants' phenotypes with regard to organ traits. In other words, the within-plant variance should be granted a descriptive value of the phenotype similar to the value traditionally conferred on the within-plant mean of organ trait values.

Acceptance of the Haldane-Roy conjecture opens the way to examining variation among characters, among species, or among populations of the same species, also from the perspective of their levels of within-individual variability. Statistically significant individual differences in variability are not universal. Real and Rathcke (1988), for example, found no differences among *Kalmia latifolia* shrubs in levels of relative variability in per-flower nectar production rate, and not every example listed in table 7.1 exhibits significant individual differences in variability. Identifying patterns and ecological correlates of interspecific variation in the magnitude of individual differences in variability will contribute to our understanding of the selective pressures ultimately responsible for that variation, as discussed in chapter 10. Patterns of geographical variation and regional phenotypic differentiation within species may also be examined from this perspective. Geographical variation in flower, fruit, or seed traits, for example, has invariably been addressed from a mean-centered perspective. Investigations of geographical variation in floral traits have traditionally proceeded by first characterizing each individual plant by its mean value for the phenotypic trait of interest (e.g., corolla length), and then examining differences between regions or populations in these plant means (e.g., Herrera et al. 2002; Herrera, Castellanos, and Medrano 2006). But recognition of the fact that within-plant variability is another trait of individuals should lead us, when investigating geographical variation, to consider the possibility of population differentiation in that trait and not only in the mean. This essentially unexplored aspect of population differentiation is illustrated in figure 7.2 for 15 southeastern Spanish populations of the shrub *Lavandula latifolia*. For these populations, Herrera, Castellanos,

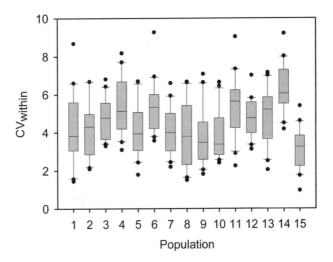

FIG. 7.2 Variation across 15 southeastern Spanish populations of the shrub *Lavandula latifolia* in within-plant variability in corolla length, as estimated with CV_{within} (defined in chapter 3). Box plots show the 10%, 25%, 75%, and 90% percentiles of the distributions of plant CV_{within}. Plant values beyond the 10–90% range are shown as dots. Populations are significantly heterogeneous with regard to their mean CV_{within} values ($\chi^2 = 75.6$, $P < 0.001$, Kruskal-Wallis ANOVA). Twenty plants were sampled per population, with 20–25 flowers measured per plant (additional information on these *L. latifolia* populations may be found in Herrera 2004; Herrera, Castellanos, and Medrano 2006).

and Medrano (2006) demonstrated significant geographical differentiation in corolla length on the basis of conventional analyses based on
plant means alone. The data plotted in figure 7.2 show, in addition to
population differences in plant means, significant differences in levels
of within-plant variability in corolla length, with each population tending to have its characteristic level of within-plant variation. Some populations are characterized by highly variable individuals (e.g., population 14),
while others are consistently made up of constant plants (e.g., population
15). Acknowledging that populations of the same species may also exhibit
geographical differences in levels of within-individual variation is a first
step toward understanding yet another facet of intraspecific geographical
differentiation. This aspect acquires particular importance in relation to
the expanded model of phenotypic selection that is proposed in chapter
10, where variability is explicitly incorporated as another trait potentially
subject to selection.

CHAPTER EIGHT

Consequences of Within-Plant Variation for Interacting Animals

Phytophagous animals' discrimination among organs of the same plant can lead to the most profitable choice but has attendant costs that may influence their overall performance and promote among-plant selectivity.

The main message conveyed by the preceding chapter was that within-plant variance in organ traits deserves a descriptive status similar to that traditionally accorded to plant means. This claim is grounded in both statistics and biology. It vindicates the use of within-plant variation, on one hand, because it is a valuable albeit traditionally neglected descriptor of an individual plant's phenotype, and on the other, because within-plant variability often has a genetic basis and therefore can respond to selection. But these are neither the sole nor the most important reasons why differences among individuals, populations, and species in levels of within-plant variation in organ traits should be examined carefully. Plant reiterated structures are the pivotal elements in most sorts of mutualistic and antagonistic plant-animal interactions, including herbivory, pollination, seed dispersal, and fruit and seed predation. Within-plant variation in the characteristics of reiterated organs may therefore influence interaction with animals in a number of significant ways.

A considerable number of empirical and theoretical studies that have accumulated in recent decades deal with population and behavioral responses of herbivorous, nectarivorous, frugivorous, and granivorous animals to patchiness in the abundance and quality of their food resources (for reviews see, e.g., Lawton 1983; Senft et al. 1987; Hunter et al. 1992; Karban 1992; Goulson 1999; Herrera and Pellmyr 2002). Emerging clearly from all this research is the notion that phytophagous animals generally interact with their food resources and exert discrimination at several nested levels of ecological resolution, namely regional systems, landscapes, plant communities, individual plants, and individual organs within plants. This is the "from-leaf-to-landscape" model of plant-animal interactions, as defined by Weisberg and Bugmann (2003). This general framework of plant-animal interactions explicitly based on ecological hierarchies of scale was originally conceived, and subsequently developed and most thoroughly tested for the case of large mammalian herbivores, which are endowed with considerable mobility and usually forage over wide areas (Senft et al. 1987; Danell et al. 1991; Weisberg and Bugmann 2003; Searle et al. 2005). Nevertheless, this hierarchical view may easily be expanded to incorporate smaller phytophagous animals that recognize and respond to patchiness in the abundance and quality of their food resources at spatial scales ranging from regional systems down to individual plants and organs within plants, as illustrated by Sallabanks (1993) for frugivorous birds, Rabasa et al. (2005) for egg-laying butterflies, and Roslin et al. (2006) for leaf miners.

At the topmost level of spatial resolution in the leaf-to-landscape model, fruit- and nectar-feeding birds and mammals and insect herbivores commonly exhibit foraging responses that involve displacements over broad geographical areas. Frugivorous birds, for example, are well known for their remarkable ability to track variable food supplies across a wide range of spatial scales by means of seasonal migrations and habitat shifts (Herrera 2002b; Saracco et al. 2004). Foraging responses of frugivores at this scale are illustrated by Australian fruit pigeons' migrations across rain forest (Crome 1975), the nomadic wandering behavior of blackcap warblers (*Sylvia atricapilla*) overwintering in southern Spanish Mediterranean shrublands (Rey 1995), and the large-scale movements of African and Asian hornbills (Bucerotidae) in response to regional variation in fruit availability (Kinnaird et al. 1996). Nectarivorous birds and mammals, in the tropics and elsewhere, also respond to spatial variation in nectar resources by habitat shifting, elevational displacements, and migratory behavior (Bertin 1982; Fleming 1992; van Schaik et al. 1993). In the case of adult insect herbivores, the ecological correlates and triggering cues of their frequent displacements at large spatial scales have been less thoroughly investigated, but both classic observational studies (Baker 1969;

Janzen 1987) and recent studies using genetic markers (Sosa-Gómez 2004; Scott et al. 2005) demonstrate that extensive migrations over wide areas are also relatively common among these small herbivores.

At the bottom end of the leaf-to-landscape hierarchy lie those foraging responses that take place within habitats, and these are the ones I will be mostly concerned with hereafter. In a given habitat, phytophagous animals are confronted with the opportunity to discriminate *both* among individual plants of the same species and among the multiplicity of reiterated organs generally borne by each of them. Very few plant-animal interactions occur entirely at the domain of plant individual by animal individual. They happen, for example, when one large mammal grazing in a meadow ingests and kills a small annual herb in a single bite, or when elephants uproot and eventually kill *Acacia* treelets in a single foraging bout (MacGregor and O'Connor 2004). In these and similar instances, discriminatory processes by very large animals eating very small plants will necessarily stop at the among-plant level, and one would expect within-plant variation in organ features to be irrelevant to the interaction between plant and consumer. All else being equal, responsiveness to within-plant variation in organ features should vary inversely with the body size of animals. This is supported, for example, by the results of Hódar and Palo's comparative investigation (1997) on patterns of twig size (a surrogate for food nutritional quality) selection by grouse (*Lagopus lagopus*), hare (*Lepus timidus*), and moose (*Alces alces*) when feeding on *Betula pubescens* trees in an arctic site in winter. As predicted from its larger body size, moose were considerably less responsive than either grouse or hares to spatial variation at several scales in chemical features of food. Since within-plant heterogeneity in chemical and size-related features of reiterated organs predominantly takes place at very small spatial scales (chapter 4), purely mechanical constraints derived from the size of the feeding apparatus will sometimes limit the capacity of large animals to respond selectively to this variation. Among mammalian browsers, for example, mouth size often constrains their ability to respond to very small-scale intraplant variation in forage quality. This is nicely illustrated by Wilson and Kerley's detailed study (2003) on the variation in bite size among six species of mammalian browsers ranging in body size from 5 kg (blue duiker, *Philantomba monticola*) to more than 1,000 kg (black rhinoceros, *Diceros bicornis*). These authors found that the mean diameter of stems bitten off ranged between 1.0 mm for blue duiker to 6.3 mm for black rhinoceros. Mean bite diameter was an increasing function of body size, being allometrically linked to body mass (M) by the exponent $M^{0.32}$. This implies that large mammalian browsers can hardly respond with selective feeding to within-plant variations in foliage quality when these occur at very small spatial scales.

Situations where the body size of the animal exceeds or is roughly comparable in magnitude to that of the plants with which it interacts are, however, relatively scarce in nature. The body size of plants generally exceeds by one or several orders of magnitude that of the majority of animals with which they interact. Simply because of a matter of scale and relative body sizes, therefore, most foraging animals will perceive individual plants as distinct habitat patches where the resources sought for (leaves, flowers, fruits, seeds) are found clustered in locally dense aggregations. Only a fraction of the resources available in each patch is generally exploited on each occasion that an animal visits the patch (i.e., the plant), which means that some within-patch foraging decisions are undertaken. This applies even to some large herbivores that browse on plants roughly equaling them in size. In arctic habitats, foraging moose (*Alces alces*) consume only a fraction of the biomass available in each of the small trees they browse upon, even though it would be theoretically feasible to completely exhaust the food resources available on each tree (Åström et al. 1990).

In most situations, therefore, the elemental ecological links that connect plants and animals ultimately revolve around individual reiterated structures rather than around individual plants. Having arrived on a plant, individual insect folivores, pollinators, or vertebrate frugivores will interact with it by feeding on only a fraction of all the leaves, flowers, or berries available. Since similar organs borne by a plant differ slightly in some important characteristics (and at times not so slightly, as documented earlier in this book), individual animal foragers may perceive this within-plant heterogeneity and develop behavioral responses to it. They may, for example, feed only on some preferred organs while leaving others aside. Through this and related mechanisms, within-plant variation in organ traits might influence the foraging patterns of the animals that feed on them, which might eventually translate into consequences for the plants themselves. The ecological relevance of considering within-plant variation in organ traits in the context of the foraging decisions of phytophagous animals has been previously suggested (Kadmon and Shmida 1992; Sallabanks 1993; Obeso and Herrera 1994; Suomela and Ayres 1994). Its ultimate implications were also aptly summarized by Feinsinger (1983, 51): "Values for biological phenomena are often condensed into means.

Theoretically, organisms dealing with those values can 'expect' the mean value and adapt for it. In reality, organisms encounter values one by one, so if variance is high the mean may be irrelevant."

In this chapter I look at the consequences to animals of within-plant variation in characteristics of reiterated organs. In the first part I consider behavioral responses taking place at the level of the individual animal. Phytophagous animals perceive the variation among organs of the same plant, and respond behaviorally by preferring some organs and rejecting others by virtue of their phenotypic traits. I review observational and experimental evidence documenting discrimination by frugivorous seed dispersers, pollinators, seed predators, and herbivores, among reiterated structures simultaneously borne by the same plant, which represents the lowest level in the hierarchy of discrimination. The second part of the chapter considers the direct ecological costs to animals arising from their within-plant discrimination and choice, which can be measured in terms of time, energy, experienced competition, performance, and mortality risk. In the third section I consider the indirect costs to animals of within-plant variation that arise from Jensen's inequality (described later in this chapter) as a consequence of the nonlinearity of relationships between consumer performance and organ trait variation. Direct or indirect costs to foraging animals arising from within-plant variation may give rise to discriminating behavior and selectivity at the among-plant level. I deal with this crucial aspect in the final section of the chapter, where I attempt to bridge the within- and between-plant selection levels by providing a general framework of the consequences of within-plant variation in terms of differential selection *among* plants by consumers. The ecological and evolutionary implications of such selection will be dealt with in chapters 9 and 10, respectively.

Behavioral Responses to Within-Plant Variation: Discrimination and Choice

With the exception of the relatively frequent work focusing on within-plant host selection patterns by phloem- and leaf-feeding insects (reviewed in Schultz 1983; Whitham et al. 1984; Gill et al. 1995; see later in this section), few studies have explicitly addressed the foraging responses of phytophagous animals to within-plant variation in the characteristics of reiterated structures (e.g., Duffield et al. 1993; Orians and Jones 2001), and most published evidence bearing on the subject is either anecdotal or indirect. Nevertheless, abundant experimental evidence demonstrates that all classes of phytophagous animals possess considerable discriminating abilities. This suggests that within-plant selection among reiterated structures differing in morphological, chemical, or nutritional properties is likely to be the rule rather than the exception. In this section I provide empirical support for this claim by reviewing evidence suggesting that within-plant variation in fruit, flower, seed, and leaf features generally elicits discriminating responses and choice *among organs* by the animals that rely on these structures for food and/or breeding sites, namely frugivores, pollinators, fruit and seed predators, and herbivores. For each of these major groups, I briefly review experimental studies documenting their discriminating competence in relation to organ features that have been shown previously to vary subindividually (chapters 2 and 3), and then I consider direct and indirect field evidence demonstrating that, under natural conditions, animals in effect respond to within-plant variation by selecting some organs over others. As a common background to this section, the reader should keep in mind two features of within-plant variation in organ traits documented earlier in this book: within-plant variation is generally comparable or even greater in magnitude than variation among plants, and most of it is simultaneous and occurs over very small spatial scales within the plant. There is thus no question that animals will routinely have ample opportunities for sensorial comparisons, discrimination, and choice among organ variants borne by the same plant.

Fruit Variation and Vertebrate Frugivores

Extensive experimental and observational evidence demonstrates that frugivorous mammals and birds are able to discern and respond to variation in a number of features of fleshy fruits (berries, drupes, and similar structures), including size, seediness, seed and pulp volume, and nutritional and chemical composition of the pulp, and that they routinely discriminate among fruits of the same or different species on the basis of variation in any of these traits (Sallabanks and Courtney 1992; Herrera 2002b; Stanley and Lill 2002; and references therein). Under controlled experimental conditions, frugivorous primates detect extremely small differences in sugar concentration in their food and use sweetness as a criterion for food choice (Laska et al. 1996). Likewise, frugivorous bats are able to discriminate among fruits differing in size, degree of ripeness, and sugar content (Korine and Kalko 2005; Nelson et al. 2005). Most evidence on feeding responses of frugivorous vertebrates to variation in fruit features, however, comes from a plethora of aviary experiments in which frugivorous birds are presented with artificial or natural fruits differing in some trait of interest but similar in other respects.

When offered a choice of otherwise similar fruits, captive avian frugivores generally discriminate among fruits and exhibit preferences based on differences in size (McPherson 1988; Avery et al. 1993; Sallabanks 1993; Rey and Gutiérrez 1997; Stanley et al. 2002), seed load (Herrera 1981b; Hegde et al. 1991; Murray et al. 1993; Stanley and Lill 2002), and nutritional content and chemical properties of pulp, including lipid and protein concentration, fatty acid composition, sugar composition, sugar concentration, and presence of secondary compounds and volatiles (Willson and Comet 1993; Avery et al. 1995; Giles and Lill 1999; Lepczyk et al. 2000; Stanley and Lill 2001; Bosque and Calchi 2003; Pierce et al. 2004; Saxton, Creasy, et al. 2004; Saxton, Hickling, et al. 2004). It seems reasonable to expect that, when confronted in natural settings with the variation in fruit features occurring within individual shrubs or trees, or even within individual infructescences, frugivorous birds will respond to it and behave as selectively at the within-plant level as they ordinarily do in experimental arrays. There are two main reasons for this expectation. First, frugivores are characterized by their ability to detect and respond to extremely small variations in fruit traits, which are considerably smaller in magnitude than those ordinarily occurring within the fruit crops of individual plants. Tanagers, a group that includes some of the commonest frugivores in the Neotropics, provide an illustrative example. These frugivores crush fruits in their bills, thereby releasing juice onto their tongues, which allows them to assess fruit chemical properties by taste before swallowing. Levey (1987) showed that three species of tanagers were able to detect differences in diets containing 8%, 10%, and 12% sugar, and four species studied by Schaefer et al. (2003) detected differences in sugar concentrations of only 1%, and differences in lipid content of only 2%. Most remarkably, blue-gray tanagers (*Thraupis episcopus*) react to differences in protein content as small as 0.09% of fresh matter (Bosque and Calchi 2003). Very low odor detection thresholds have also been reported for some frugivorous birds (Clark 1991), and particularly low discrimination thresholds for sugar concentration seem to be a distinguishing feature of frugivorous primates and bats (Laska et al. 1996; Herrera et al. 2000). Second, most natural within-plant variation in fruit traits typically occurs at very small spatial scales within plants, generally within the same branch or infructescence (chapter 4). Frugivorous birds tend to be most selective among simultaneously available, alternative fruit types when these latter are highly accessible and located near each other (Moermond and Denslow 1983; Levey et al. 1984; Willson and Comet 1993).

A few field studies provide direct empirical support for the expectation that frugivorous birds respond selectively to within-plant variation in fruit features. In Costa Rican tropical montane forest, seeds of the lauraceous tree *Beilschmiedia costaricensis* are dispersed only by the few large frugivorous birds that are capable of swallowing the large drupes (mean diameter = 23 mm) characteristic of the species (Wheelwright 1985). The tree also exhibits large within-plant variation in fruit diameter, which may range between 17 and 25 mm within a single tree. By comparing the frequency distribution of diameters of *B. costaricensis* seeds regurgitated (and, by inference, fruits consumed) below an isolated tree by emerald toucanets (*Aulacorhynchus prasinus*) with the distribution of seed diameters of fruits dropped uneaten by birds below the same tree, Wheelwright (1985) demonstrated that, when feeding on that tree, toucanets fed preferentially on the smaller fruits and rejected the larger ones. Sallabanks (1993) investigated patterns of within-plant fruit selection by American robins (*Turdus migratorius*) foraging in hawthorn (*Crataegus monogyna*) shrubs. Fruits picked by robins but subsequently dropped beneath 14 focal shrubs were identified in the field by their characteristic beak marks (two or more parallel or pointed lines left on the surface of the fruit by the edge of the bird beak holding it), and their features were compared with those of intact fruits picked from the same shrubs using a paired design. All the fruit traits measured (fruit diameter, fruit mass, pulp mass, seed mass, and pulp-to-seed ratio) were significantly greater for bird-dropped fruits compared with untouched fruits remaining on shrubs, which denoted significant size-based, within-plant selectivity by foraging robins. I replicated this part of Sallabanks's study at four southeastern Spanish populations of *C. monogyna* by sampling available and rejected, beak-marked fruits underneath a total of 55 trees (C. M. Herrera, unpublished data). In these populations, *C. monogyna* fruits were eaten by several *Turdus* (thrush) species, and within-tree variation accounted on average for 61% of population-wide variance in fruit width (table 3.3). As in Sallabanks's study, southeastern Spanish thrushes consistently exerted within-plant discrimination based on fruit size when feeding on these variable fruit crops. The mean diameter of rejected, beak-marked fruits found underneath plants was larger than the mean diameter of the fruits borne by the same plant in 53 out of the 55 trees studied (fig. 8.1). Similar size-based discrimination by frugivorous birds among the fruits produced by individual plants has also been reported for other fleshy-fruited Mediterranean plants such as *Osyris lanceolata* (Herrera 1988) and *Prunus mahaleb* (Jordano 1995b).

All the preceding examples refer to within-plant discrimination by avian frugivores in the field, based on differences in fruit size. This is probably because size is the single fruit trait most easily measured and the one for which most quantitative information on intraspecific variation down to the within-plant level exists (chapter 3). Nevertheless, I would expect frugivores to respond selectively also to within-plant variation in less apparent fruit traits such as chemical composition of pulp (chapter 2). Field and aviary investigations have revealed that frugivorous birds possess considerable discriminating competence, which allows them to select among simultaneously available fruits on the basis of subtle characteristics that are not easily perceived visually. For example, blackbirds (*Turdus merula*) foraging on *C. monogyna* shrubs are able to distinguish, and

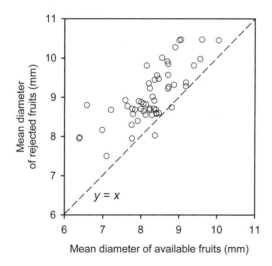

FIG. 8.1 Frugivorous thrushes (*Turdus* spp.) consistently exert within-tree discrimination on fruit size while foraging on *Crataegus monogyna* trees in southeastern Spain. The mean diameter of rejected, beak-marked fruits lying on the ground underneath individual plants is plotted against the mean diameter of fruits borne by the same plant. Each symbol corresponds to a different tree (*N* = 55). Except in two trees, rejected fruits were larger on average than those available on the plant. The *y* = *x* dashed line represents no discrimination. C. M. Herrera, unpublished data.

reject before swallowing, fruits whose seeds contain insect larvae, despite the fact that infestation is not readily apparent from visual inspection of the fruit (Manzur and Courtney 1984). Similar fine-tuned discrimination based on cryptic fruit infestation by microbes or insect pests has been demonstrated for other species of avian frugivores feeding on a variety of fruit types (Jordano 1987; Krischik et al. 1989; Valburg 1992a, 1992b; Traveset et al. 1995; García et al. 1999), as well as for small fruit-eating tropical bats in the family Phyllostomidae (Korine and Kalko 2005).

Flower Variation and Pollinators

Experimental evidence has accumulated over more than a century proving that animal pollinators are characterized by remarkable visual and olfactory discriminatory abilities in relation to a long list of morphological and chemical floral features (Clements and Long 1923). (In this section, the word *pollinator* is used loosely to mean animals adapted to visiting flowers even if they are not high-quality pollinators of the plant they are visiting.) Most information comes from studies of social bees, notably honeybees (*Apis mellifera*) and bumblebees (*Bombus*). For these, investigations with captive individuals have documented a considerable capacity to detect and respond to subtle variations in features of artificial model flowers (reviews in Goulson 1999; Chittka and Thomson 2001). In addition to responding to variation in nectar concentration, volume, and secretion pattern (Waddington 1980; Waddington et al. 1981; Cnaani et al. 2006), honeybees and bumblebees also discriminate among individual flowers on the basis of differences in structural and chemical characteristics, including size (Møller 1995; Blarer et al. 2002), orientation (Ushimaru and Hyodo 2005), symmetry (Lehrer et al. 1995; Møller 1995; Rodríguez et al. 2004), scent (Raguso 2001 and references therein), and the presence or concentration of amino acids and secondary compounds in the nectar (Inouye and Waller 1984; Singaravelan et al. 2005).

Outstanding discriminatory competence, however, is not by any means exclusive to social bees. Although much less frequently investigated, insect pollinators such as lepidopterans, beetles, and flies are also endowed with the capacity to respond to slight variations in floral traits. Flower-visiting butterflies and the day-flying hawk moth *Macroglossum stellatarum* discriminate among artificial nectars with different sugar and amino acid composition (Erhardt 1991; Kelber 2003), and individuals of the hawk moth *Manduca sexta* foraging in a laboratory flight cage selected those artificial flowers that emitted the highest levels of CO_2, a characteristic of the newly opened, rewarding flowers of their *Datura wrightii* food plants (Thom et al. 2004). In a field experiment, Møller and Sorci (1998) demonstrated that floral visitors belonging to several families of Coleoptera and Diptera discriminated among artificial flower models differing in degree of symmetry.

Among vertebrate floral visitors, hummingbird responses to variation in floral features have been most frequently studied. These nectarivorous birds are able to discriminate among flowers differing in nectar volume and concentration (Hainsworth and Wolf 1976; Tamm and Gass 1986; Blem et al. 2000), nectar sugar composition (Stromberg and Johnsen 1990), and nectar amino acids, vitamins, and minerals (Hainsworth and Wolf 1976; Carroll and Moore 1993; Bouchard et al. 2000). Other nectarivorous birds, including species of African Nectariniidae, can also discriminate among floral nectars differing in sugar composition (Jackson et al. 1998). Among mammalian floral visitors, nectar-feeding glossophagine bats have recently been shown to possess the remarkable capacity of discriminating among simultaneously available flowers of the same species that differ slightly in morphology on the basis of their different "echo fingerprints" (von Helversen and von Helversen 2003).

The preceding examples, although incomplete, show that pollinators exhibit considerable discriminating capacity among flowers or artificial flower surrogates when tested under controlled experimental conditions. On this basis, one would predict that they also respond selectively to within-plant variation in floral features occurring in the field, and this is supported by the direct and indirect field evidence presented below. Before considering it, however, it is important to acknowledge an inherent difficulty associated with the interpretation of direct field observations on flower selection by pollinators. Pollinators differ from other phytophagous animals (e.g., frugivores, seed predators) in the way they "sample" the reiterated structures simultaneously available on the same plant. For example, individual frugivores foraging on a tree or shrub sample the available fruits without replacement, chosen fruits being taken away from the plant and becoming subsequently unavailable to that frugivore or any other. In contrast, floral visitors often sample with replacement the available flowers on a plant. Flowers of most species last for more than one day (Ashman and Schoen 1996), and even the shortest-lived ones may receive many visits over their lifetime (Kato 1988; Jones et al. 1998); hence a flower visited by a pollinator will remain exposed to further visits by others for some time afterwards. This means that, at any given time in the life of individual flowers, some of its features that may influence the behavioral responses of pollinators, particularly the amount of pollen presented or nectar content, reflect not only intrinsic properties of the flower (pollen and nectar production), but also the vagaries of its prior visitation history. This history generates complex patterns of within-plant variation in "pollinator-modifiable" floral traits that may be only weakly related to intrinsic patterns of variation, as exemplified by weak correlations across flowers between rates of nectar production and instantaneous nectar standing crops (Zimmerman 1988), and by the extensive within-plant variation in nectar sugar composition in the field, but not in the greenhouse, reported for flowers of two species of *Aquilegia* (Canto et al. 2007). Furthermore, mounting evidence indicates that pollinators may actively "manipulate" the probability of individual flowers being subsequently visited, thus additionally contributing to decouple intrinsic floral traits and the probability of visitation. Several groups of insect pollinators, including social and solitary bees, deposit short-lived repellent odors on the corolla when collecting nectar or pollen, and subsequent visitors of the same or different species tend to avoid these recently depleted flowers (Giurfa and Núñez 1992; Goulson et al. 1998; Stout and Goulson 2001; Hrncir et al. 2004; Gawleta et al. 2005; Reader et al. 2005). Pollinator-induced decoupling between observable and intrinsic floral traits thus greatly limits the use of field observations to assess whether within-plant flower discrimination by pollinators, when it occurs, is actually related to intrinsic within-plant variation in floral features or to differences among flowers originated by contrasting histories of pollinator visitation. Keeping this caveat in mind, however, field evidence relating differences among flowers of the same plants to differential responses by floral visitors may still be interpreted as indicative of within-plant flower choice by pollinators.

After approaching a plant, individual pollinators eventually visit only some of the flowers available (range 5–70%; references in Harder et al. 2004). This aspect of plant-pollinator interactions has been examined frequently by pollination ecologists, because variations in the proportion of flowers visited per pollinator approach may translate into variations in the likelihood of geitonogamous pollinations and, consequently, in the proportion of selfed progeny (Hessing 1988; Karron et al. 2004; Mitchell et al. 2004). Despite the interest elicited by the widespread phenomenon of "nonexhaustive foraging" by pollinators, relatively little effort has been devoted to investigate the proximate mechanisms and discriminatory processes involved in within-plant flower selection. The available evidence, however, suggests that the subset of flowers visited per approach will rarely be, if ever, a random subsample of those simultaneously borne on the plant. This has been well investigated in plants with flowers linearly arrayed on simple vertical inflorescences, where pollinators often tend to forage selectively on available flowers according to well-defined directional patterns. Usually, approaching bees commence foraging at the bottom flowers, work predominantly upward, and leave inflorescences before visiting all flowers. These patterns of within-plant flower selection may reflect discrimination of pollinators in favor of the largest or most rewarding flowers in the array, as well as intrinsic preferences for foraging on certain specific locations in the inflorescence (Pyke 1978a; Corbet et al. 1981; Dreisig 1985; McKone et al. 1995). Regardless of inflorescence type, some bee species may also prefer those flowers of a plant that are located at some particular height above the ground, and this form of within-plant selection has been reported for species ranging from low herbs to tall trees (Levin and Kerster 1973; Frankie and Coville 1979; Roubik et al. 1982; Kadmon et al. 1991; Rinderer et al. 1996). Rinderer et al. (1996), for example, found that each of several species of *Apis* and *Trigona* bees foraging on a single individual of the tree *Peltophorum pterocarpum* in a Bornean forest tended to select flowers located at different heights above the ground. Kadmon et al. (1991) studied flower visitation by anthophorid bees (*Anthophora* and *Eucera*) on a single plant of the annual herb *Anchusa strigosa* and found that the probability of approaches from another plant varied significantly between individual flowers, being negatively correlated with the height of the flower above the ground surface. Individual flowers located at 20–30 cm above the ground received, on average, twice the number of external approaches per unit time than those at 60–70 cm. As there was no consistent relationship between flower height and nectar production, their findings reflect intrinsic preferences of bees for flowers located at lower positions on the plant.

More generally, and regardless of plant size or type of inflorescence, it is a common observation that individual pollinators frequently approach individual flowers and reject them without landing or attempting to probe them for nectar or pollen (e.g., Kadmon et al. 1991), behavior that clearly denotes discrimination among available flowers on the same plant. In the case of bees, within-plant discrimination of this sort generally results in foragers probing the most rewarding flowers and skipping the least rewarding ones (Heinrich 1979; Wetherwax 1986; Kato 1988; Kadmon 1992; Duffield et al. 1993). This is accomplished by assessing within-plant variation in rewards, either directly (evaluating nectar reward itself), by association (using some correlated floral cue), or by rejecting flowers upon which previous visitors have left scent marks (Thorp et al. 1975; Marden 1984b; Goulson et al. 1998; Goulson et al. 2001). These studies demonstrate within-plant selectivity by pollinators, although it is difficult to place them properly in the context of pollinator responses to within-plant variation in *intrinsic* floral features because, as noted above, floral traits involved in the discrimination can vary dynamically as a consequence of previous pollinator visitation.

Field observations of flower selection by pollinators in response to variation in floral traits not susceptible to modification by prior visitation, such as corolla size or degree of symmetry, provide better evidence of within-plant discrimination. Duffield et al. (1993) investigated the choice of flowers at two hierarchical levels by honeybees foraging on plants of the Mediterranean shrub *Lavandula stoechas*. Within plants, bees landed preferentially on inflorescences with more of their flowers open, while tending to ignore or actively reject inflorescences with relatively fewer flowers. Once on an inflorescence, bees usually probed only a small proportion of the open flowers, tending to probe more often those flowers with the longer and wider corollas. Møller (1995) used a method based on the comparison of the characteristics of flowers in nearest-neighbor pairs on the same ramet to study the relationship of pollinator visitation with the size and degree of symmetry of flowers of bumblebee-pollinated *Epilobium angustifolium*. The degree of petal asymmetry in flowers visited by a bumblebee was smaller than that of the nearest neighboring flower on the same ramet that was ignored by the visitor. First-visited flowers were also larger than the nearest neighboring flower that was not visited first by a bumblebee. Møller and Eriksson (1995) did a similar study in three different sites in Spain, Denmark, and Sweden, involving ten species pollinated by bees, beetles, or flies. In all the species studied, the petals of insect-visited flowers were significantly more symmetrical than those of the unvisited nearest neighboring flower in the pair. Furthermore, in seven of the species, petals were significantly longer in the insect-visited flowers than in the neighboring unvisited flowers. Using a similar approach, Martin (2004) likewise found that honeybees foraging on *Mimulus guttatus* were found consistently more often in the larger flower of the nearest-neighbor pair. In contrast to Møller's work (1995), these two later studies compared the traits of a visited flower with that of the nearest unvisited one on a different plant. Provided that the variance in floral traits occurring within-plants may be similar or even greater than the variance occurring among plants (chapter 3), results of these studies support within-plant discrimination of flowers by pollinators in response to variation in size and degree of symmetry of flowers.

Fruit and Seed Predators

Ovipositing females of invertebrate fruit and seed predators whose sessile larvae spend a protracted developmental period confined within a single structure should be particularly selective when exposed to the fruit or seed variants occurring in a single plant's crop. In these cases, oviposition choices will ultimately determine the amount and quality of the food available to larvae during development, and strong selective pressures should favor subtle discriminating abilities and marked preferences for those fruits or seeds whose traits are conducive to more numerous and viable progeny. A large number of experimental investigations and field studies clearly support this expectation.

Because of the frugivorous habits of their larvae, many species of the true fruit fly family Tephritidae have become major damaging agents in fruit orchards, and they figure prominently among the world's most damaging agricultural pests (e.g., *Anastrepha, Ceratitis, Dacus, Rhagoletis*). Their economic impact has prompted many studies on the ecological, behavioral, and sensorial aspects involved in host choice and fruit discrimination by these insects (reviews in, e.g., Boller and Prokopy 1976; Fletcher 1987; Sallabanks and Courtney 1992). These studies have shown that female fruit flies use a two-step fruit selection procedure. After selecting individual host plants on the basis of foliage color, shape, size, and odor emissions, ovipositing fruit fly females of most species discriminate visually among the developing fruits simultaneously available on the chosen plant. Shape, size, and contrast against background are the fruit characteristics most commonly used by ovipositing flies in this second step to discriminate among fruits within plants. For example, ovipositing flies of different species of *Rhagoletis* discriminate visually among inanimate spherical models used as surrogates of host fruits on the basis of differences in diameter (Prokopy 1969, 1977; Prokopy and Bush 1973). Fruit discrimination and preferences by ovipositing females based on size differences have frequently been reported for other tephritid species (*Anastrepha*, Sugayama et al. 1997; *Dacus*, Katsoyannos and Pittara 1983). A visual discrimination mechanism of this sort based on differences in fruit size may also be responsible for the nonrandom distribution of tephritid fly larvae among berries of the shrub *Berberis hispanica* in southeastern Spain (Herrera 1984b). In fruit crops of this species, the estimated per-fruit probability of infestation by fly larvae varied within plants and was related to variation in the number of seeds per fruit, increasing from 0.31 through 0.62 to 0.88 for fruits enclosing one, two, and three seeds, respectively. In this example, the within-plant preference of ovipositing flies for larger fruits enclosing more seeds was clearly related to increasing larval survival with increasing number of seeds (Herrera 1984b).

Plants with multiovulate carpels often exhibit considerable within-plant variation in fruit size and seediness, which generally occurs at very restricted spatial scales (chapters 3 and 4). Furthermore, in these species fruit size and seediness are generally correlated across fruits of the same plant (C. M. Herrera, unpublished data). Size-based within-plant discrim-ination of fruits similar to that exhibited by tephritid flies feeding inside fleshy fruits is therefore expected to be widespread among other sessile invertebrate predators whose larvae spend their whole lives feeding on the seed content of a single fruit. Results of a study of the incidence of the seed-eating larvae of a bruchid beetle on the ripe fruits of *Guazuma ulmi-folia*, a tropical dry forest tree, are consistent with this suggestion (Herrera 1989c, and unpublished data). Ovipositing females of *Amblycerus cisteli-nus* lay a single egg on each *G. ulmifolia* fruit, and there the larva com-pletes its development feeding on most or all of the 20–80 enclosed seeds until pupation, which also takes place inside the fruit. Within individual trees, infested fruits are consistently larger (fig. 8.2) and contain more seeds, on average, than uninfested ones, which reveals size-based discrimi-nation and egg-laying preference by ovipositing *A. cistelinus* females for the larger, more profitable fruits at the within-tree level. Similar results were reported by McClure et al. (1998) for two species of *Strobilomyia* flies feeding within the seed cones of *Larix laricina* trees. These seed pred-ators tended to select for oviposition the longer cones in each tree.

Examples abound of small predispersal seed predators (Curculioni-dae and Bruchidae in the Coleoptera; Eurytomidae and Torymidae in the Hymenoptera) whose larvae develop inside single seeds (Janzen 1969). In these species, the resources available to the larva are determined by the quality of the seed on which the egg was deposited. After entering a seed, the larvae's principal source of mortality may be failure to complete development and reach the pupation stage. In *Bruchus brachialis* feed-ing on *Vicia villosa* seeds, 23.5% of larvae died within the seeds (Dicka-son 1960). Strong selective pressures are therefore expected on the egg-laying behavior of females, and within-plant discrimination by ovipositing

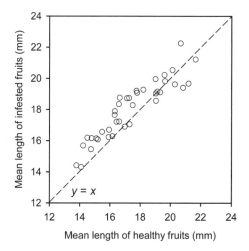

FIG. 8.2 Within individual fruit crops of *Guazuma ulmifolia*, a tropical deciduous forest tree, the fruits infested by larvae of the bruchid beetle *Amblycerus cistelinus* tend to be larger than the uninfested ones. The graph plots the mean length of fruits infested by larvae against the mean length of uninfested fruits on the same plant for a Costa Rican *G. ulmifolia* population. Each symbol corresponds to a different tree ($N = 40$). The $y = x$ dashed line represents no discrimination. C. M. Herrera, unpublished data.

females should proceed one step beyond selection among fruits and involve selection among individual seeds. Nalepa and Grissell (1993) did a detailed study of within-plant variation in seed size in a single plant of *Rosa multiflora*, and how it related to variation in adult size, emergence, and morphology of a torymid wasp seed predator. *Megastigmus aculeatus* is a small (thorax width 1.0–1.5 mm) chalcid wasp that feeds internally in the seeds of wild and cultivated species of *Rosa*, where it consumes the embryo and the endosperm. Only one larva attains full growth in a single seed. Nalepa and Grissell classified more than 2,600 seeds into three size categories, and used X-rays to nondestructively assess the presence of *M. aculeatus* larvae inside individual seeds. Larvae were nonrandomly distributed among seed-size classes. The proportion of infested seeds increased significantly from 16% in size class 1 (<2.0 mm), to 37% in class 2 (2.0–2.8 mm), to 44% in class 3 (>2.8 mm), which clearly denoted seed-size-based discriminating capacity by the wasps and an active oviposition preference for the largest seeds available on the single shrub studied. Similar size-based selection of *Rosa rugosa* seeds by ovipositing *M. aculeatus* wasps was documented by Gillan and Richardson (1997). At their two study sites, between 79 and 95% of the achenes with wasp exit holes were more than 2.0 mm, while infestation was only 5–21% among the smaller achenes

(1.4–2.0 mm). Further evidence of within-plant selection by chalcid seed predators is provided by Chung and Waller's work on patterns of seed predation by the chalcid wasp *Eurytoma seminis* in clones of the shrub *Rhus glabra* (1986). The incidence of chalcid wasps on seeds varied widely among stems and infructescences of the same clone, variation being considerably greater within than among clones (Chung and Waller 1986, fig. 2). Within-clone variation in chalcid wasp incidence was related to variation in seed size, with seed length being positively correlated with predation rate in four out of nine clones. Subindividual variation in seed size thus partly accounted for intraplant variation in predation rates through its effects on the host-selection behavior of the predator.

Although few field investigations have approached the study of invertebrate seed predation and its correlates from the perspective of their within-plant variation, I would predict that patterns similar to those revealed by the studies of the *Amblycerus-Guazuma*, *Megastigmus-Rosa*, and *Eurytoma-Rhus* systems should be widespread whenever (1) a predictable relationship exists between progeny fitness and some seed or fruit trait, and (2) ovipositing females are able to assess the within-plant distribution of seed or fruit quality (e.g., size, seediness) and select the best seed or fruit. This combination of conditions is probably the rule in nature, at least judging from the results of numerous experimental and observational studies conducted on seed-eating beetles of the families Bruchidae and Curculionidae. These include Dickason (1960) on *Bruchus brachialis* feeding on seeds of *Vicia villosa*; Mitchell (1975) on *Callosobruchus maculatus* feeding on mung beans (*Vigna aureus*); Mitchell (1976) on egg-laying *Mimosestes amicus* on *Cercidium floridum* pods; Bradford and Smith (1977) on *Caryobruchus buscki* and *Scheelea rostrata* seeds; Fox and Mousseau (1995) on *Stator beali* and *Chloroleucon ebano*; Moegenburg (1996) on *Caryobruchus gleditsiae* and *Sabal palmetto*; Redmon et al. (2000) for *Bruchidius villosus* ovipositing on *Cytisus scoparius* pods; Cope and Fox (2003) on *Callosobruchus maculatus* and *Vigna unguiculata*; Campbell (2002) on *Sitophilus oryzae* on *Triticum aestivum* seeds; and Koo et al. (2003) on *Mechoris ursulus* ovipositing on *Quercus serrata* and *Q. mongolica* acorns. These studies have shown that egg-laying females of seed-eating beetles discriminate among simultaneously available seeds on the basis of their size and select relatively larger seeds for oviposition, and that the body size of offspring, and thus their prospective fitness, increases with increasing size of the host seed. Within-plant seed-size selection by these small, sedentary, seed-eating invertebrates should be the rule under natural conditions.

Some invertebrate seed predators whose larvae feed inside developing fruits lay eggs on flowers rather than on the developing ovaries. In these cases, ovipositing females should not only find an appropriate host plant, but, once it has been found, they should select for oviposition those flowers with a higher probability of setting seed. One would therefore expect ovipositing females in these cases to exert active within-plant selection based on floral characters likely to enhance maternal pollination success, a prediction supported by some field evidence. In the hummingbird-pollinated *Ipomopsis aggregata*, Brody (1992a, 1992b) and Brody and Waser (1995) found that flowers chosen by ovipositing females of the seed predator *Hylemya* tended to have longer corollas and a higher probability of setting seed than flowers that were not chosen. In the bumblebee-pollinated *Monotropastrum globosum*, the developing fruits are infested by microlepidopteran larvae. Within individual plants, fruits originating from flowers with comparatively larger petals are infested significantly more often than fruit from flowers with smaller petals (Ushimaru and Imamura 2002). These two examples reveal that predispersal fruit predators may also exert discrimination within plants by preferring some flowers over others for oviposition.

Vertebrate fruit and seed predators may also exert within-plant selection of fruits and seeds, although their food-selection patterns have been examined from this perspective even less frequently than for insect predators. Small mammals are important predispersal seed predators of many species, and in some cases exert within-plant discrimination among seeds and fruits. In the Iberian Peninsula, the long-tailed field mouse *Apodemus sylvaticus* is the main predispersal seed predator of the perennial herb *Helleborus foetidus* (Fedriani 2005). Experiments with captive animals have shown that, after climbing the infructescence of a plant, mice discriminate among the available fruits and feed preferentially on those with the greater number of follicles (J. M. Fedriani, personal communication). Granivorous birds that behave as predispersal seed predators may also exert within-plant discrimination on the basis of differences in fruit or seed characteristics, as demonstrated for some fringillid finches. *Carduelis chloris* is a major predator of the seeds of the Mediterranean shrub *Pistacia lentiscus*, a species that produces fruit crops composed of a mixture of filled and unfilled seeds (Jordano 1989). In an experimental and field study designed to evaluate the fruit- and seed-selection behavior of *C. chloris*, Jordano (1990) demonstrated that birds were able to discriminate between *P. lentiscus* fruits containing filled and unfilled seeds by bill-weighing, and actively rejected fruits enclosing unfilled seeds without having to crack them in the bill. Similar abilities to discriminate among simultaneously available conspecific seeds that differ in size, soundness, weight, hardness, or some combination of these features, along with distinct preferences for seeds possessing certain traits, have been frequently reported for other granivorous birds in the families Corvidae, Fringillidae, Paridae, and Estrildidae (Hespenheide 1966; Ligon and Martin 1974; Senar 1983; Greig-Smith and Crocker 1986; Johnson et al. 1987; Heinrich et al. 1997; van der Meij and Bout 2000). In an experimental study of food selection by Clark's nutcrackers (*Nucifraga columbiana*) feeding on *Pinus edulis* seeds, Christensen et al. (1991) found that nutcrackers first discriminated among trees based on differences in mean seed size, and then discriminated among cones within a tree by selecting the longer cones with more numerous and proportionally more viable seeds. Captive bullfinches (*Pyrrhula pyrrhula*) presented with single sunflower fruits, pairs of fruits, and batches of fruits, selected better when fruits could be directly compared with one another, and choices were based on their relative features (Greig-Smith and Crocker 1986).

Further proof of the remarkable discriminating competence of avian seed predators is provided by observations indicating that they can distinguish externally between fruits or seeds based on whether they are infested by larval seed predators. When feeding on the infructescences of *Banksia attenuata*, the cockatoo *Calyptorhynchus funereus* seeks and eats the larvae of seed-eating weevils; the birds forage within-plants in a selective manner, tending to feed preferentially on infested infructescences (Scott and Black 1981). Blue jays (*Cyanocitta cristata*) feeding on *Quercus palustris* acorns discriminate among nuts infested and uninfested by weevils, and they handle, open, and consume uninfested nuts significantly more often than infested ones (Dixon et al. 1997). The capacity to discriminate among simultaneously available seeds illustrated by these examples, in combination with the remarkably high variances in seed traits commonly occurring within single plants (chapter 3), leads to the prediction that granivorous birds that pick seeds one at a time from dehiscent fruits or exposed infructescences (e.g., grass panicles) should ordinarily engage in within-plant seed discrimination.

Leaf Variation and Herbivores

"Variation is the rule" in host plant selection by herbivores, as expressed by the title of a chapter in Schoonhoven et al.'s book on insect-plant biology (2005). It is thus not surprising that the causes and consequences of variation in plant traits in relation to herbivores have been examined in considerable detail by innumerable studies done on a plethora of species. Reviews may be found in Karban (1992), Hare (1992), Hoy et al. (1998), and Schoonhoven et al. (2005). These reviews have generally highlighted patterns of variation in herbivory-related plant traits among species and among individuals or populations of the same species, but they have generally ignored or considerably played down the significance of variation within individual plants. Nevertheless, the many ways whereby within-plant variation in foliage may influence food selection by herbivores have been known for a long time, and examples in the primary ecological literature are not as rare as their neglect by recent reviews of plant-herbivore interactions would seem to suggest. For example, more than 20 years ago Whitham (1978, 1980) and Zucker (1982) demonstrated clearly that variation in leaf size and quality within *Populus angustifolia* tree crowns strongly affected oviposition choices by phloem-feeding *Pemphigus* aphids. Schultz (1983) reviewed what was known at the time about the influence of tree heterogeneity on the foraging behavior of larval lepidopterans, and concluded that within-tree heterogeneity had a number of important consequences for both the populations and the communities of forest insects. In a detailed study of host leaf selection by the miner *Stilbosis juvantis* on *Quercus emoryi* trees, Faeth (1985) demonstrated that leaf selection by ovipositing females took place at the level of individual leaves rather than individual trees, with females first locating a tree of the correct host species and then making oviposition decisions within that tree. These and other early studies documenting the ecological significance of within-plant foliage variation in relation to host selection by herbivores were considered in a series of comprehensive reviews by Whitham and associates in the early eighties (Whitham 1981, 1983; Whitham et al. 1984). The main purpose of these reviews was to provide support for Whitham and Slobodchikoff's original theory (1981) that within-plant genetic mosaics caused by somatic mutations might have direct ecological implications and adaptive significance because of their role in the defense of long-lived plants against short-lived herbivores (the "genetic mosaic theory of plant defense," GMT hereafter). I suspect that the neglect in recent reviews of the within-plant level of variation in relation to herbivory is a direct consequence of the dismissal of the GMT, which no longer seems to be perceived by ecologists as a worthwhile approach to the study of plant-herbivore coevolutionary interactions. Even if the overall judgment is right that genetic mosaicism is not a principal contributor to plant adaptation to herbivores, I contend that rejection of the GMT should not lead us to automatically dismiss, without further scrutiny, the significance of within-plant variation in relation to herbivory.

An explicit dissection of the GMT into its four main conceptual constituents is useful to tease apart the variable degrees of support for the different parts of the theory, to identify its critical weak point, and to ascertain which of its elements should still be considered valid and used in our current views of plant-herbivore interactions (even if the GMT as such is poorly supported). The GMT, as well as the variant developed by Gill (Gill and Halverson 1984; Gill 1986), was originally framed around four conceptual elements: (1) the common *observation* that individual plants often are internally heterogeneous with regard to some important phenotypic traits that influence the acceptability of plant parts as food for herbivores (e.g., secondary compounds, nutritional composition); (2) the *claim* that this observed within-plant phenotypic heterogeneity is the observable consequence of accumulated somatic mutations; (3) the frequent *observation* that herbivores respond to within-plant variation and discriminate among plant parts that differ in nutritional or defensive traits; and (4) the *inference* that, because of 2 and 3, genetic mosaicism played a crucial role in the evolution of defensive strategies of long-lived plants against herbivores (Gill et al. 1995). In retrospect, element 2 emerges as the weak element in the GMT. As discussed in detail in chapter 5, genetic mosaicism most likely plays only a minor role as a cause of the extensive within-plant variation in phenotypic traits of reiterated structures commonly exhibited by *wild plants*. And the evidence advanced by early studies in support of the role of genetic mosaicism as a cause of observed within-plant variation in phenotypic traits often turned out inconclusive. Too often, within-plant phenotypic heterogeneity was automatically equated with genetic mosaicism without considering plausible alternative causes, perhaps because the notion that a single genotype can produce a range of phenotypes was less readily accepted two decades ago than it is now. Illustrative examples of unsupported conflations of within-plant phenotypic variation with genetic mosaicism are found in Gill and Halverson 1984, 114; Niemelä et al. 1984; and Edwards et al. 1990. In these and other examples, phenotypic mosaicism was automatically interpreted as a sign of genetic mosaicism, without an empirical test. Recognition of both the rarity and unlikelihood of genetic mosaicism as a cause of within-plant variation in wild plants, along with the weak or insubstantial inferences that originally linked within-plant phenotypic heterogeneity and genetic mosaicism, unavoidably discredits element 4 of the GMT and hence the whole theory as originally formulated. It is also important to note that the GMT was largely devised to answer the question of why short-lived pathogens and herbivores don't break the defenses of their long-lived host plants (Whitham 1983). As this question was prompted by the disparity between pathogens and their host plants in generation times and recombination potential, the GMT mostly emphasized within-plant variation in trees, the growth form where genetic mosaics generated by accumulated somatic mutations were expected to occur most often (chapter 5). Nevertheless, as shown by the numerous examples provided in preceding chapters, within-plant variation in phenotypic traits of reiterated structures is neither more ubiquitous nor quantitatively more important among long-lived, woody plants. Annual plants, where any role of genetic mosaicism derived from somatic mutations in generating subindividual variation can be confidently ruled out, can exhibit extensive within-plant variation in many traits, including chemical defenses, an observation that clearly militates against the GMT. In short, therefore, the main flaw underlying the GMT lay in its purported relationship of causality linking genetic and phenotypic within-plant mosaicism, element 2 above.

Elements 1 and 3 of the GMT received strong support from the reviews mentioned above and from numerous subsequent investigations. Regardless of size, growth form, and life history, most plants are internally heterogeneous with regard to organ traits that may be highly influential on host selection by herbivores, such as structural features, nutritional value, and concentration of defensive secondary compounds of leaves. This conclusion has been reinforced by the evidence summarized in chapters 2–4. Furthermore, the early reviews by Whitham and associates (and Raupp and Denno 1983; Schultz 1983; Gill 1986; Gill et al. 1995) succeeded in putting together a solid body of observational and experimental evidence showing that many invertebrate herbivores perceive and respond unambiguously to within-plant variation in traits that likely play defensive roles. I present below an updated summary of what is currently known on the foraging responses of herbivores to within-plant variation in leaf traits. I return in chapter 9 to the ecological and evolutionary significance of the two "surviving" elements 1 and 3 of the GMT in the context of plant-herbivore interactions.

Herbivores typically discriminate among conspecific individuals (Karban 1992). In addition, they do not treat individual plants as homogeneous food resources. Herbivores discriminate and respond behaviorally to within-plant variation in foliage traits, as shown by the frequent observation that the eggs, larvae, and adults of phytophagous insects are neither regularly nor randomly distributed among the leaves of a given plant, and that such patchiness is often related to variation in measurable leaf attributes. I have gathered a selection of examples in table 8.1, which includes observational and experimental investigations of leaf miners, sap feeders, and foliage chewers and involves a variety of insect orders (Homoptera, Lepidoptera, Coleoptera, and Hymenoptera) and host-plant growth forms. Most examples come from tree-feeding species. This reflects the fact that within-plant heterogeneity for herbivores has been particularly sought after in trees, which by virtue of their large size are expected to exhibit greater variability than other plants and, hence, to provide greater opportunities for selection by herbivores. Nevertheless, table 8.1 also includes examples of within-plant patchiness in herbivore distribution in herbs, which is at odds with the notion that small plants provide a narrower range of feeding conditions and reduced opportunities for within-plant selectivity by insect herbivores. Chapter 3 showed that subindividual variability is not necessarily smaller in small plants. Examples in table 8.1 involving within-plant selectivity by invertebrates feeding on herbs provide confirmation that within-plant variation, and the selective responses by herbivores to such variation, is not exclusive to trees. Within-plant selectivity by invertebrate herbivores feeding on small plants was early on documented in detail by Thompson (1983a, 1983b) for the larvae of species of *Depressaria multifidae* and *D. leptotaeniae* (Oecophoridae) feeding on *Lomatium grayi* and *L. dissectum*, respectively.

The patchy distribution and food selection of insect herbivores within individual plants is often predictably related to some extrinsic (e.g., height above ground, orientation, incident solar radiation) or intrinsic (e.g., leaf nodal position) plant gradient, as illustrated by some examples in table 8.1. In some cases, within-plant patchiness in herbivore distribution probably reflects a direct response to variation in the physical (e.g., microclimate) or biotic (e.g., parasitoids) environment rather than to patchiness in foliage attributes itself. This applies, for example, to the orientation-dependent distribution of *Yponomeuta mahalebella* caterpillars within *Prunus mahaleb* trees described by Alonso (1997a; see also Moore et al. 1988 for *Malacosoma californicum*); the oviposition preferences exhibited by *Papilio glaucus* on several tree species (Grossmueller and Lederhouse 1985); and the height-dependent distribution of *Popillia japonica* beetles on *Tilia cordata* trees (Rowe and Potter 1996). In this last case, beetles did not discriminate among foliage from different canopy zones in laboratory

TABLE 8.1 **A selection of observational and experimental studies documenting correlates of within-plant heterogeneity in the distribution of feeding or oviposition sites of leaf-miner, sap-feeding, and chewer invertebrate herbivores.**

Herbivore	Host plant	Factors accounting for within-plant distribution	Reference
Leaf miners			
Cameraria sp. (Gracillariidae)	*Quercus emoryi*	Leaf size, sun-shade regions of tree crown, presence of endophytic fungi	Faeth 1990, 1991; Wilson and Faeth 2001
Phyllocnistis populiella (Gracillariidae)	*Populus tremuloides*	Leaf height above ground	Condrashoff 1964
Phyllonorycter blancardella, P. crataegella (Gracillariidae)	Cultivated apple trees	Leaf height above ground, compass orientation	Barrett 1994
Stilbosis juvantis (Cosmopterigidae)	*Quercus emoryi*	Prior damage to leaves	Faeth 1985
Tuta absoluta (Gelechiidae)	Cultivated tomato plants	Leaf nodal position	Torres et al. 2001
Sap feeders			
Adelges cooleyi (Adelgidae)	*Picea engelmanni*	Whorl height, nodal position	Fay and Whitham 1990
Aonidiella aurantii (Diaspididae)	Cultivated orange trees	Leaf and fruit height above ground, orientation	Carroll and Luck 1984
Bemisia argentifolii (Aleyrodidae)	Cultivated tomato plants	Leaf age, nodal position	Schuster 1998
Bemisia argentifolii (Aleyrodidae)	Cultivated *Euphorbia pulcherrima*	Leaf age, nitrogen content	Bentz et al. 1995
Bemisia tabaci (Aleyrodidae)	Cultivated cotton plants	Leaf nodal position	Naranjo and Flint 1995
Bemisia tabaci (Aleyrodidae)	Cultivated tomato and pepper plants	Leaf age, nodal position	Muñiz et al. 2002
Ctenarytaina eucalypti, C. spatulata (Psyllidae)	*Eucalyptus globulus*	Leaf morph class (juvenile vs. adult)	Brennan et al. 2001
Euceraphis betulae (Drepanosiphinae)	*Betula pendula*	Branch vigor, pathogenic stress	Johnson et al. 2003
Hormaphis hamamelidis (Hormaphididae)	*Hamamelis virginiana*	Leaf size, nodal position	Rehill and Schultz 2001
Metopolophium dirhodum (Aphididae)	Cultivated oak and barley plants	Leaf size, chlorophyll content, nodal position	Honěk and Martinková 2002
Pemphigus betae (Aphididae)	*Populus angustifolia*	Leaf size, nodal position on shoot, phenol concentration	Whitham 1978; Zucker 1982
Rhopalosiphum padi Aphididae	Cultivated wheat plants	Leaf height above ground	Gianoli 1999
Chewers			
Arcte coerulea (Noctuidae)	*Boehmeria nipononivea*	Shoot length and height above ground	Ide 2006

(continued)

TABLE 8.1 *(continued)*

Herbivore	Host plant	Factors accounting for within-plant distribution	Reference
Atta colombica (Formicidae)	*Piper marginatum*	Water, proline, and nonstructural carbohydrate concentration	Meyer et al. 2006
Choristoneura pinus (Tortricidae)	*Pinus banksiana*	Leaf height above ground	Wallin and Raffa 1998
Chrysopharta agricola (Chrysomelidae)	*Eucalyptus nitens*	Leaf morph class (juvenile vs. adult)	Nahrung and Allen 2003
Coelocephalapion aculeatum (Apionidae)	*Mimosa pigra*	Inflorescence stage	Heard 1995
Coleophora fuscedinella (Coleophoridae)	*Betula papyrifera*	Leaf height above ground, depth into tree crown	Raske and Bryant 1977
Frankliniella occidentalis (Thripidae)	*Cucumis sativus*	Leaf nodal position	de Kogel et al. 1997
Hypsipyla robusta (Pyralidae)	*Toona australis*	Shoot length and basal diameter	Mo et al. 1997
Malacosoma disstria (Lasiocampidae)	*Acer saccharum*	Leaf height above ground	Fortin and Mauffette 2002
Manduca sexta, M. quinquemaculata (Sphingidae)	Cultivated tobacco plants	Leaf nodal position, nicotine concentration	Kester et al. 2002
Mnesampela privata (Geometridae)	*Eucalyptus globulus, E. dunnii*	Leaf morph class (juvenile vs. adult), monoterpenes and cuticular waxes	Steinbauer 2002; Steinbauer et al. 2004
Neodiprion abietis (Diprionidae)	*Abies balsamea*	Leaf height above ground	Anstey et al. 2002
Papilio glaucus (Papilionidae)	*Liriodendron tulipifera, Prunus serotina*	Insolation, height above ground, orientation	Grossmueller and Lederhouse 1985
Popillia japonica (Scarabeidae)	*Tilia cordata*	Leaf height above ground	Rowe and Potter 1996
Spodoptera exigua, Mamestra brassicae (Noctuidae)	*Senecio jacobaea*	Leaf nodal position, pyrrolizidine alkaloid concentration	de Boer 1999
Stenoma aff. assignata (Oecophoridae)	*Copaifera langsdorfii*	Leaf sesquiterpene concentration	Macedo and Langenheim 1989
Thoressa varia (Hesperiidae)	*Sassa veitchii*	Leaf age and coloration	Ide 2004
Trigonophora flammea, Noctua janthe (Noctuidae)	*Daphne laureola*	Branching order of leaf whorl, stem length, number of leaves	Alonso and Herrera 1996
Yponomeuta mahalebella (Yponomeutidae)	*Prunus mahaleb*	Branch orientation	Alonso 1997a

Note: Only studies focusing on within-plant food selection by single herbivore species have been considered. The large number of studies documenting exclusively the effects of leaf-age variation on within-plant herbivore distribution have not been included (see Raupp and Denno 1983 for a review).

choice tests, which demonstrated that their within-plant selectivity in the field was mediated by factors other than variation in food quality. In other cases, however, the relationship between within-plant herbivore distribution and plant gradients probably reflects the behavioral responses of herbivores to variation in leaf traits that were predictably correlated with nodal position, height above ground, orientation, or insolation. For example, within-plant variation in leaf size and nitrogen content is generally related to changes in the light environment (chapter 4); thus the preference of *Malacosoma disstria* larvae for the upper canopy leaves of *Acer saccharum* trees is most likely a consequence of their preference for food with higher nitrogen content (Fortin and Mauffette 2002; see also Yamasaki and Kikuzawa 2003).

In some of the examples in table 8.1, within-plant selection of feeding sites by herbivores was not correlated with any major extrinsic or intrinsic plant gradient, but it was explained by within-plant patchiness in leaf characteristics including size, concentration of secondary compounds (alkaloids, phenols, terpenes, waxes), presence of fungal endophytes, pathogenic stress, prior herbivory, and chlorophyll content. These relationships have convincingly been proven, for example, for *Pemphigus* aphids on *Populus* trees (Whitham 1978; Zucker 1982), *Euceraphis* aphids on *Betula* (Johnson et al. 2003), and noctuid and geometrid larvae feeding on leaves of *Eucalyptus globulus*, *Senecio jacobaea*, and *Daphne laureola* (Alonso and Herrera 1996; de Boer 1999; Steinbauer et al. 2004). In addition to studies based on correlative evidence, table 8.1 includes some experimental studies showing that, when leaves collected from different parts of plants are offered to herbivores under controlled conditions, they discriminate among them and exhibit clear preferences (Brennan and Weinbaum 2001; Brennan et al. 2001; Fortin and Mauffette 2002). These findings strengthen the interpretation that correlations between within-plant herbivore distribution and intrinsic or extrinsic gradients in leaf traits actually reflect within-plant selectivity by herbivores.

Within-plant patchiness in the distribution of single herbivore species will commonly arise as a response to intraplant variation in the quality of the food or the microenvironment. In spite of this, however, regular or random within-plant distribution of herbivores could still be possible in plants attacked by multiple herbivores, each of which responds differently to within-plant heterogeneity. This possibility is not supported by studies examining within-plant distributions of complete, multispecific herbivore assemblages. These studies have generally found significant patchiness in the cumulative incidence of all herbivores on single plants, thus denoting a nonrandom cumulative outcome of the behavioral responses of all herbivores that concur on a plant. Within-plant patchiness in the cumulative effects of diverse herbivore guilds has been found related to differences among branches (Marquis 1988; Hochwender et al. 2003) and variation in insolation level (Alliende 1989), depth within the crown (Wilkens et al. 2005), and leaf size (Shibata et al. 2001).

Preferences of insect herbivores for particular host-plant species or individuals are generally correlated with differential performance, either of the adults or of their progeny (reviews in Thompson 1988; Schoonhoven et al. 2005). Similar preference-performance associations may also hold at the within-plant level. Some herbivores select for feeding and/or egg-laying those leaves or branches of individual plants that eventually result in improved fitness of adults and/or their progeny via enhanced growth, survival, or fecundity. Behavioral responses of herbivorous insects to within-plant heterogeneity, and the ensuing patchiness of their distribution within plants, should therefore be most parsimoniously interpreted as consequences of intraplant variation in food quality. Whitham's outstanding study of within-plant selection by *Pemphigus* aphids (1978; see also Zucker 1982) provided an early elegant demonstration that discrimination by a herbivore among parts of the same plant with different characteristics is a critical aspect in explaining its nonrandom distribution, and that such within-plant selection has measurable fitness consequences for the insects. Similar evidence linking preference and performance of insect herbivores at the within-plant level has been obtained for, among others, the larvae of *Coleophora fuscedinella* feeding on *Betula papyrifera* (Raske and Bryant 1977); *Stilbosis juvantis* miners on *Quercus emoryi* leaves (Faeth 1985); *Papilio glaucus* larvae on *Prunus serotina* (Grossmueller and Lederhouse 1985); the spruce gall aphid *Adelges cooleyi* on *Picea engelmanni* (Fay and Whitham 1990); *Bemisia argentifolii* whiteflies on *Euphorbia pulcherrima* (Bentz et al. 1995); larvae of the weevil *Coelocephalapion aculeatum* on *Mimosa pigra* inflorescences (Heard 1995); *Malacosoma disstria* larvae on *Acer saccharum* leaves (Fortin and Mauffette 2002); *Mnesampela privata* caterpillars and *Ctenarytaina* psyllids on different leaf morphs of heterophyllous *Eucalyptus dunnii* and *E. globulus* (Brennan and Weinbaum 2001; Steinbauer 2002); and *Manduca sexta* and *M. quinquemaculata* larvae on tobacco plants (Kester et al. 2002). All these studies found that herbivores tended to select preferentially those parts of individual plants that eventually resulted in greater performance. Exceptions to this prevailing pattern are relatively infrequent, as suggested by the small number of studies failing to document correlations between preference and performance at the within-plant scale (Faeth 1990; Rowe and Potter 1996; Nahrung and Allen 2003).

Differences in performance among conspecific insect herbivores feeding on different leaves of the same plant can be far from negligible. This was thoroughly documented in a comprehensive series of field and laboratory experiments on *Epirrita autumnata* larvae feeding on leaves of *Betula pubescens* (Suomela and Nilson 1994; Suomela 1996; Suomela, Ossipov, and Haukioja 1995; Suomela, Kaitaniemi, and Nilson 1995). Leaves of the same trees that differ in size, orientation on the tree, and height above the ground induced large differences in larval growth rate and pupal mass that were of the same magnitude as the differences exhibited by larvae feeding on leaves from different trees. Suomela and associates also studied among-tree and within-tree variation in larval growth rate, using a series of hierarchically nested sampling levels: among trees (genets), among ramets within trees, among branches within ramets, among shoots within branches, and among leaves within shoots. Differences among ramets, branches, shoots, and leaves of the same tree accounted together for 27% of total variance in larval growth rate, while variation among trees accounted for only 18%. Larval growth rate was 11–32% lower on the worst ramet than on the best ramet within trees, 8–18% lower on the worst branch than on the best branch within ramets, and 12–30% lower on the worst shoot than on the best shoot within branches. Large within-tree variance in larval growth rate was mainly the outcome of within-plant variation in leaf quality (water content, specific weight, toughness, nitrogen content), inducing variation in food-utilization efficiency by the larvae.

Similarly high within-plant variance in insect performance has been reported for other plant-herbivore systems. For *Pemphigus betae* feeding on *Populus angustifolia* trees, Whitham (1978) found that the proportion of aborted galls due to death of stem mothers (fundatrices) ranged between 0 and 80%, the number of aphids per gall between 30 and 200, and fundatrix weight between 0.24 and 0.40 mg, variation in all these parameters being closely related to within-plant variation in leaf size. Within trees of *Picea engelmanni*, the realized fitness of fundatrices of the spruce gall aphid *Adelges cooleyi* (computed as the product of survivorship × fecundity × growth) ranged between 0 and 2.8, and a nearly perfect correlation existed between the spatial location of fundatrices within spruce trees and their fitness (Fay and Whitham 1990). Gripenberg and Roslin (2005) performed a detailed investigation of the relative importance of different sources of variation in larval mortality for the oak-specific leaf miner *Tischeria ekebladella* feeding on *Quercus robur* foliage. By using a hierarchical design, they partitioned total variance in larval mortality at the habitat level into components due to differences among trees, shoots within trees, and leaves within shoots. Moth survival varied both among trees and among different parts of the same tree. Only about 30% of total variance in survival occurred among individual trees, while the rest of variance was accounted for by variation among shoots of the same tree (10%) and, principally, among individual leaves within the same shoot (60%). In this system, therefore, variation among trees in average foliage quality was of secondary importance as a determinant of larval survival in comparison to variation among different parts of the same tree (Gripenberg and Roslin 2005). Whitham (1981) and Zucker (1982) provided further examples showing that the magnitude of variation in herbivore performance among different parts of the same plant can be as large or even larger than the variation occurring among conspecific plants.

Direct Costs to Animals of Within-Plant Variation

In the preceding section I have shown that all major classes of animal consumers relying on plant reiterated structures for food are similar in that they do not treat individual plants as homogeneous food sources, and discriminate among organs borne on the same plant on the basis of differences in quality. Under most circumstances, it will be advantageous to animals to discriminate and select among the variable organs of individual plants, and natural selection will ultimately favor those individuals whose within-plant choices result in greater survival, or more numerous and better progeny. This possibility was emphasized by Suomela and Nilson (1994, 45) in relation to *Epirrita autumnata* larvae feeding on birch trees, noting that while "among-tree variation should select for discrimination by ovipositing females and dispersing larvae, within-tree variation should select also for optimal foraging behaviour of larvae." From the perspective of zoologists, therefore, the main consequence of within-plant heterogeneity is that it provides a distinct opportunity for natural selection to enhance the animals' discriminating abilities and develop behavioral responses allowing them to identify and exploit the most profitable items. Whenever the organs simultaneously available on individual plants vary in their contribution to the fitness of the animals that consume them, a mutation conferring improved abilities to choose the flowers, fruits, seeds, or leaves in a plant that yield greater fitness returns would quickly spread in an animal population made up of nondiscriminating individuals that select organs at random within plants. For *Pemphigus* aphids feeding on *Populus angustifolia*, for example, Whitham (1978) estimated that within-tree leaf selection by fundatrices doubled their fitness in relation to that expected if leaves were selected at random. Within-plant variation, therefore, provides ample opportunities for the evolution of adaptive behavioral responses among animal consumers.

Within-plant choice, however, has several potential short-term costs to foraging animals; considering these costs is essential to a proper understanding of the possible mechanisms linking within- and between-plant selectivity. I consider in the following sections three major classes of costs resulting from within-plant discrimination by animals: augmented competition, time and energy costs, and constraints on optimal foraging.

Augmented Competition

The attendant costs of choice in response to within-plant heterogeneity are ecologically most significant for sedentary or scarcely mobile animals such as leaf miners, sap feeders, and seed predators, all of which will often experience enhanced competition in preferred patches within-plants. If different organs borne by the same plant have different food value (and thus presumably different fitness returns), and individuals of the same or different species share the same preferences, then aggregate distributions will result, and competition will become most intense in those plant parts or organs that possess the preferred characteristics.

One example of within-plant heterogeneity generating localized competition is provided by insect herbivores with sedentary larvae and ovipositing adults behaving selectively with respect to leaf traits. In these cases, spatial variation in leaf traits within plants will lead to aggregation of larval populations in patches with preferred leaf characteristics, with the consequence that intraspecific competition will become most intense at these places. A situation of this sort was described by Faeth (1990, 1991) for *Cameraria* leaf miners feeding on *Quercus emoryi* leaves. In this system within-plant variation in leaf size, combined with the behavioral responses of ovipositing females to such variation, gives rise to broad within-plant variation in the degree of aggregation of larvae and, consequently, to variation in the competitive environment and larval survival prospects. Most larvae tended to occur in the largest, preferred leaves, and survival was lowest there because of stronger competition. Similar results have been obtained by studies of patterns of within-plant distribution, fecundity, and survival of sap-feeding herbivores. Fundatrices of the gall-making aphid *Hormaphis hamamelidis* colonizing *Hamamelis virginiana* plants preferentially select the distal leaves of buds, which grow more than the proximal leaves. This leads to an aggregated distribution of galls among leaves and increased mortality of fundatrices in multiply-galled leaves (Rehill and Schultz 2001). As shown by these examples, herbivore aggregation on preferred places within-plants frequently leads to decreased individual performance as a result of increased local competition.

Seed and fruit predators whose larvae spend most or all of their lives in single seeds or fruits are also expected to show increased aggregation, and thus to experience intensified competition, in patches within single plants bearing seeds or fruits with preferred features. As commonly found in other beetle seed predators (see earlier in this chapter), Campbell (2002) reported that the body size of *Sitophilus oryzae* adults was directly related to the size of the single *Triticum aestivum* seed where they grew up, and that ovipositing females preferred larger seeds for oviposition. Females not only laid eggs more frequently in the larger seeds, but also tended to lay significantly more eggs per seed as seed size increased. Despite variation in number of eggs deposited per seed, however, the proportion of seeds producing at least one adult beetle remained fairly constant across seed-size classes. This happened because the probability of an individual egg eventually producing an adult declined with increasing seed size, presumably because competition among larvae increased with increasing number of larvae per seed. Aggregated distribution of eggs on seeds, greater initial density of eggs on larger seeds, and severe competition when more than one larva share the same seed, have been demonstrated for a number of bruchid and curculionid beetle seed predators infesting the seeds of wild and cultivated plants. These include *Bruchus brachialis* infesting *Vicia villosa* seeds (Dickason 1960), *Kytorhinus sharpianus* infesting *Sophola flavescens* (Ishihara and Shimada 1993), *Stator limbatus* infesting *Acacia greggii* (Fox et al. 1996), *Bruchidius dorsalis* infesting *Gleditsia japonica* (Shimada et al. 2001), *Acanthoscelides obtectus* infesting *Phaseolus vulgaris* (Szentesi 2003), *Callosobruchus maculatus* infesting *Vigna unguiculata* (Credland et al. 1986; Cope and Fox 2003), and *Revena rubiginosa* infesting *Syagrus romanzoffiana* (Alves-Costa and Knogge 2005). Although some of these studies did not explicitly report a direct relationship between seed size and the intensity of intraspecific larval competition, they did show that some seeds received single eggs while others received multiple ovipositions consisting of variable egg numbers. If, as seems reasonable, this pattern of variable larval aggregation was related to within-plant variation in seed size, results of these studies are consistent with the hypothesis that foraging responses by seed consumers to within-plant variation in seed size will create patchiness in the competitive environment and, particularly, will give rise to competitive hot spots in those within-plant patches that possess the preferred characteristics from the viewpoint of the consumers.

Time and Energy Costs

Animals that discriminate among similar plant organs by visual, tactile, or olfactory means upon arrival at an individual plant have to invest some time in assessing the phenotypic diversity of the available organs, as well as in deciding whether to accept or reject particular organs for feeding or oviposition. This added time represents an extra foraging cost in relation to a theoretical baseline whereby all organs on the plant are identical or, if variable, are chosen fully at random. This cost to animals can be considered as arising directly from within-plant variation. The magnitude of the extra time required to materialize within-plant selection will be directly related to the "grain size" of within-plant variation (chapter 4) and the animal's choosiness, and inversely related to its mobility and sensorial ability to assess organ quality quickly (e.g., visually from a distance). Such added-time costs should be greatest for very choosy animals endowed with little mobility, foraging on organs exhibiting coarse-grained within-plant variation, and needing to probe organ quality one-by-one at close range. In contrast, added-time costs are expected to be smallest for highly mobile animals, foraging on organs that exhibit fine-grained variation, and being able to assess and compare the quality of individual organs without direct probing at close range.

Only a few studies have directly addressed the issue of the foraging costs derived from within-plant discrimination by phytophagous animals; hence quantitative estimates of these costs are scarce. The limited evidence available, however, indicates that the added-time costs of within-plant selection can sometimes be substantial. Most evidence I was able to locate is related to seed- or fruit-eating birds foraging on fruit crops of fleshy-fruited species. Jordano's study of fruit selection by the finch *Carduelis chloris* feeding on filled and empty drupes of the Mediterranean shrub *Pistacia lentiscus* (1990) revealed that foragers were very selective, cracking only sound fruits containing filled seeds. Unfilled seeds were detected by bill weighing and rejected. Jordano estimated that, as a consequence of the handling and rejection of rewardless fruits (i.e., of exerting discrimination among fruit classes), finches would have to spend up to 48% more time in plants than if all fruits contained only filled seeds and birds did not need to perform within-plant discrimination. Fruit crops of *Pistacia terebinthus*, a deciduous Mediterranean treelet, contain a mixture of green high-lipid drupes along with red low-lipid ones. Frugivorous birds may consume both types of fruits, but when both are available they definitely prefer, and actively seek, the highly rewarding green fruits. By means of a controlled experiment, Fuentes (1995) demonstrated that the presence of unpreferred red fruits in infructescences significantly reduced the ingestion rate of the preferred green fruits by birds, the effect being due to red fruits reducing the visibility and accessibility of the preferred fruits. Individuals of *Turdus merula* feeding on the fruits of *Crataegus monogyna* in England exert discrimination among the fruits simultaneously available on the same bush, selecting healthy fruits and rejecting those with the seed infested by larval *Blastodacna hellerella* (Manzur and Courtney 1984). As the percentage of insect-infested fruits in individual plants increased, and thus more thorough within-plant discrimination was necessary to locate preferred fruits, the birds made significantly more mistakes, had to spend proportionally more time per healthy fruit ingested, and thus eventually foraged less efficiently.

As noted earlier in this chapter, animal pollinators often approach some flowers and reject them without landing. Apparently they are discriminating among the flowers simultaneously available on a plant. Although I failed to find in the abundant literature on pollinator foraging behavior any direct quantitative assessment of the magnitude of the extra time attributable to this widespread behavior, indirect evidence suggests that it may be quite substantial in some instances. Anthophorid bees foraging on *Anchusa strigosa* plants rejected 6% of approached flowers without attempting to probe them (Kadmon et al. 1991), and *Bombus terricola* foraging on *Trifolium repens* rejected up to 27% of flowering heads in areas that had been previously exposed to pollinators (Heinrich 1979). Regardless of the actual figures, however, it is obvious that spending any time to approach and hover in front of flowers that will not be subsequently probed for reward necessarily exacts some time costs on pollinators.

Time itself is a limiting fitness-related commodity in some cases, particularly in short-lived insect herbivores or seed predators whose females have limited time to mate, find appropriate host plants, and, within these, discriminate among available organs and distribute eggs among them. To these animals, the additional time spent in within-plant discrimination represents a cost insofar as it competes with the time devoted to other critical activities like mating or egg laying. In insect parasitoids, which typically spend most of the time seeking suitable hosts, the importance of the constraints imposed on time budgets by extended search and discriminating time has been well established by theoretical models (Iwasa et al. 1984; Cullen Speirs et al. 1991). Conclusions of these models could easily be extrapolated to short-lived insects that exert systematic within-plant discrimination among simultaneously available leaves, fruits, or seeds.

More generally, and even if time is not a seriously limiting commodity, the additional time spent by foraging animals in individual plants as a consequence of within-plant discrimination translates into increased exposure to parasitoids and predators and reduced overall foraging efficiency, both of which in turn translate into associated fitness costs. In the case of small frugivorous vertebrates, the risk of predation is higher in fruiting plants than in nearby areas, since predators may be attracted to trees that prey species visit regularly (Howe 1979). Sapir et al. (2004) found that predator avoidance was the main factor explaining microhabitat selection by three species of heavily frugivorous *Sylvia* warblers migrating through Israel, which preferentially selected densely foliated trees for feeding where detectability by predators was reduced. In a review of information on the frequency of predator attacks on frugivorous birds while feeding at several species of fruiting trees and bushes in western Europe, Guitián et al. (1994) found attack rates that ranged between 0.02 and 0.35 attacks per hour of observation time. These figures mean that frugivores may have to withstand a predator attack once every 3–50 hours of stay on a fruiting plant, which is far from a trivial risk. Even in human-altered habitats, the threat from predators experienced by frugivorous vertebrates may be surprisingly high, as shown by the frequent attacks by raptors on frugivorous birds recorded by Snow and Snow (1988, 228) while the birds were feeding on fruiting trees and shrubs in the English countryside. To individual frugivores, therefore, staying on fruiting plants for extra time because of the added-time costs imposed by within-plant discrimination is expected to consistently increase the likelihood of suffering predation.

Reduced foraging efficiency is probably the most common consequence for phytophagous animals of the added-time costs derived from within-plant discrimination. This effect will be particularly marked under stressful environmental conditions and for organisms characterized by severe energetic constraints. Manzur and Courtney (1984) emphasized that the overall decline in foraging efficiency of *Turdus merula* with increasing proportions of unpreferred fruits in *Crataegus monogyna* crops should be particularly critical in winter, when alternative food is scarce and there is a marked decrease in both temperature and day length in which to feed. In the case of pollinators that reject approached flowers without visiting them, the added-time costs derived from such within-plant selectivity will impair their energetic balance. The importance of this effect will presumably be most pronounced among endothermic insects such as bumblebees, which among pollinators have been considered "the most extravagant utilizers of energy on a weight-specific basis," and for which energetic constraints seem to have molded most aspects of their foraging behavior (Heinrich 1975).

Constraints on Optimal Foraging

Within-plant variation in features of reiterated organs impose yet another foraging cost on phytophagous animals that feed or oviposit on them, namely, setting limits on the optimality of foraging decisions. Since the early days of the formulation of the theory of optimal foraging by MacArthur and Pianka (1966), the issue of how animals partition foraging time among and within feeding sites to harvest resources most efficiently has received much attention (Pyke 1984; Goulson 1999). As noted above, individual plants represent distinct feeding patches characterized by locally dense aggregations of the resources sought by phytophagous animals for food or oviposition (flowers, seeds, fruits, leaves). One of the aspects of optimal foraging that has perhaps been considered most frequently by theoretical models is the evolution of the rules that should govern the decision of foraging animals to depart from a feeding patch in order to maximize energy intake, the so-called departure rules. Models predicting different optimal departure rules have been developed and tested for flower-visiting nectarivores (Charnov 1976; Pyke 1978b; Pleasants 1989; Kadmon and Shmida 1992), but the underlying concepts and assumptions can be extrapolated to other phytophagous animals that feed or oviposit on reiterated plant organs other than flowers.

The decision of a foraging animal to depart from a particular plant may become considerably more complicated than originally envisaged by departure-rule models if organ quality varies widely and is patchily distributed within individual plants, two features that seem nearly universal in view of the evidence summarized in earlier chapters. Under such circumstances, a forager encountering one or a few poor-quality organs upon arrival at a plant has to decide whether it is in a low-quality plant (i.e., a plant with mean organ quality inferior to the population mean) or in a low-quality patch within that particular plant (Kadmon and Shmida 1992). In the case of pollinators, Boose (1997, 497) remarked that "high within-plant variation could make it more difficult for a pollinator to differentiate among plants based on mean nectar production rates, because any subset of flowers visited on a plant would only provide a rough estimate of the mean reward value of the plant as a whole." In the absence of well-defined, predictable spatial patterns of within-plant variation (chapter 4), the likelihood of making a wrong foraging decision depends on the relative magnitudes of within- and among-plant variance in organ quality in the population. In the hypothetical case where plants in a population differed in mean organ quality from the perspective of the animal but were otherwise similar in having zero within-plant variance (i.e., all organs borne by the same plant have identical quality), then the finding of just a few low-quality organs in a plant would provide an unambiguous indication to the animal forager that it is in a low-quality plant. At the opposite extreme, if all plants in a population were nearly identical in mean organ quality and most population-level variance occurred within plants, then finding a few low-quality organs in a plant would be uninformative as to the plant's average quality, and wrong decisions would occur frequently. These two extreme examples illustrate that the capacity of animals to develop and practice optimal departure rules from plants will be substantially hampered, or even precluded, when the organ traits relevant to the interaction (e.g., corolla length, nectar volume, fruit width, seed size) are characterized by comparatively large within-plant variance. In particular, the maximum possible accuracy (relative to a hypothetical optimum) of departure decisions is expected to decline steadily whenever $\%Var_{within} > \%Var_{among}$. Given the large values of $\%Var_{within}$ occurring in nature for many organ traits relevant to animal foragers (chapter 3), one would expect within-plant variation in organ traits to consistently limit the opportunities of phytophagous animals for developing optimal plant departure rules under natural field conditions.

The "grain size" of within-plant variation (chapter 4) will also affect the opportunities for developing optimal foraging strategies within plants (patches). For pollinators, Kadmon and Shmida (1992) found an inverse relationship between the magnitude of the reward in the last visited flower and the flight distance to the next flower within the same plant. As shown in chapter 4, within-plant variation in organ trait values is often very fine-grained, and spatial autocorrelation, when it occurs, takes place only at very small distances within the plant. Under these circumstances, the behavioral rule documented by Kadmon and Shmida for pollinators will not necessarily enhance the probability of finding a high-quality organ after a visit to either a poor-quality or a high-quality one.

Indirect Costs: Implications of Jensen's Inequality

In general, within-plant selectivity will only reduce, rather than completely eliminate, the within-plant variance in organ traits experienced by animal foragers. This may be due to sensorial thresholds or behavioral constraints limiting the perception of organ differences and hence the accuracy of discrimination (e.g., detection thresholds for size or chemical differences), but also to compromise behaviors arising from trade-offs between the gains and costs derived from selectivity. Regardless of its causes, the phenomenon that I want to emphasize here is that, even if most phytophagous animals exert some within-plant discrimination, they will eventually interact (i.e., feed, oviposit) with an array of organs on each plant that are different from each other, and whose value differs from the animals' viewpoint. Possibly the best-known and most thoroughly investigated example is that of nectar-feeding pollinators sequentially visiting flowers on the same plant that contain widely different nectar rewards (Pleasants 1981; Biernaskie et al. 2002; Biernaskie and Cartar 2004). Similar phenomena, however, are also expected to affect other classes of animal consumers that oviposit or feed on reiterated plant organs. Ovipositing females of insect leaf miners or seed predators lay eggs on a number of leaves or seeds of the same plant, which will differ slightly in some phenotypic characteristics affecting their quality as larval food (e.g., size, nutritional traits). In the course of their lives, caterpillars staying on the same shrub or tree crown feed on a number of leaves of the same plant, which will almost certainly differ in nutritional quality or chemical defense. Individual frugivorous birds coming to a fruiting plant ingest fruits that differ in traits affecting their energy and nutrient efficiency, including size, seediness, and pulp nutritional quality. In short, upon arrival at a plant and despite exerting some discrimination and choice, individual animals will eventually feed or oviposit on an array of variable organs differing in phenotypic traits that influence animals' energy budget, nutritional condition, fecundity, survival, or, in more general terms, fitness prospects. Such "after-choice" variance experienced by animals can have consequential effects derived from Jensen's inequality.

Jensen's inequality is a mathematical property inherent to nonlinear functions. The inequality states that for a nonlinear function $f(x)$, and a set of x_i values with a mean of x_{mean} and a variance greater than zero, the average result of $f(x_i)$, Mean$[f(x_i)]$, does not equal the result of the function for the average x_i, $f(x_{mean})$ (Ruel and Ayres 1999). This effect is depicted in figure 8.3. When $f(x_i)$ is decelerating (i.e., a concave-down function with second derivative negative), Mean$[f(x_i)]$ is less than $f(x_{mean})$ (fig. 8.3a). When $f(x_i)$ is accelerating (i.e., a concave-up function with second derivative positive), Mean$[f(x_i)]$ is greater than $f(x_{mean})$ (fig. 8.3b). The difference between Mean$[f(x_i)]$ and $f(x_{mean})$ may be termed Jensen's effect. Its sign depends only on the form of the function (decelerating vs. accelerating), while its absolute value depends on both the shape of $f(x)$ and the probability distribution of x_i. In general, the magnitude of Jensen's

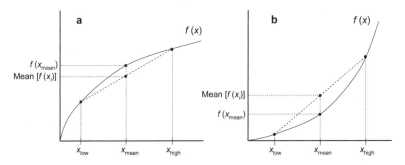

FIG. 8.3 Jensen's inequality describes how variance in an independent or driving variable depresses the mean response variable in decelerating, concave-down nonlinear functions (a), and elevates the mean response variable in accelerating, concave-up nonlinear functions (b) (Ruel and Ayres 1999). In concave-down functions (a), the mean of the function is greater than the function of the mean, while the reverse is true for concave-up functions (b). For simplicity, graphs depict situations where the probability distribution of x over the interval $x_{low} - x_{high}$ is uniform, but the general idea applies to distributions of any shape (see text).

effect increases with increasing nonlinearity (i.e., absolute value of second derivative) and increasing variance in the driving, or independent, variable x_i. For simplicity, the graphs in figure 8.3 depict situations where the probability distribution of x_i over the interval $x_{low} - x_{high}$ is symmetrical and uniform ($x_{mean} = (x_{low} + x_{high})/2$), but the general idea applies for x_i distributions of any shape. For a given $f(x)$ and x_{mean}, distributions of x_i differing in variance, skew, or kurtosis (i.e., the higher moments of the distribution) may give rise to Jensen's effects of variable magnitudes. Although Jensen's inequality is generally cited because of its effects on arithmetic means, it is also valid for a wider class of statistical location parameters, including the geometric mean, median, trimmed mean, and midrange (Burnside 1975; Spiegelman 1985).

Jensen's inequality has relevance to any area of biology that includes nonlinear functions, including enzymatic reactions, life-history evolution, plant biomass allocation strategies, ecosystem and community ecology, population dynamics, and animal behavior (Gillespie 1977; Salisbury and Ross 1992; Smallwood 1996; Ruel and Ayres 1999; Drake 2005; Inouye 2005; Laird and Aarssen 2005). I focus here on the implications of Jensen's inequality to phytophagous animals that feed or oviposit on the variable organs of single plants. In this context, Jensen's inequality leads to the relatively straightforward prediction that within-plant variance in organ trait values will influence the *mean* quality of individual plants from an animal's perspective whenever the intrinsic quality or the perceived value of individual organs is functionally related to organ trait values by some nonlinear function. This prediction implies, for example, that individual plants with identical means but different variances for some organ character relevant to their interaction with animals (e.g., flower corolla length, nectar sugar concentration, seed size, fruit seediness, leaf nutrient content) will differ in their mean quality from the viewpoint of animal consumers whenever the shape of the quality–trait value relationship is not linear. Similar predictions can be advanced that involve differences in higher moments of organ character distributions other than the variance. Since the magnitude of a Jensen's effect depends on the shape of the probability distribution of x_i, individual plants may also differ in their *mean* value to animals in situations where $f(x)$, x_{mean}, and the variance of x_i are all identical, but the shape of within-plant trait distributions differs among plants (e.g., by differing in skew or kurtosis, as illustrated by examples in chapter 7). It must be stressed that Jensen's inequality is not a biological phenomenon per se but a mathematical consequence of nonlinear functions; hence the above predictions should apply generally regardless of the type of plant-animal interaction, organ, or trait involved. The only crucial requisite for these predictions to be valid is that the nutritional, energetic, performance, or fitness value of individual organs to animals be nonlinearly related to the organ trait under consideration. As shown below, nonlinear relationships of this kind are most likely the rule rather than the exception in nature, which provides compelling evidence that animals feeding or ovipositing on variable plant organs will nearly always be subject to Jensen's effects.

In the context of the interaction between animals and variable plant organs, two main classes of nonlinear relationships may be distinguished that can give rise to Jensen's effects. The first group, "value-cue relationships," includes those nonlinear relationships linking variation among organs in food or fitness value to animals with variation in traits used as cues to discriminate among organs of different value. Consider, for example, the relationship linking flower nectar content to flower size. As shown earlier in this chapter, nectarivores often discriminate among flowers simultaneously available on a plant on the basis of the size of their corollas. Corolla size, however, plays only the role of a proximate cue furnishing information on the truly important floral character from the perspective of the animal foragers, namely nectar content. Other value-cue relationships relevant to plant-animal interactions include those linking pulp mass (value) and cross diameter (cue) of single fleshy fruits; nutrient content (value) and area (cue) of individual leaves; and number or mass of enclosed seeds (value) and fruit linear dimensions (cue). As illustrated by these examples, cue- and value-related organ traits often differ in the dimensionality of the variables describing them. While cue-related traits are often described by one- or two-dimension variables (length, area), value-related traits are more frequently described by three-dimensional variables (biomass, volume). Plant traits differing in dimensionality are generally linked by nonlinear allometric relationships described by exponential and power functions (Niklas 1994); hence for this reason alone, one would expect nonlinear relationships between value and cue organ traits to be commonplace in nature.

The form of the relationship between different organ traits has been examined rather infrequently in intraspecific contexts (in contrast to the large number of studies considering such variation at the between-species level; see, e.g., Niklas 1994), but the evidence available confirms the expectation of frequent nonlinearities of value-cue relationships. For the single-seeded drupes of the bird-dispersed tropical tree *Ocotea tenera*, Wheelwright (1993, fig. 2) illustrated an increasing, accelerating (i.e., concave-up) nonlinear relationship between pulp mass and cross-sectional diameter of individual fruits. A similar relationship holds for the fleshy fruits of southern Spanish species of bird-dispersed shrubs and trees, including *Crataegus monogyna*, *Daphne laureola*, *Osyris lanceolata*, and *Phillyrea latifolia* (C. M. Herrera, unpublished data). In the last species, the proportional contribution of pulp mass to total fruit mass is an increasing, decelerating (concave-down) function of fruit diameter (fig. 8.4a). The preceding examples refer to single-seeded drupes. In species with multiseeded berries, variation in the number of seeds per fruit will likewise induce nonlinear variations in the food value of fruits to frugivores. For the multiseeded berries of *Solanum pubescens*, which contain a widely variable number of seeds (range = 1–30), Hegde et al. (1991, fig. 1) demonstrated that the relationship between pulp/seed-mass ratio and the number of seeds per fruit was best described by a declining, decelerating (concave-up) power function. In *Smilax aspera*, the proportion of fruit mass contributed by the pulp is a decreasing, decelerating (concave-up) function of seed number per fruit (fig. 8.4b). Fruit pulp is the nutritious reward obtained from fruits by frugivores, and indigestible seeds represent indigestible ballast. Since vertebrate frugivores often use fruit size as a proximate criterion for fruit selection (see earlier in this chapter), the preceding examples of nonlinear value-cue relationships suggest that, through their feeding on fruits differing in size and amount of ballast, frugivorous animals will often be subject to Jensen's effects.

Nonlinear value-cue relationships between organ traits should also be frequent in the case of flower, leaf, and seed features important for the interaction with pollinators, herbivores, and seed predators, respectively. In the bumblebee-pollinated perennial herb *Helleborus foetidus*, nectary length, a surrogate for the maximum volume of nectar potentially available to floral visitors, is linked to flower size by an increasing, decelerating nonlinear function (fig. 8.4c). In the tropical tree *Guazuma ulmifolia*, the total mass of seeds per fruit is an increasing, decelerating function of fruit size (C. M. Herrera, unpublished data), a cue used by the beetle seed predator *Amblycerus cistelinus* to select among simultaneously available fruits for oviposition (fig. 8.2). Midgley et al. (1991) found that, within single shrubs of *Hakea sericea* and *H. drupacea*, the mass of individual seeds was nonlinearly related to the mass of the follicle within which it

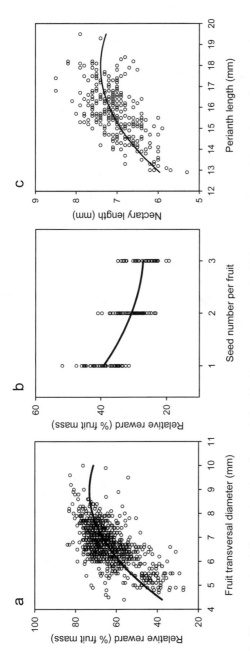

FIG. 8.4 Three examples of allometry-based, nonlinear relationships between parameters describing an organ's potential food value from the viewpoint of animal consumers (vertical axes) and organ phenotypic traits that typically vary within plants (horizontal axes). *a* and *b*, relationship between the percentage of the whole ripe fruit represented by the pericarp (on a dry mass basis; an index of a fruit's inherent food value for frugivorous birds, Herrera 1981b) and fruit diameter and seediness for *Phillyrea latifolia* (*a*) and *Smilax aspera* (*b*), respectively, two southern Spanish species of passerine-dispersed, fleshy-fruited plants. *c*, relationship between nectary length, a surrogate for the maximum volume of nectar available per nectary to insect visitors, and perianth length in *Helleborus foetidus*, a bumblebee-pollinated perennial herb. In all graphs each symbol corresponds to a different flower or fruit, and lines are the least-squares-fitted second-order polynomials. Based on data from Herrera 1981b; Herrera et al. 1994; Herrera et al. 2002.

was enclosed. In the boreal trees *Betula nana*, *B. pendula*, and *B. pubescens*, intraspecific variation in foliar nitrogen concentration is linked to variation in leaf area by nonlinear relationships (Niinemets et al. 2002). These examples suggest that, as in the case of frugivores, the nonlinear relationships linking value- and cue-related organ traits will often give rise to Jensen's effects among nectar eaters, seed predators, and larval herbivores, and that within-plant variance in organ traits used as feeding or oviposition cues by these animals will influence the mean value of organs from the animals' perspective.

The second and presumably most influential class of nonlinear relationships inducing Jensen's effects are those directly linking behavioral, physiological, energetic, or developmental responses of animals to variation in the characteristics of plant organs. Representative examples of this category are listed in table 8.2, which summarizes the mathematical form of responses by vertebrate and invertebrate frugivores, nectar eaters, herbivores, and seed predators, to naturally occurring or experimentally induced intraspecific variation in plant food characteristics or in organ traits that are important for the animal's interaction with plants. The driving variables involved in these nonlinear relationships have to do with nutritional (e.g., sugar, nitrogen, secondary metabolite concentration) and morphological (e.g., fruit and seed size, corolla length) organ features. Response variables include a variety of parameters related to food and energy intake rate, handling time, assimilation efficiency, growth rate, survival, and body size of the phytophagous animals involved in each case. Out of a total of 37 examples of nonlinear relationships gathered in table 8.2, 10 cases correspond to concave-up functions, while the remaining 27 instances correspond to concave-down functions. Relationships where the dependent variable involves some kind of cost to animals (e.g., handling time, failure rate, deterrence) are predominantly concave-up functions (87.5%). In contrast, relationships where the dependent variable represents some kind of gain, benefit, or advantage to animals (e.g., intake rate, energy gain, body mass) are almost exclusively concave-down functions (89.7%), regardless of the food trait, organism, and response variable involved.

If the examples compiled in table 8.2 are not too seriously biased as to generate spurious patterns, then the unequal distribution of concave-up and concave-down functions among cost- and benefit-related classes of response variables suggests that different signs of Jensen's effects (Mean$[f(x_i)] - f(x_{mean})$; fig. 8.3) are expected for cost-trait relationships than for benefit-trait ones: predominantly positive Jensen's effects in the case of cost-related functions, as exemplified by figure 8.3b, and predominantly negative in the case of gain-related functions, as in figure 8.3a. Interestingly, however, these two trends are like two sides of the same coin, as they actually point to the same conclusion, namely that within-plant variance in organ trait (x_j) will predominantly have negative effects on the foraging animals experiencing it, either through an increase in the per-plant mean cost (positive Jensen's effect) or through a decrease in the per-plant mean benefit (negative Jensen's effect) of exploiting a mixture of variable organs. This indicates that within-plant variation in organ traits that are influential on any aspect of the performance of foraging animals will generally act to reduce the mean per-organ value characteristic of each plant or, in other words, that variance in the energy or food value of plant organs used by animals will generally depress the latter's performance. This prediction is nicely corroborated by the experimental results of Stockhoff (1993) for larvae of the polyphagous gypsy moth (*Lymantria dispar*) reared on three artificial diet treatments with identical mean nitrogen content but differing in their variance (constant, low variance, high variance). Despite equal relative consumption rates and nitrogen consumption rates across all treatments, larvae experiencing variation in the diet suffered a reduction in pupal mass and extended development time. The existence of measurable, specific costs to larvae associated with variation in diet quality was interpreted as a consequence of nonlinearity in the relationship between nitrogen and food utilization, and this example was subsequently brought forward by Ruel and Ayres (1999) as representative of the ecological consequences of Jensen's effects. The findings of Stockhoff and other studies showing that variance in food resource quality depresses the overall performance of animals (Real 1981; Waddington et al. 1981; Miner and Vonesh 2004) have important implications for the interaction between phytophagous animals and variable plants, as shown in the next section.

Bridging Within- and Among-Plant Levels of Selection

Some evolutionary implications of within-plant variation in organ traits should be anticipated in the context of plant-animal interactions if, in addition to influencing the behavioral responses of animals at the within-plant level, such variation also explained foraging decisions that implicate discrimination *among* plants (i.e., genotypes). As shown in chapter 7, conspecific plants usually vary widely in both the magnitude and shape of

TABLE 8.2 **Representative examples of nonlinear relationships between responses of animals to intraspecific variation in food characteristics or organ traits relevant to their interaction with plants.**

Consumer organism	Response variable (y)	Food constituent or organ trait (x)	Shape of the nonlinear relation $y = f(x)$	Reference
Frugivorous birds and mammals				
Bombycilla cedrorum	Fruit-handling time	Fruit diameter	Increasing, concave-up	Avery et al. 1993
Bombycilla cedrorum	Intestinal sucrase activity	Sucrose concentration	Increasing, concave-down	Martínez del Río et al. 1989
Bombycilla cedrorum, Turdus migratorius, Hylocichla mustelina	Dry matter intake, weight change, sugar and energy assimilation	Glucose concentration	Increasing, concave-down	Witmer 1998
Caluromys philander, Microcebus murinus	Solution intake rate	Fructose concentration	Increasing, concave-up	Simmen et al. 1999
Carpodacus mexicanus	Sugar intake rate	Hexose and sucrose concentration	Increasing, concave-down	Avery et al. 1999
Pycnonotus cafer	Fruit-handling time, percent failure to swallow	Fruit diameter	Increasing, concave-up	Hegde et al. 1991
Sylvia atricapilla	Body mass	Tannic acid concentration	Decreasing, concave-down	Bairlein 1996
Sylvia atricapilla, Turdus philomelos	Fruit-handling time	Fruit diameter	Increasing, concave-up	Rey and Gutiérrez 1997
Turdus migratorius	Intestinal glucose uptake	Glucose concentration	Increasing, concave-down	Levey and Karasov 1992
Zosterops pallidus	Solution intake rate	Xylose concentration	Decreasing, concave-up	Franke et al. 1998
Nectar-feeding ants, butterflies, moths, and bees				
Agraulis vanillae, Phoebis sennae	Energy intake rate	Sucrose concentration	Increasing, concave-down	May 1985
Bombus appositus	Probe time per flower	Amount of nectar per flower	Increasing, concave-down	Hodges and Wolf 1981
Bombus occidentalis	Energy intake rate	Nectar volume per flower	Increasing, concave-down	Cartar and Dill 1990
Three bumblebee species	Nectar intake rate	Sucrose concentration	Decreasing, concave-down	Harder 1986
Camponotus mus, C. rufipes, Pachycondyla villosa	Sugar and fluid intake rate, relative crop filling	Sucrose concentration	Increasing then decreasing, concave-down	Josens et al. 1998; Paul and Roces 2003
Macroglossum stellatarum	Nectar intake rate	Sucrose concentration	Decreasing, concave-down	Josens and Farina 2001
Thymelicus lineola	Sugar intake rate	Sucrose concentration	Increasing then decreasing, concave-down	Pivnick and McNeil 1985
Nectar-feeding birds				
Acanthorhynchus tenuirostris	Sugar intake rate	Sugar concentration	Increasing then decreasing, concave-down	Mitchell and Paton 1990

Species	Response variable	Factor	Relationship	Reference
Amazilia rutila	Handling time per flower	Corolla length	Increasing, concave-up	Montgomerie 1984
Anthochaera carunculata, Phylidonyris novaehollandiae	Sugar intake rate	Sugar concentration	Increasing, concave-down	Mitchell and Paton 1990
Archilochus colubris	Handling time per flower	Corolla length	Increasing, concave-up	Temeles 1996
Coereba flaveola	Sugar intake rate	Sucrose concentration	Increasing, concave-down	Mata and Bosque 2004
Eugenes fulgens, Diglossa baritula	Nectar intake rate	Hexose and sucrose concentration	Decreasing, concave-down	Schondube and Martínez del Río 2003
Selasphorus rufus	Energy intake rate	Sucrose concentration	Increasing then decreasing, concave-down	Tamm and Gass 1986
Selasphorus rufus	Handling time per flower	Corolla length	Increasing, concave-up	Temeles and Roberts 1993
Sephanoides sephanoides	Daily net energy gain	Sucrose concentration	Increasing then decreasing, concave-down	López-Calleja et al. 1997
Insect herbivores				
Anticarsia gemmatalis	Larval survival, pupal mass	Caffeine concentration	Decreasing, concave-down	Slansky and Wheeler 1992
Choristoneura occidentalis	Larval survival, pupal mass	Nitrogen content	Increasing, concave-down	Clancy 1992
Lymantria dispar	Larval growth rate and food utilization efficiency	Nitrogen content	Increasing, concave-down	Stockhoff 1993
Manduca sexta	Pupal weight	Nicotine concentration	Decreasing, concave-down	Parr and Thurston 1972
Spodoptera eridania	Larval mass and survival	Nitrogen content	Increasing, concave-down	Karowe and Martin 1989
Spodoptera exigua	Relative food intake	Pyrrolizidine alkaloid concentration	Decreasing, concave-up	van Dam et al. 1996
Spodoptera littoralis	Feeding deterrence index	Furanocoumarin content	Increasing, concave-down	Calcagno et al. 2002
Trichoplusia ni	Larval duration, mortality	β-Nitropropionic acid concentration	Increasing, concave-up	Byers et al. 1977
Insect fruit and seed predators				
Amblycerus cistelinus	Body mass of adult beetle	Fruit transversal diameter	Increasing, concave-down	C. M. Herrera unpubl.
Bruchus brachialis	Body length of adult beetle	Seed size	Increasing, concave-down	Dickason 1960[a]
Rhagoletis pomonella	Oviposition frequency	Fruit diameter	Increasing then decreasing, concave-down	Prokopy and Bush 1973; Prokopy 1977

[a]Nonlinearity of the relationship tested on digitized data from figure 1 of the original publication.

within-plant variability in organ traits, and these individual differences can have a genetic basis. Therefore, any mechanism that can predictably connect within-plant variation to among-plant differences in their interaction with animals will be apt to generate opportunities for animal selection on the characteristics of subindividual variability. This would happen, for instance, if animals forage differentially among individual plants in response to their differences in within-plant organ variability, that is, if they exhibited variance-sensitive behaviors. As shown below, the direct (e.g., suboptimal foraging, predation risk) and indirect (Jensen's effects) costs to animals of within-plant variation described in the two preceding sections may combine to bridge the within- and among-plant selectivity levels by stimulating variance-sensitive behaviors among phytophagous animals.

Costs of Within-Plant Variation and Variance-Sensitive Behaviors

The fact that subindividual organ variability entails both direct costs and benefits to foraging animals suggests that trade-offs will frequently occur, with within-plant selectivity being neither so meticulous as to incur large costs nor so sloppy as to use available items on a plant in a fully random fashion. The net profitability to an animal of using the variable organs of a given plant eventually depends on the relative magnitude of costs and benefits associated with that particular plant's inherent variability. The functions relating costs and benefits to within-plant variation in a particular plant-animal system are difficult to obtain, and are expected to differ across animal and plant species, types of interaction, and environments. For example, the fitness returns accrued from the selective exploitation of available organs on a variable plant should be directly related to the strength of the correlation between organ trait values and the performance differential (via variation in fecundity or survival) accrued to animals from using them for feeding or oviposition. The direct costs derived from increased predation risks, for example, will be directly related to the animal's susceptibility to predators and, all else being equal, would be expected to decline with increasing body size, mobility, and defensive competence. Energetic costs, or the constraints imposed by the inability to reach optimal plant departure rules, are expected to be most exacting on energy-limited animals or those living in ecological scenarios where restricted time to feed and adverse abiotic conditions set limits to energy acquisition, and energetic efficiency is at a premium. These hypothetical examples illustrate that the sign and magnitude of the net effect on animals of within-plant variation may vary widely depending on the particularities of each plant-animal system. Unfortunately, the comparative magnitudes of direct costs and benefits of within-plant variation do not seem to have been explored so far for any class of phytophagous animal, and only some rather simple predictions are possible at this stage. High direct costs will tend to favor the appearance of among-plant selectivity by animals, with the least variable individual plants tending to be preferred over the most variable ones. This situation may arise when predation-prone, time- and energy-limited animals forage on highly variable organs, such as ovipositing leaf miners or small insect seed predators. Low direct costs, in contrast, will often lead to animals selecting plants regardless of their inherent subindividual variability levels. This would be expected, for example, in large, long-lived, mobile vertebrates feeding on temporarily superabundant fruit crops. Regardless of the actual frequency of high- and low-cost scenarios in nature, the aspect I wish to emphasize here is that, under most circumstances, direct costs to animals derived from within-plant variation are expected to cause among-plant selectivity related to plant differences in variability levels.

Theory predicts that indirect costs of within-plant variability derived from Jensen's effects will generally favor variance-sensitive behaviors and among-plant selectivity by animals. Behavioral models incorporating the effects of nonlinear responses and explicitly framed in terms of Jensen's inequality predict that animals should evolve variance-sensitive (or "risk-sensitive," as they are most frequently termed) behaviors that will ultimately translate into discernible foraging responses to among-patch differences in the variance of resource quality (Real and Caraco 1986; Smallwood 1996; Kacelnik and Bateson 1997). In particular, when variance in resource quality is involved (as opposed to variance in resource acquisition delay; Kacelnik and Bateson 1997), these models predict that variance-averse ("risk-averse") behaviors should develop whenever the Jensen's effects involved are large and positive, and variance-indifferent behaviors should arise when they are small or nonexistent (Smallwood 1996). Other theoretical models not explicitly invoking Jensen's effects have similarly predicted variance-sensitive behaviors from consideration of the uncertainty in fitness returns associated with different behaviors. Real (1980) formulated a behavioral model that explicitly considered the responses of animals to resource uncertainty, and predicted that under a "law of diminishing returns" organisms should tend to minimize uncertainty. The vast majority of nonlinear relationships linking animal performance with plant food traits shown in table 8.2 actually represent instances of diminishing returns, either in the form of increasing, concave-down gain functions or through its complementary version, that is, increasing, concave-up cost functions. Since within-plant variation in the food or survival value of individual organs is just a particular class of the uncertainty or stochasticity experienced by animal foragers while visiting different feeding or oviposition patches, Real's pioneering model may also be used to predict that phytophagous animals should generally tend to be variance-averse and avoid the most uncertain plant individuals, that is, the ones with the largest variances in per-organ food reward or fitness value. This prediction was also supported by more elaborated models proposed by Real and Caraco (1986). As shown in the next section, empirical data largely corroborate these expectations and support the prediction that within-plant variation in organ traits may ultimately induce among-plant discrimination by phytophagous animals.

Evidence for Variance-Sensitive Behavior

Experimental inquiries about variance-sensitive foraging have generally been based on manipulating the reward distribution available to an animal. One class of experiments examines preference in relation to reward variance when mean rewards are fixed, while a second class tests trade-offs between mean and variance by simultaneously changing both parameters (Real and Caraco 1986; Kacelnik and Bateson 1996). I am concerned here exclusively with the first class of evidence because it is the most directly relevant to the subject dealt with in this section; this does not imply disregard for the importance of considering the joint influence of means and variances on foraging responses (see chapter 10).

Studies of the behavioral consequences of the exposure of phytophagous animals to variance in resource quality have been largely concerned with nectarivorous animals foraging on flowers with variable nectar sugar rewards. This bias probably is attributable to historical rather than biological reasons, since the first attempts to incorporate the effects of resource variance into optimal foraging models mainly focused on pollinators (Real 1980, 1981; Waddington et al. 1981; Harder and Real 1987; but see Caraco et al. 1980) because their often strict energy requirements led to the expectation that behavioral adaptations for successfully meeting these requirements should be commonplace (Heinrich 1975; Real 1980).

Subsequent investigations of variance-sensitive behavior in animals have mostly followed the trail of earlier researches on pollinators, and this long tradition has resulted in the accumulation of considerable empirical evidence demonstrating that variance-sensitive behaviors commonly occur among both vertebrate and invertebrate nectarivores, as illustrated by some recent reviews (Kacelnik and Bateson 1996; Perez and Waddington 1996; Shafir 2000).

A representative selection of studies examining the behavioral responses of insect and vertebrate nectarivores is shown in table 8.3. With only a single exception, all studies demonstrate that when nectarivores are offered a choice between feeding patches with contrasting variance in food reward (volume and concentration of sugar solution), they develop variance-averse behaviors and tend to prefer the least variable patches. In these studies, variance aversion may be expressed by differential visitation rates to constant and variable flower patches, as reported for example by Waser and McRobert (1998) for *Selasphorus* hummingbirds feeding on arrays of *Ipomopsis aggregata* flowers. Birds paid significantly fewer visits to high-variance arrays than to low-variance ones. In other cases, variance-averse foraging may be expressed as patch-departure preferences, as found by Biernaskie et al. (2002) in a study of hummingbirds and

TABLE 8.3 **Experiments on the responses of nectarivorous animals to variance in food quality.**

Species	Quantity that is variable (or "risky")	Behavioral response to variance	Reference
Invertebrates			
Apis mellifera	Sugar solution volume	Variance-averse	Shafir et al. 1999
Bombus sandersoni	Sugar solution volume	Variance-averse	Real 1981
Bombus edwardsii	Sugar solution volume	Variance-averse	Waddington et al. 1981
Bombus flavifrons	Sugar solution volume	Variance-averse	Biernaskie et al. 2002
Vespula maculifrons	Sugar solution volume	Variance-averse	Real 1981
Xylocopa micans	Sugar solution volume and concentration	Variance-indifferent	Perez and Waddington 1996
Vertebrates			
Coereba flaveola	Sugar solution volume and concentration	Variance-averse	Wunderle and O'Brien 1985
Selasphorus rufus	Sugar solution volume and concentration	Variance-averse	Hurly and Oseen 1999; Bateson 2002
Selasphorus rufus	Sugar solution volume	Variance-averse	Biernaskie et al. 2002
Selasphorus rufus *S. platycercus*	Sugar solution volume	Variance-averse	Waser and McRobert 1998

Note: Most studies shown were based on binary choices among food patches, either natural (e.g., inflorescences) or experimental (e.g., arrays of artificial flowers), characterized by high and low variances in reward to animal foragers.

bumblebees foraging on artificial inflorescences differing in variance of nectar reward. When they were allowed to assess resource variability by concurrent sampling, birds and bumblebees visited fewer flowers on variable inflorescences than on constant ones.

Information on variance-sensitive behavior by phytophagous animals other than nectarivores is scarce. The available evidence, albeit limited, suggests that granivores, frugivores, and herbivores may also exhibit variance-averse behavior in response to among-patch differences in the variance of resource quality. Variance-averse behavior in response to patch differences in seed-number variability was shown by Caraco et al. (1980) and Caraco (1982) for the granivorous birds *Junco phaeonotus* and *Zonotrichia leucophrys*. Variance-sensitive foraging has also been reported for European starlings (*Sturnus vulgaris*), an occasional frugivore (Brito e Abreu and Kacelnik 1999). Collared lemmings (*Dicrostonyx groenlandicus*) consistently minimize the variance in encounter rates with food rewards by choosing feeding patches that offer the least variation in food reward rather than those that maximize food intake rates (Searle et al. 2006). Despite this scarcity of data, however, it seems safe to predict that when the amount of reward is variable, variance aversion should be as common among nonnectarivorous phytophages as it is among nectarivorous ones or among nonphytophagous animals in general (Kacelnik and Bateson 1996).

If variance sensitivity is to influence behavior in a significant way, animals must discriminate among, and strategically use, different variable rewards, and not only be able to make choices between a constant reward and a variable one (Real and Caraco 1986). That the degree of variance sensitivity expressed by animals may effectively depend on the characteristics of the reward distributions has been shown by Shafir et al. (1999) for honeybees feeding on sugar solutions. On the basis of the results of their study, they also suggested that, when evaluating variability, animals may be most sensitive to relative measures, such as the CV, than to any single parameter such as the absolute variance in reward. These suggestions were subsequently tested and verified by Shafir (2000) by analyzing a large set of published experiments (see also Weber et al. 2004 for some related analyses). In the case of nectarivores, the level of variance sensitivity depends on the perception of variability, for it is linearly related to the relative magnitude of variability, as measured by the CV (fig. 8.5), but not to absolute measurements such as the variance or standard deviation (Shafir 2000). The higher the CV, the stronger the variance aversion.

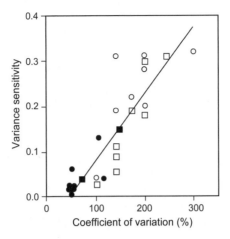

FIG. 8.5 Variance sensitivity of invertebrate (*circles*) and vertebrate (*squares*) nectarivores to variable reward distribution in experiments where animals were offered a choice between constant and variable reward (nectar volume or nectar concentration). *Open symbols* denote experiments in which the variable reward distribution included empty rewards; *closed symbols* denote experiments in which all options were rewarding. The strength of variance sensitivity was defined as the absolute value of the difference between 0.5 and the mean proportional preference for the constant reward. Redrawn from Shafir 2000.

Results of Shafir's review (2000) are important for two reasons. In the first place, they show that it is relative, rather than absolute, measurements of variability in resource quality that animals are sensitive to. Using CV_{within} to evaluate the extent of within-plant variation, as done in chapter 3, thus acquires biological sense in addition to being a statistically sound procedure. Second, Shafir showed that there is a gradient of responses to resource variability rather than the simple dichotomy (variance-sensitive vs. variance-indifferent behavior) often implied in the interpretation of binary-choice experiments (e.g., Kacelnik and Bateson 1996). This result suggests that, all else being equal, the extent of among-plant discrimination exerted by phytophagous animals in the field should be directly related to the magnitude of *relative* within-plant variability in traits that influence food or fitness returns. Given the broad differences existing among plant organ types and organ traits in levels of relative within-plant variability (CV_{within}, chapter 3), one would expect, for example, that variance sensitivity and among-plant discrimination should be most evident when animals are foraging for highly variable organs such as leaves or seeds (e.g., leaf miners, insect fruit and seed predators), and when the interaction is mediated by organ traits characterized by extreme levels of within-plant variability, such as nectar volume per flower or individual seed mass.

While the abundant experimental evidence on variance-sensitive animals considered in this section clearly supports the hypothesis that within-plant variation in organ traits may lead to among-plant selection by consumers, two important caveats should be explicitly acknowledged. Most theoretical models and experimental results for variance-sensitive behaviors are specifically concerned with responses to variability in energetic rewards, to a large extent because such models initially arose as an outgrowth of optimal foraging theory (Real 1980; Real and Caraco 1986). Theoretical models and predictions relating animal variance sensitivity to heterogeneity in other currencies are still in their infancy (e.g., Karban et al. 1997; Shelton 2000, 2004; Tenhumberg et al. 2000). These models strongly suggest, however, that insofar as nonlinearities, the law of diminishing returns, and Jensen's effects are involved, variance-sensitive behaviors will also develop in response to variability in plant organ traits that, although not directly affecting the energetic rewards of animals, may affect other components of their fitness or short-term performance (e.g., variability in leaf nitrogen and secondary metabolite concentration, corolla dimensions, nutritional composition of fruit pulp). Another limitation of experimental studies demonstrating variance-sensitive behaviors is that they have invariably been conducted on artificially manipulated resource patches. No field study so far seems to have documented variance-sensitive foraging responses by phytophagous animals to *natural* variation among feeding patches (e.g., individual plants) in food-reward variability. While the experimental evidence summarized in this section strongly suggests that such responses to variability will also occur under natural conditions, empirical studies are needed that explicitly look for animal responses to naturally occurring variation in the field. Some indirect evidence presented in the next chapter, showing that differences among conspecific plants in organ variability levels often correlate with individual variation in the strength of interaction with animals, does suggest that animal responses to within-plant variation in natural conditions would be frequently revealed if appropriately sought after.

Fitness Consequences of Subindividual Variability in Organ Traits for Plants

Subindividual variation in the characteristics of reiterated organs may influence the fecundity or vegetative performance of plants, and through this mechanism, individual fitness differences may arise as a consequence of variation in the extent and organization of variability.

M yriad mechanisms are possible whereby within-plant variation in leaf, flower, fruit, or seed characteristics influences diverse aspects of plant vegetative (e.g., growth rate, carbon assimilation) and reproductive (e.g., fecundity) performance, thus eventually affecting the fitness of individuals. This chapter presents an overview of such mechanisms and, in so doing, argues by way of example that subindividual variability in organ traits might in the long run have some nontrivial evolutionary consequences to the plants, an aspect that is explicitly dealt with in chapter 10.

In some cases, the performance consequences of within-plant variability are unrelated to the interaction of plants with animals, and they are most parsimoniously explained by considering aspects of the abiotic environment alone. This is the case of the relationship between within-plant leaf variation and within-plant light gradients (Field 1983; Givnish 1988; Hollinger 1996). In other instances, however, the putative fitness implications of plant variability are the direct or indirect outcome of diverse interactions with animals. This may apply, for example, to within-plant variation in certain floral features (Waser and Price 1984; Ott et al. 1985). For the sake of completeness, and to broaden the ecological basis for the discussion on the evolutionary significance of within-plant variation in chapter 10, I consider in the present chapter consequences both related and unrelated to animals. Continuous as well as discontinuous within-plant variation can have fitness consequences for plants. Several excellent reviews have recently focused on the implications of discontinuous variation in leaf, flower, fruit, and seed attributes (heterophylly, dimorphic cleistogamy, heterocarpy, and seed heteromorphism; Mandák 1997; Winn 1999a; Wells and Pigliucci 2000; Imbert 2002; Matilla et al. 2005; Culley and Klooster 2007). This relieves me of presenting a detailed account of the implications of these relatively well-investigated topics. Only succinct summaries are presented, aimed at drawing some parallels with continuous within-plant variation and pointing out some weaknesses in our current knowledge. I am mainly concerned in this chapter with enumerating the variety of possible mechanisms linking continuous subindividual variation to individual fitness through the action of animals. This aspect remains largely unexplored in the literature, and there have been almost no attempts to directly look for effects of continuous within-plant variation on individual performance or reproductive success. For this reason, much of the evidence presented below is admittedly circumstantial.

To ascertain the effects on individual plants of subindividual variability in a given organ trait, one should ideally compare some functional performance (e.g., growth rate, photosynthesis, carbon balance, nutrient status) or life-history measurement (e.g., fecundity, survival, longevity) for conspecifics that differ in levels of subindividual variation in the trait but are identical in every other respect, including trait means. This kind of comparison can hardly be done on plants growing naturally in the field, since trait means usually vary among individuals. An alternative, albeit imperfect, method consists of simply looking for relationships across individuals between trait variability and some measure of fitness or performance, regardless of whether they differ in means or not. One obvious problem with this approach, however, is that plant means tend to covary with the corresponding within-plant variances (fig. 3.1), which may confound the effects of trait variance on individual fitness or performance with the effects of the mean. This caveat was highlighted by Real and Rathcke (1988, 734) when considering the possible responses of pollinators to individual differences in levels of within-plant variability in nectar abundance per flower, and they noted that "the existence of a strong correlation between

the arithmetic mean and variance in nectar volume distribution compli-
cates our ability to infer pollinator preferences under conditions of uncertainty (risk) based on their observed behavior." This complication should
therefore be kept in mind when interpreting examples presented in this
chapter in which fitness or performance measurements are related to
subindividual variability without previously filtering out, or statistically
accounting for, the possible "contamination" of the data by concomitant
variation in plant means. It must be emphasized, however, that the problem of confounding mean and variance effects does not exclusively affect
assessments of the fitness consequences of variability. By exactly the same
token, assessments of the fitness correlates of differences in plant trait
means, routinely done in plant phenotypic selection studies, may also be
confounded to an undetermined extent by the possible effect on fitness of
individual differences in trait variances, as shown by examples presented
in chapter 10.

Discontinuous (Discrete) Variation

As noted in chapters 2 and 3, there is a long tradition of research on discrete within-plant polymorphisms affecting leaves, flowers, fruits, and
seeds, and a voluminous literature has accumulated over more than a century, examining the functional significance and possible fitness implications of the phenomenon. Regardless of the obvious functional differences
among the distinct types of structures involved, the common theme unit-
ing all modalities of discrete within-plant variation is that they exemplify
cases of "multiple strategies" of variation (sensu Lloyd 1984), in which
one plant simultaneously operates distinct types of structures that perform the same function. If the environment predictably favors different
variants of the same structure at different times or in different locations
(e.g., leaves on different parts of the same plant, as with submerged and
floating parts of amphibious plants; Sculthorpe 1967), then the production of a mixture of variants may enhance the fitness of individual plants
(Lloyd 1984; Winn 1999a). This canonical explanation for the evolution
of discrete within-plant variation in organ traits, or multiple strategies of
variation in Lloyd's terminology (1984), rests on two central postulates:
(1) Each alternative form of a structure performs differently in different
locations or at different times or, in other words, represents a unique, nonoverlapping pathway of contributing to the fitness of its bearer. I refer to

this postulate as the "multiroute" condition throughout this section. (2) Individuals producing a mixture of variants are consistently superior in fitness to those producing exclusively one of the variants. I refer to this postulate as the "mixed-superiority" condition. Both conditions have to be met to properly interpret instances of multiple strategies as mechanisms enhancing the fitness of individuals (Winn 1999a). The empirical support available for each of these two conditions is, however, markedly different. As shown below, there is ample evidence supporting the multiroute condition, but remarkably little substantiating the mixed-superiority condition.

Leaves

Heterophylly has been most thoroughly investigated in the case of aquatic plants that potentially possess two or more leaf types (submerged, floating, and emergent) differing in shape, anatomy, chemical profile, and chlorophyll concentration. In species of *Potamogeton*, for example, these differences are correlated with contrasting rates of photosynthesis. Floating leaves achieve much higher rates of photosynthesis than submerged leaves in air, and when compared under water, rates of photosynthesis increase in submerged leaves and decrease in floating leaves relative to the performance of the same leaf type in air (Wells and Pigliucci 2000). Thus each leaf type not only exhibits higher rates of photosynthesis in its respective environment, but also is better suited to function in that environment than the alternative leaf, a finding in direct accordance with the multiroute condition. Wells and Pigliucci (2000) present further examples similarly showing that the respective characteristics of distinct leaf types produced by aquatic heterophyllous plants are best suited to their particular environment. Equivalent information is much scarcer for terrestrial heterophyllous plants, although some evidence suggests also that functional differences between leaf types match the characteristics of their respective environments. Individuals of *Viola septemloba* can produce both leaves that are entire and cordate and leaves that are deeply lobed (fig. 2.1d). Winn (1999a) examined the possibility that lobed leaves dissipate heat better than cordate leaves, a scenario that would be consistent with an advantage of bearing lobed leaves in the hot summer months, which is the pattern observed in *V. septemloba*. As predicted, the temperature of lobed leaves was significantly below that of cordate leaves on the same plant. Since high leaf temperatures may result in tissue damage, the ability of lobed leaves to maintain lower temperatures could represent an advantage during summer months.

In addition to the advantages derived from "partitioning" the abiotic environmental gradient by alternative leaf forms, heterophylly may also contribute to the fitness of individual plants by reducing damage from herbivores. Developmental heterophylly, in which different leaf morphs are associated with different ontogenetic stages in the same plant (juvenile and adult shoots), is particularly frequent on oceanic islands (Friedmann and Cadet 1976). A few large, ground-living browser species are or have been abundant in many archipelagos (e.g., giant tortoises, flightless birds), and Givnish et al. (1994) hypothesized that the high incidence of developmental heterophylly on islands represents a convergent response of plants to herbivory by these visually oriented herbivores. Under this hypothesis, juvenile-morph leaves located closer to the ground would be better defended and less palatable to browsers than adult-morph ones located higher, and the former would advertise their condition to browsers by their distinctive shapes. Eskildsen et al. (2004) have recently tested and verified a crucial element of this hypothesis. In a choice experiment, they recorded the feeding response of Aldabra giant tortoises to ten species of Mauritian plants, of which seven were heterophyllous and three homophyllous (i.e., lacking leaf dimorphism). Foliage of homophyllous species was preferred over that of heterophyllous ones, and among heterophyllous species adult leaves were preferred to juvenile ones. Differences between leaf morphs in acceptability to tortoises thus corresponded with the differential risk of damage derived from their distinct positions relative to ground, which is in accordance with the multiroute condition. Differential acceptability of leaf morphs is also in accordance with the optimal defense theory, which predicts that, to maximize growth individual plants should allocate defenses among leaves in direct proportion to the leaf's photosynthetic value (Iwasa et al. 1996; van Dam et al. 1996). Developmental heterophylly has also been shown to aid in resistance against invertebrate herbivory in *Eucalyptus* and *Populus* trees (Kearsley and Whitham 1997; Brennan and Weinbaum 2001; Brennan et al. 2001). These studies, and the frequent observation of chemical differences between juvenile and adult leaf morphs (Stein and Fosket 1969; Les and Sheridan 1990; Murray and Hackett 1991), suggest that developmental heterophylly often represents a mechanism of developmental resistance to herbivory, in which resistance undergoes changes along the "developmental stream" of individual plants (Kearsley and Whitham 1989, 1997).

A pattern of heterophylly that corresponds to differences in optimum leaf phenotype among environments (multiroute condition) provides circumstantial evidence that heterophylly is advantageous to individual plants. A direct test, however, requires a comparison of the fitnesses of individuals that exhibit different patterns of heterophylly (mixed-superiority condition), as emphasized by Winn (1999a). In some species, individual plants differ in the extent of heterophylly, and these differences have a genetic basis (Wells and Pigliucci 2000); hence the mixed-superiority condition could be tested quite simply by looking for correlations between fitness and extent of heterophylly across individuals. As remarked by Wells and Pigliucci (2000), examples of this test in the literature on heterophylly are very rare, and the few data available do not provide a clear indication that the magnitude of heterophylly affects the fitness of individual plants. In *Nuphar lutea*, for example, Kouki (1993) showed that experimental variation in the amount of herbivory experienced by floating leaves resulted in modifications of the proportions of submerged and floating leaves subsequently produced by the plants. These induced alterations in the extent of heterophylly, however, had no measurable effects on short-term flower production by the plants.

Flowers

Dimorphic cleistogamy, in which a plant produces cleistogamous (CL), automatically self-pollinated flowers in addition to chasmogamous (CH) ones, may have a number of fitness advantages, which have been thoroughly reviewed by Culley and Klooster (2007). Among these, reproductive assurance and preservation of adapted gene complexes seem most important. CL flowers offer reproductive assurance when pollinators are rare or absent, and in some species CL flowers actually increase seed production when CH flowers remain unpollinated. CL selfing may prevent the disruption of locally adapted gene complexes by avoiding the recombination that frequently accompanies outcrossing. In addition, CL seeds do not disperse very far in some species, which contributes further to preserving locally adapted complexes. CH flowers, on the other hand, generally produce more genetically variable progeny, and CH seeds are often dispersed farther from the maternal plant, which may alleviate parent-offspring and sibling competition and enhance the possibilities of colonizing favorable microsites.

Theoretically, the production of a mixture of CH and CL flowers would be advantageous to individual plants in either fine-grained (at the within-plant scale) or coarse-grained (across an area or season) heterogeneous environments. When environmental heterogeneity is fine-grained, the fitness of a phenotype that produces both CH and CL flowers at a single time would be enhanced if each flower type is produced in the environment for which it is best suited, as when flowers at different spatial positions experience varying numbers of pollinator visits. If environmental heterogeneity is coarse-grained, then the fitness of a phenotype would increase by the production of a different flower type within each environment to maximize reproductive success (Lloyd 1984; Schoen and Lloyd 1984). Judging from the exhaustive review of Culley and Klooster (2007), it does not seem that the whole-plant advantages of producing a mixture of CL and CH flowers implied by these and other (Masuda et al. 2001) theoretical models have ever been examined empirically, and the mixed-superiority condition thus appears to remain untested for dimorphic cleistogamy. Indeed, Culley and Klooster included among their suggestions for future research the need for empirical data to test the main postulates of theoretical models, including quantitative information on the variation of floral types within and between individuals.

Fruits and Seeds

Seed heteromorphism represents the production of discontinuous types of seeds by a single plant (chapter 2). The differentiation may affect either the whole diaspore (heterocarpy) or exclusively the seed (heterospermy). Both phenomena are frequently associated (Mandák 1997; Imbert 2002), and their consequences for plants are difficult to separate; hence I treat them jointly in this section.

Functional differences between the distinct morphs produced by heterospermous and heterocarpous plants are usually well-defined and ecologically important, thus unambiguously fulfilling the multiroute condition. The distinct types of seeds or diaspores typically differ in size, dispersal ability, dormancy, germination requirements, and seedling emergence, survival, and growth (Mandák 1997; Imbert 2002; Matilla et al. 2005). Most often, differences between morphs involve several of these aspects simultaneously. In the annual herb *Spergularia marina*, winged seeds are longer and heavier, and disperse farther than unwinged ones (Telenius and Torstensson 1989). In *Leontodon longirrostris*, the achenes occupying the peripheral positions on the fruiting head are heavier, lack a pappus, are dispersed at short range, and exhibit extended dormancy and slow germination, while the central achenes are much lighter, possess a well-developed pappus, are dispersed by wind, and germinate faster under a broad spectrum of conditions (Ruiz de Clavijo 2001). Similar differences have been reported for many other heterospermous and heterocarpous species. Consequently, spreading offspring in time and space as a consequence of variation in germination time, diaspore morphology, and dispersal method seem the two ecologically most significant consequences of seed heteromorphism to individual plants (Lloyd 1984; Imbert 2002). In highly variable and unpredictable habitats, spreading offspring in time may reduce temporal variance in realized fecundity, which may eventually result in increased individual fitness (Gillespie 1977). Likewise, within-plant variation in seed size, morphology, and dispersal agent generally contributes to enlarging the seed shadow of individual plants, and hence to spreading offspring in space (Augspurger and Franson 1987, 1993). The same arguments discussed later in this chapter in relation to the more general situation of continuous fruit and seed variation apply also here to the case of discontinuously varying diaspores. Individual differences in the extent of seed heteromorphism are expected to translate into variations in spatial and temporal characteristics of seed shadows, which may then result in individual fitness differences.

Despite the large number of studies of seed heteromorphism, variation among individuals in the extent (proportions of the different diaspore types) and characteristics (functional contrasts between diaspore types) of heteromorphism, and the possible fitness consequences of such variation, remain two poorly explored topics. Individual variation in heteromorphism can be extensive in some species, which provides opportunities for simple tests of the mixed-superiority condition. In two populations of the heterocarpous *Thymelaea velutina* studied by de la Bandera and Traveset (2006), some individuals produced both dry fruit (achenes) and fleshy fruit (drupes), thus effectively qualifying as heterocarpous, while others produced exclusively fleshy fruits. In addition, truly heterocarpous individuals varied widely in the relative proportions of fruit types produced, the percentage of fleshy fruits ranging between 40 and 95%. It is most likely that such a broad spectrum of within-plant variation in fruit features results in fitness differences among plants, but this aspect remained unmeasured in that study.

There is almost no information on the possible responses of seed and fruit predators to variation in seed and fruit characteristics associated with heterospermy. Mandák's exhaustive review (1997) mentioned only a study of *Atriplex sagittata* reporting preferential damage of one type of fruit by insects. Imbert (2002) did not report any example, and emphasized the need for data concerning differences in predation rate among seed morphs. Scarcity of empirical data notwithstanding, the different types of fruit and seeds produced by heterospermous plants will most likely differ in risk of pre- and postdispersal damage by invertebrates and vertebrates, since these generally exhibit discriminatory behavior and feeding responses to variation in seed traits that are ordinarily involved in heterocarpy and heterospermy, such as size and shape (chapter 8). Individual variation in the proportion of different diaspore types produced is therefore expected to translate frequently into differential seed-survival prospects, but this possible mechanism does not seem to have been explored so far for any species exhibiting seed heteromorphism.

Continuous Variation: Leaves

Continuous within-plant variation in leaf features can have three major classes of consequences at the whole-plant level: those related to the spatial and temporal heterogeneity in the abiotic environment; the cascading consequences that result from induced variation in other organs, such as flowers, fruits, and seeds; and the food selection responses by foliage-eating animals. I consider each of these three categories in turn.

Consequences Related to the Abiotic Environment

Ontogenetic changes in the size and shape of plants, and in the spatial arrangement of their parts, contribute to spatial and temporal heterogeneity in the abiotic environment around plant parts. The influence of neighbors also adds to this heterogeneity. The different leaves of the same plant are thus produced in, and become successively exposed to, different phylloclimates (chapter 6), and a significant fraction of continuous intraplant variation in leaf traits (size, shape, photosynthetic features) typically occurs along intrinsic and extrinsic environmental gradients (chapter 4). As noted above for heterophyllous plants that produce discontinuous leaf types, the most direct implication of continuous intraplant variation in structural and functional leaf features is the efficient partitioning of spatial and temporal abiotic gradients (Winn 1999a).

By producing leaves of different nitrogen concentration, specific area, and photosynthetic capacity at different microsites along within-plant light gradients, individual plants can maximize overall photosynthetic performance in the face of environmental heterogeneity. Models predict that total carbon gain of individual plants will increase if leaf spatial and physical properties vary so that light is more uniformly intercepted, and leaf nitrogen is apportioned among leaves proportionally to the amount of incident light (Field 1983; Hirose and Werger 1987; Givnish 1988; Hollinger 1989). In particular, Field (1983) predicted that, in plants where some microsites are heavily shaded and others are rarely shaded, carbon gain for the whole plant should be maximized when leaf nitrogen is distributed such that the leaves in the microenvironments receiving the highest amount of photosynthetically active radiation have the highest nitrogen contents. In contrast, in plants characterized by uniformly illuminated microsites with frequently saturating light intensities, whole-plant carbon gain should be maximized when the available leaf nitrogen is distributed uniformly among leaves. These early predictions have generally been upheld by subsequent empirical studies and simulations (Hollinger 1996; Le Roux et al. 1999; Casella and Ceulemans 2002; Kull 2002). For example, simulations by Hollinger (1996) showed that the observed pattern of nitrogen allocation among leaves of the broad-leaved tree *Nothofagus fusca* results in a greater canopy carbon gain than would be expected if the same total quantity of nitrogen were allocated randomly or equally among microsites. In that species, assuming a uniform distribution of nitrogen among leaves may underestimate plant-level assimilation by up to 10%, or in other words, the unequal distribution of nitrogen among leaves accounts for an additional photosynthetic gain at the whole-plant level of around 10%.

Although the plant-level consequences of within-plant variation in leaf features have been investigated most frequently for forest canopy trees, they apply to smaller, herbaceous species as well. The importance of between-leaf photosynthetic resource partitioning is well established for herbs growing in dense stands, where vertical gradients in leaf nitrogen content result in improved whole-plant-level photosynthetic performance (Hirose and Werger 1987; Lemaire et al. 1991; Connor et al. 1995; Pons and Anten 2004). For example, the actual (i.e., uneven) distribution of nitrogen among leaves of densely growing plants of *Solidago altissima* achieved over 20% more photosynthesis than that under an uniform distribution (Hirose and Werger 1987). In experimental stands of *Lysi-*

machia vulgaris, Pons and Anten (2004) found that whole-plant carbon gain was reduced by 19.6% when plants were given the relatively shallow within-plant gradient in leaf nitrogen content characteristic of uniformly lighted plants grown in the open. The results of their study clearly show that varying levels of within-plant variation in leaf nitrogen content will give rise to individual differences in overall photosynthetic performance and carbon gain. Although there have been few empirical connections between plant functional and population biology features (Ackerly and Monson 2003), it is expected that consistent differences between individuals in overall photosynthetic performance will in the long run translate into variation in fecundity or progeny quality (e.g., seed size). This effect should be most apparent in annual plants, as they depend exclusively on short-term, current-season photosynthesis for reproduction. In a comparative study of two winter wheat (*Triticum aestivum*) cultivars, Wang et al. (2005) found that steeper vertical gradients in nitrogen distribution among leaves tended to be associated with greater seed quality, as measured by gluten and protein content of grains. The corollary to the preceding studies is that, whenever the leaves of a plant are immersed in a spatially heterogeneous, patchy light environment, producing variable leaves whose photosynthetic features (e.g., nitrogen content) match the spatial distribution of incident light will be intrinsically superior to producing identical leaves throughout the plant. In other words, inequality among leaves in their photosynthetic capacity will be directly advantageous for the whole plant if it matches differences in the amount of light received.

Leaves produced by the same plant at different times in the seasonal cycle may differ in size, shape, anatomy, chemistry, water-use efficiency, and photosynthetic capacity (Westman 1981; Mulkey et al. 1992; Kitajima et al. 1997; Winn 1999a, 1999b; Palá-Paúl et al. 2003). These and other studies have demonstrated or inferred a good correspondence between alternative leaf types and the environmental conditions prevailing at the time each type is produced. For example, in the tropical understory shrub *Psychotria marginata*, leaves produced just prior to the dry season have higher specific mass and, during drought, have lower stomatal conductance and higher water-use efficiencies than leaves produced at the beginning of the wet season (Mulkey et al. 1992). These and similar observations (Westman 1981; Kitajima et al. 1997) suggest that the seasonal component of within-plant leaf variation may contribute to enhance physiological efficiency at the whole-plant level in the face of temporal heterogeneity in the light, temperature, and water environment (Winn 1999a, 1999b). Nevertheless, in contrast to the relatively well-investigated consequences of spatial within-plant heterogeneity in leaf traits, the whole-plant consequences of seasonality in leaf traits have been investigated directly on very few occasions. In the only well-worked example known to me, Winn (1999b) tested in the field whether seasonal variation in the size, thickness, and stomatal density of leaves of the annual *Dicerandra linearifolia* was advantageous to individual plants. Regression of individual dry mass (a proxy for fitness) on leaf traits revealed no evidence supporting the hypothesis that different leaf phenotypes are favored in different seasons. Individuals with large and thick leaves in both winter and summer consistently had a size advantage during the two years of the study, a result that suggests within-plant variation in leaf traits is not intrinsically advantageous to *D. linearifolia*. It is possible, however, that the whole-plant effect of leaf variation could have been detected if other response variables such as plant longevity, number of seeds produced, or seed size had been considered.

Cascading Consequences for Heterogeneity of Flowers, Fruits, and Seeds

Flowers, fruits, and seeds generally behave as net photosynthate sinks that are principally supplied by their local source leaves, and their growth is largely determined by local photosynthetic rates, as discussed in chapter 6. In sectorial plants, within-plant heterogeneity in the photosynthetic yield of single leaves will therefore be partly responsible for within-plant variation in certain characteristics of flowers, fruits, and seeds, and this variation will in turn have multifarious consequences to individual plants, as shown in later sections of this chapter. Consequently, the distribution and strength of local photosynthate sources within a plant may define a sort of initial template from which heterogeneity in the characteristics of other reiterated organs, with its associated host of ramifying consequences to individual plants, will arise in a developmental cascade. The congruence between the initial template defined by variation in the sources and the map of variation eventually exhibited by the sinks will depend on a complex combination of internal and external factors (chapter 6). A few predictions may be tentatively advanced at this point, however, implying that the cascading effects of leaf variation on whole-plant fecundity or performance via its effects on the variation of other organs will depend on habitat type and the ecological characteristics of species.

Within-plant heterogeneity in the photosynthetic capacity of individual leaves is greater in species that ordinarily grow in dense stands or whose branch architecture leads to high within-crown patchiness in incident light intensity (Field 1983; Hirose and Werger 1987). Furthermore, shade-tolerant and shade-intolerant species differ in levels of within-plant variation in leaf morphology and physiology, with shade-intolerant trees being less variable than shade-tolerant ones (Niinemets et al. 1998; Uemura et al. 2006; and references therein). It may therefore be predicted that, for a given level of plant sectoriality, the cascading effects of leaf heterogeneity on within-plant variation in those flower, fruit, and seed traits that are susceptible to variations in photosynthate import (e.g., size, nectar sugar secretion, nutritional composition) will be more important in shade-tolerant species than in shade-intolerant species. Such effects will also be more important in dense shady forest with steep light gradients and strong overshadowing effects from neighbors (e.g., tropical and temperate forests) than in sunny open habitats with widely spaced, uniformly lit individuals (e.g., open woodlands and shrublands in deserts and Mediterranean environments).

Consequences Related to Herbivory

The possible plant-level consequences of within-plant variation in leaf features due to interactions with folivores have received considerably less attention than the implications related to the heterogeneity in the abiotic environment referred to above. Proponents of the genetic mosaic theory of plant defense (GMT), discussed in chapter 8, frequently alluded to the potential implications for individual plants of becoming mosaics of resistance to herbivores, but direct supporting evidence was scarce at the time, and it was admitted that "the ecological implications [of within-plant variation] for the plant are unclear" (Whitham et al. 1984, 39). Recent theoretical models and the limited evidence available confirm these earlier suggestions, as they reveal a variety of ecological mechanisms whereby intraplant heterogeneity in leaf characteristics, via its effects on herbivores, may translate into individual differences in growth or fecundity. Some of these effects follow indirectly from patterns of within-plant food selection by herbivores, while others stem directly from among-plant choices by herbivores driven by individual differences in leaf variability itself.

Folivores discriminate and respond selectively to within-plant variation in leaf nutritional and defensive characteristics (chapter 8), which accounts for the common observation that some leaves of a plant are eaten more thoroughly than others and some parts of the canopy lose more leaf area than others (see, e.g., Whitham 1983, fig. 3). From a plant's perspective, one of the most direct consequences of this variation influencing acceptability by herbivores may be an increase in the spatial dispersion of damage over the plant (Suomela and Ayres 1994). The magnitude and spatial scale over which this heterogeneity-driven dispersion takes place (e.g., among parts of the same leaf, among leaves of the same branch, or among branches) will depend on the size, mobility, foraging behavior, and choosiness of the herbivore, and on how leaf variability is spatially patterned within the plant (chapter 4); hence meaningful generalizations or predictions are risky without carefully taking into consideration each of these aspects. It is relatively well established, however, that the distribution of damage over a canopy can affect plant recovery from herbivory, and that certain spatial patterns of damage may be more detrimental to plant fitness than others, depending apparently on the rate of movement of nutrients, hormones, wound-induced signals, and defensive compounds among and within integrated physiological units of the plant (Marquis 1996; Avila-Sakar and Stephenson 2006). Sectorial plants should be able to compensate better for small amounts of local damage dispersed over many sectors than for more intense damage concentrated on just one or a few sectors. Dispersed damage should thus be less detrimental to individual plants than concentrated damage, and this pattern has been generally confirmed by most studies specifically addressing this prediction (Marquis 1996; Meyer 1998; Avila-Sakar et al. 2003). Consequently, insofar as within-plant variation in leaf features enhances the dispersion of the damage caused by herbivores, it will also contribute to reducing the overall impact of herbivory on the growth and reproduction of individual plants possessing sectorial physiological organization.

Optimal defense theory predicts that, in order to minimize loses to herbivores and maximize growth, individual plants should allocate defenses among leaves in direct proportion to the latter's photosynthetic value, and the frequent observation that leaves of the same plant differ in the concentration of defensive compounds has thus been interpreted in terms of optimal defense allocation (Iwasa et al. 1996; van Dam et al. 1996; Anderson and Agrell 2005). Although theoretical models were mainly derived for the case of variations in photosynthetic value associated with differences in leaf age, their general predictions may also apply to the case of predictable variations in leaf photosynthetic value due to other factors, such as position within the canopy. Because herbivores typically respond to within-plant variation in leaf defense levels by avoiding the most heavily defended ones (chapter 8), one proximate consequence for individual plants of better defending those leaves with greater productive prospects will be a net reduction in the cumulative impact of herbivory. This has been documented by de Boer (1999) for *Senecio jacobaea*, among others. In plants of this species, pyrrolizidine alkaloid and nitrogen leaf content both declined steadily from top to bottom positions on the stem, in parallel with increasing leaf age. Larvae of *Spodoptera exigua* and *Mamestra brassicae*, two generalist herbivores, exhibited a strong preference for the oldest, least-defended leaves at the basal positions. Photosynthetic losses to individual plants were thus minimized as a consequence of the unequal distribution of defensive metabolites among leaves, since the damage principally affected leaves that were near the end of their life span and thus possessed the least photosynthetic value to the plant.

I have considered so far only those consequences to plants of intraplant variation in leaf traits that stem from the within-plant level of selection by herbivores. Other mechanisms involve the among-plant level of host selection. One of these mechanisms applies particularly to the interaction between long-lived plants and specialist sedentary herbivores. Specialist phytophagous insects with limited dispersal capacity may become locally adapted to particular individuals of long-lived host species, forming local demes on individual plants that gradually become adapted to the host's characteristics over time. This "adaptive deme formation" (ADF) hypothesis was first formulated by Edmunds and Alstad (1978), and has since received unequal support from empirical tests (Van Zandt and Mopper 1998). It has recently been suggested that the high within-plant heterogeneity in the nutritional and defensive characteristics of leaves relative to among-tree variation may prevent tree-level specialization by monophagous, sedentary herbivores as envisaged by the ADF hypothesis, because from the perspective of these herbivores the largest source of variation in food quality and survival prospects will occur within rather than among plants (Gripenberg and Roslin 2005; Roslin et al. 2006; see Whitham 1983 for a forerunner to this idea). According to this reasoning, and provided that the within-plant component tends to be the main source of variance in leaf characteristics in woody plant populations (chapter 3), it is perhaps not surprising that only 5 out of 12 studies included in a meta-analysis by Van Zandt and Mopper (1998) supported the ADF hypothesis. The ADF hypothesis is mainly herbivore-centered, and I am not aware of any theoretical or empirical exploration of the lifetime consequences for long-lived individuals of becoming the objects of specialization by one or more demes of species-specific herbivores. Nevertheless, it seems intuitively reasonable to postulate that, everything else being equal, it will be better for a tree's fitness not to become the host of an adapted deme of its sedentary herbivore, and that in a hypothetical population of trees made up of individuals with and without associated demes, those lacking them will have an advantage. By preventing the formation of long-lasting, adapted demes of specialized herbivores on individual plants, therefore, extensive within-plant variation in foliage defensive properties may have important fitness consequences for long-lived plants, a circumstance emphasized long ago by Whitham (1981, 1983) in relation to the GMT.

Within-plant heterogeneity in leaf defense levels may act as a defensive property of the individual, reducing the overall damage inflicted by either specialist or generalist herbivores (Whitham 1981, 1983; Karban et al. 1997; Shelton 2000). Shelton designed a dynamic state variable model that compared the consequences of herbivory by generalists for two types of plants. One type produced a constant level of some defensive toxin in all its parts, whereas the other type produced the same average amount of toxin but distributed unevenly throughout its parts. In Shelton's model, a herbivore feeding on a constant-toxin plant would always consume plant tissue with a determinate amount of toxin, whereas a herbivore that feeds on a variable-toxin plant would consume tissue with a random toxin level selected from a distribution function. The two principal assumptions of the model were that the curve relating the food benefit for the herbivore to the toxin level was a decreasing, concave-down function (following Karban et al. 1997; see examples in table 8.2), and that herbivores can distinguish between constant-toxin and variable-toxin plants. In addition to investigating how herbivores respond to variability in plant defense, the model examined whether variability benefited the plant by increasing herbivore avoidance. The model predicted that, when herbivores had intermediate energy levels, they would become variance averse and reject variable-toxin plants, as long as there were constant-toxin plants in the environment and the benefit curve was sufficiently concave, that is, the magnitude of Jensen's effect was sufficiently large. Shelton's model thus implies that within-plant variability in leaf defenses per se, regardless of whether these are constitutive or induced, will favor herbivore avoidance of variable plants, which will have enhanced their defenses by increasing the uncertainty experienced by herbivores. This general conclusion is similar to that reached by Karban et al. (1997) for within-plant variation in induced defense levels, and by Shelton (2004) in a series of models examining the implications of variation in defense levels at several hierarchically nested scales of variation (among plants, among leaves within-plants, and among parts within leaves). The significance for plant fitness of very small-scale, spatially random variation in induced defenses has also been discussed in detail by Shelton (2005), who emphasized the need for studies examining the patterns and magnitude of within-plant chemical variation at scales that are important to herbivorous insects.

Continuous Variation: Flowers

Continuous within-plant variation in floral traits may induce three major, not necessarily mutually exclusive, types of effects: within-plant heterogeneity in per-flower pollination success, among-plant differences in overall pollination success, and cascading consequences arising from the influence of pollination heterogeneity on the within-plant variability of fruit and seed crops.

Direct Consequences: Within-Plant Heterogeneity in Pollination Success

In animal-pollinated species, the most immediate effect of within-plant variation in floral traits is an increase in the variance of the cumulative pollinator service received by individual flowers of the same plant, as a consequence of within-plant discrimination and choice exhibited by pollinators (chapter 8). This effect will cause inequalities in the total amount of pollen received and exported by flowers of the same plant over their life spans. Several lines of evidence support this view.

In the ordinary case of species whose flowers last for one or a few days (Ashman and Schoen 1996), each flower may be visited by pollinators multiple times over its life span. If individual flowers of the same plant differ in their attractiveness to pollinators by virtue of their different characteristics (e.g., corolla size, nectar production rate; Mitchell and Waser 1992; Conner and Rush 1996), then within-plant variation in floral traits may give rise to within-plant heterogeneity in the cumulative number of visits received by individual flowers over their life spans. For example, in experimental arrays consisting of conspecific flowers of different sizes, those with the largest corollas are generally visited more times by pollinators (Stanton and Preston 1988; Conner and Rush 1996). Pollination ecologists, however, have traditionally been more concerned with studying how many different flowers on the same plant are probed in succession by approaching pollinators, than with determining how many separate visitation events are received by individual flowers. Surprisingly few studies have quantified the visitation history of individual flowers exposed to natural pollination in the field, as stressed recently by Karron et al. (2006). Consequently, we know next to nothing about what the shape of the frequency distribution of total number of pollinator probes received per flower looks like for any species in its natural setting. The limited available evidence suggests that flowers on the same plant may differ widely in the cumulative pollinator service received over their life spans (Kato 1988; Kadmon et al. 1991; Kadmon 1992; Hodges 1995; Jones et al. 1998; Karron et al. 2006), and that such differences may give rise to heterogeneity in pollination success that is attributable, at least in part, to within-plant variation in floral characteristics.

Kato (1988) did a detailed field study of patterns of bumblebee visits to individual flowers of *Impatiens textori* plants growing close together in several small patches in a Japanese locality. Foragers actively avoided visiting flowers that had recently been visited by another individual, which might theoretically have equalized the distribution of visits among available flowers. Despite this, however, individual flowers at all sites differed greatly in the cumulative number of visits received from sunrise to sunset, which spanned nearly one order of magnitude (fig. 9.1). In two wild individuals of *Anchusa strigosa* systematically watched by Kadmon et al. (1991), considerable variation was also found among flowers of each plant in the visitation rate by anthophorid bees, the species' principal pollinators. In experimental arrays of *Antirrhinum majus* and *Mimulus ringens* exposed to natural visitation by pollinators in the field, Jones et al. (1998) and Karron et al. (2006), respectively, found that individual flowers differed widely in the total number of pollinator visits received.

Male and female pollination success of individual flowers generally increases with the number of pollinator probes (Mitchell and Waser 1992; Hodges 1995; Aigner 2005; Huang et al. 2006; Karron et al. 2006). Differences among flowers of the same plant in pollinator visitation are therefore expected to generate within-plant heterogeneity in cumulative pollen receipt, pollen export, or both. The few studies that have quantified within-plant variation in measurements of per-flower pollination success consistently support this expectation. Table 9.1 shows that, regard-

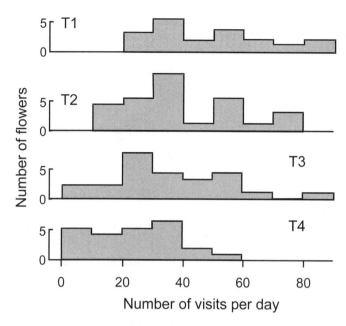

FIG. 9.1 Individual flowers simultaneously exposed to pollinators on the same plant may differ widely in the cumulative number of pollinator visits received over their life spans. Histograms depict the frequency distributions of the total number of bumblebee visits received from sunrise to sunset by individual flowers of *Impatiens textori* blooming at four small patches (T1–T4). Modified from Kato 1988.

less of plant species or pollinator type, within-plant variation in per-flower maternal pollination success (pollen grains per stigma, pollen tubes per style) is extensive, by far exceeding the among-plant component of variation. In the examples presented in table 9.1, $\%Var_{within}$ (defined in chapter 3) for measurements of maternal pollination success ranges between 60 and 100%, falling mostly in the range 80–90%. When populations of the same species were sampled over several reproductive episodes, overt predominance of the within-plant component of variance in maternal pollination success persisted over years (Herrera 2004). Partitions of variance in per-flower paternal pollination success have been quantified on even fewer occasions. The only figures available, obtained using the number of pollen grains exported per flower as a surrogate for paternal pollination success, likewise reveal a predominance of the within-plant component ($\%Var_{within} = 82\%$; table 9.1). Taken together, the figures shown in table 9.1 reveal that within-plant variation in maternal and paternal pollination success are as extensive as would be predicted if pollinator visitation

TABLE 9.1 **Proportion of population-level variance in measurements of maternal and paternal pollination success of individual flowers accounted for by differences among flowers of the same plant.**

Pollination success measurement	Species	Main pollinators	%Var$_{within}$[a]	Reference
Pollen grains received per stigma	*Daphne laureola*[b]	Small nitidulid beetles		C. Alonso unpubl.
	Hermaphrodites		85.5–98.4	
	Females		83.1–100	
	Lavandula latifolia	Honeybees, megachilids, nymphalid butterflies	78.9	C. M. Herrera unpubl.
Pollen tubes per style	*Ballota hirsuta*	Bumblebees, honeybees, large anthophorid bees	67.9	Herrera 2004
	Daphne laureola[b]	Small nitidulid beetles		C. Alonso unpubl.
	Hermaphrodites		72.6–100	
	Females		78.9–100	
	Helleborus foetidus	Bumblebees	62.0–73.3	Herrera 2002a
	Helleborus viridis	Bumblebees	60.0–85.0	C. M. Herrera and J. Guitián unpubl.
	Lavandula latifolia	Honeybees, megachilids, nymphalid butterflies	75.7–98.7	Herrera 2004
	Lindera benzoin	Small dipterans and hymenopterans	85.4	Niesenbaum 1994
	Marrubium supinum	Bumblebees, anthophorid bees	90.7	Herrera 2004
	Phlomis lychnitis	Bumblebees	84.9	Herrera 2004
	Rosmarinus officinalis	Bumblebees, honeybees, solitary bees	77.8	Herrera 2004
	Teucrium rotundifolium	Large anthophorid bees	83.2	Herrera 2004
Pollen grains exported per flower	*Lavandula latifolia*	Honeybees, megachilids, nymphalid butterflies	81.8	C. M. Herrera unpubl.

[a]Ranges of variation are given when data refer to several populations or several years for the same population. %Var$_{within}$ defined in chapter 3.
[b]A gynodioecious species; data are shown separately for female and hermaphrodite individuals.

varied as widely among flowers of the same plant as suggested by the results of Kato (1988) and other authors mentioned above.

A certain fraction of within-plant variance in per-flower pollination success is doubtless due to stochastic factors, rather than to the responses of pollinators to variation in floral characteristics. Dissecting the causal components of within-plant variance in per-flower pollination success has never been attempted, which is hardly surprising in view of the fact that intraplant variation in pollination success itself has been rarely acknowledged, not to mention quantified (Herrera 2004). An indirect approach to tackling this issue consists of looking for statistical relationships between floral features and pollination success across flowers of the same plant.

I looked for this type of relationship in a large data set containing morphological and pollination success data for more than 6,000 individual flowers from 300 *Lavandula latifolia* shrubs (this was the same data set analyzed by Herrera, Castellanos, and Medrano [2006]). Separate rank correlations were run for each plant between the number of pollen grains in the stigma and the length of the flower's upper corolla lip, a floral trait subject to selection by pollinators (Herrera, Castellanos, and Medrano 2006). Correlations were positive in 96% of the plants, a proportion evidently far beyond the 50% expected from the null hypothesis that floral features and pollinator service are unrelated across flowers of the same plant. Flowers of the same plant with longer, showier upper corolla lips thus predictably received more pollinator service than those with shorter lips. At least for one of the species listed in table 9.1, therefore, extensive within-plant variation in pollination success may safely be attributed to intraplant variability in floral features.

Direct Consequences: Among-Plant Differences in Pollination Success

In animal-pollinated species, within-plant variation in floral features may directly influence the overall pollination success or reproductive performance of individual plants through effects on the composition and diversity of the pollinator assemblage visiting each plant, the overall attractiveness of plants to pollinators, and the frequency of geitonogamous pollinations experienced by each plant, which presumably affects the extent of pollen carryover originating from the plant.

POLLINATOR COMPOSITION With relatively few exceptions, animal-pollinated plants are pollinated by taxonomically diverse arrays of pollinators (Herrera 1996; Waser et al. 1996). It has long been known that different pollinators differ in their floral preferences and exploit different subsets of the flowers simultaneously available at a given locality (Müller 1883). Because of its role in promoting reproductive isolation and enhancing floral diversification in angiosperms, the contrasting preferences of pollinators for floral features have been traditionally related to the morphological and functional diversity exhibited by different species (Fægri and van der Pijl 1966; Stebbins 1970). Nevertheless, pollinators are responsive to the much narrower variation in floral features that occur within species (chapter 8, and preceding section), and different pollinators may express contrasting preferences when facing intraspecific floral variation. This may lead to the "partitioning" of the available pollinator spectrum by conspecific flowers that differ in key characteristics. As discussed below, such partitioning may contribute to variation among individual plants in the diversity and species composition of pollinator assemblages (Herrera 1995a, 2005).

Partitioning of pollinators will be easier to detect in the case of clear-cut, discontinuous intraspecific floral variation involving, for instance, corolla color or sexual expression. Innumerable examples of differential responses by insect species to experimental alterations of the shape, size, or color of flowers may be found in Clements and Long's classic monograph on experimental pollination (1923). More recent examples are Kay (1976, 1982) and Stanton (1987), who reported differences among species of bees, flies, and butterflies in their responses to natural intraspecific variation in corolla color and sexual morph of a variety of flowers (*Cirsium arvense*, *Chrysanthemum coronarium*, *Raphanus raphanistrum*, *R. sativus*, *Silene dioica*, *Succisa pratensis*). Contrasting responses of pollinator species to continuous intraspecific variation in floral traits may be more cryptic and harder to detect, but I suspect that they would have been reported much more frequently if researchers had looked for them more often. Conner and Rush (1996), in a study of flower-size selectivity by the insect pollinators of *Raphanus raphanistrum*, found that syrphid flies consistently preferred flowers with larger corollas, while small bees either exhibited inconsistent preferences or were indifferent to corolla size variation. Ushimaru and Hyodo (2005) reported that syrphid flies, but not bumblebees, responded with differential visitation to experimental modifications in the orientation of the bilaterally symmetrical flowers of *Commelina communis*. In a detailed study of individual flower choice, Harder (1988) found that different bumblebee species, and even different castes of the same species, forage nonrandomly on flowers of the same species. In his study, pollinator partitioning by different-sized flowers was evidenced by positive correlations between the tongue lengths of foraging bees and the corolla depth of the flowers they visited. Likewise, Morse (1978) found a close direct correlation between proboscis length of individual *Bombus vagans* workers and the depth of the corolla of *Vicia cracca* florets they visited, despite the very narrow range of variation exhibited by the depth (6.1–6.6 mm). He interpreted the finding as denoting that individuals of varying size tended to forage on florets of differing corolla lengths (see Plowright and Plowright 1997 for similar results).

Pollinator partitioning should be expected to be most evident in species pollinated by taxonomically very diverse assemblages. This is exemplified by the summer-flowering shrub *Lavandula latifolia*, which is pollinated by dozens of species of bees, butterflies, and flies (Herrera 1987a, 2005). The different species of insects pollinating *L. latifolia* differ in the mean corolla size of visited flowers. Some species exhibit significant preferences for either the longer or shorter corollas available, while others do not forage selectively (fig. 9.2). As a consequence, individual *L. latifolia* flowers that differ in corolla length will tend to be visited by different sets of pollinator species. Differences among pollinators in mean corolla size of visited flowers were of similar magnitude at the population (fig. 9.2a) and within-plant (fig. 9.2b) levels. In *L. latifolia*, therefore, pollinator partitioning based on flower-size variation occurring at the population level is basically the outcome of a within-plant phenomenon, whereby each pollinator tends to visit slightly different subsets of the phenotypically distinct flowers borne on each plant. Partitioning of pollinators by different-sized *L. latifolia* flowers illustrated in figure 9.2 suggests the general prediction that, all else being equal, the broader the range of floral traits occurring within a plant, the greater the taxonomic diversity of the pollinators servicing it. Because pollinators generally differ in the quantity and quality of their pollinating services (e.g., pollen deposition and removal per

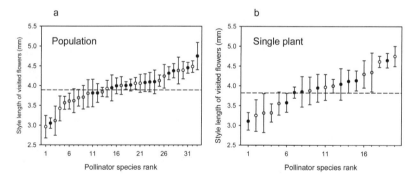

FIG. 9.2 Individual flowers of the Mediterranean shrub *Lavandula latifolia* that differ slightly in corolla size tend to be visited by different species of insect pollinators, as illustrated by differences among insects in the mean style length (a trait correlated with corolla length) of probed flowers. Corolla-size partitioning by pollinators at the population level (*a*, data from 7 plants and 33 species of pollinators) mainly reflects floral partitioning at the within-plant level (*b*, data for a representative plant visited by 20 species of pollinators). Vertical segments denote ±1 SE of the mean (*dot*), and horizontal *dashed lines* indicate the mean style length available to pollinators, as estimated from independent random samples of flowers. Each symbol corresponds to a different pollinator species: Diptera, *white dots*; Hymenoptera, *black dots*; Lepidoptera, *shaded dots*. Based on data from Herrera 1987a.

visit; Schemske and Horvitz 1984; Herrera 1987a, 1989b; Wilson and Thomson 1991), individual differences in levels of within-plant variation in floral traits may eventually translate into variation in per-flower pollen export and import average rates via effects on pollinator diversity and composition.

ATTRACTIVENESS TO POLLINATORS   Another mechanism whereby within-plant variation in floral features may influence the overall pollination success or reproductive performance of individual plants is related to the variance-sensitive behavior of pollinators. As argued in chapter 8, variance-sensitive pollinators are expected to respond to individual differences in within-plant variability of those floral features that can affect their foraging efficiency and time or energy budgets. Since the vast majority of pollinator species so far investigated are variance-averse (table 8.3), one would expect plants in a population with less variable flowers to be visited more often than plants with more variable flowers (Biernaskie et al. 2002). Everything else being equal, and if all pollinators tend to have concordant choices, an inverse relationship should arise across plants between average per-flower pollinator service received and within-plant variability in floral traits affecting the foraging efficiency of pollinators. Despite the attention received by the variance-averse behavior of pollinators in response to floral-reward variation, and the frequent suggestions that such behavior may have important ecological and evolutionary implications for the reproductive strategies of plants (Biernaskie et al. 2002; Shafir et al. 2003; Biernaskie and Cartar 2004), studies directly quantifying the impact of variance aversion on pollinator visitation to whole plants or inflorescences under natural field conditions are scarce. I am aware of only two studies comparing the number of pollinator visits to artificial inflorescences with equal means but different variances in nectar reward per flower. One of these did support the prediction that variable-reward inflorescences were visited on fewer occasions than constant-reward ones (Waser and McRobert 1998), while the other obtained inconsistent results (Biernaskie et al. 2002). No study seems to have looked for relationships between within-plant variation in per-flower nectar reward and pollinator visitation to whole plants in the field. Patterns of within-plant variability in nectar reward in the field, however, are complicated by too many factors in addition to intrinsic individual differences in nectar secretion (Zimmerman and Pyke 1986; Real and Rathcke 1988; and references in chapters 2 and 8) to reasonably expect consistent relations between nectar-reward variability and pollinator visitation to individual plants.

Although the causal connection between within-plant variability in floral features and discrimination among plants by pollinators has usually been examined in relation to variation in floral nectar production or instantaneous availability, it should not necessarily be restricted to this particularly inconstant floral feature. Dukas and Real (1993) showed that when bumblebees were exposed to a variable number of flower types having identical nectar rewards but differing in color, their learning rates and overall foraging efficiency declined as the number of offered floral types increased. Other studies have also suggested that switching between flower types may weaken or even erase the motor patterns learned by bees for handling flowers (review in Chittka et al. 1999). In addition, there are reasons to suspect that the same cognitive constraints responsible for the floral constancy of pollinators in interspecific contexts might also apply, under some circumstances, to the narrower range of variation faced by pollinators while foraging on flowers of the same species (Dukas 1998; Chittka et al. 1999). Laverty (1994) found that foraging bumblebees may be less constant to plant species with simple floral morphologies than to species with complex floral morphologies. In the latter case, but not in the former, he found evidence of increased handling times and error frequencies following switches between species. Under standardized experimental conditions, Gegear and Laverty (2001) were able to induce floral constancy by bee foragers by increasing the variation in several floral traits, including color, complexity, and size. The more flower types were offered, the more constant the bees became. Since perceptual systems have limited capacity to simultaneously process a large variety of sensory information, the nature of neural processing may bias organisms toward restricted host ranges in sensory-rich environments (Bernays and Wcislo 1994). More generally, the ability of animals to choose well may decline as the number of possible choices exceeds some threshold, and aversion to excessive choices seems to have evolved in some animal groups (Hutchinson 2005). Overall, these studies support the idea that within-plant variability in floral features other than amount of nectar might also sometimes influence among-plant discrimination by pollinators through effects on learning performance, handling efficiency, and ease of choice or willingness to choose. Floral features whose within-plant variability could induce foraging responses would be related, for example, to accessibility of the floral reward (e.g., corolla-tube depth or aperture width), nectar composition (e.g., amino acid concentration, proportions of different sugars), floral signals (e.g., scent, perianth color, degree of symmetry), or structural complexity (e.g., number and disposition of petals and nectaries). Within-plant variability in any of these structural, "nonmodifiable" morphological floral traits (in contrast to "modifiable" traits such as the amount of floral nectar) may create among-plant variation in pollinator visitation. If per-flower handling time and/or energetic returns are nonlinearly related to these morphological traits (as shown, e.g., by Harder [1988] for the relation between bumblebee net energy uptake and corolla depth), then Jensen's inequality predicts that pollinators should exhibit variance-averse responses to variation in such floral features, tending to reject plants with morphologically more variable flowers.

The intriguing possibility of an inverse relationship between morphological variability level and pollination success of individual plants does not seem to have been addressed previously. It is supported by a reanalysis of my data on floral morphology and pollination success of individual plants of *Lavandula latifolia* and *Viola cazorlensis*. The two species differ widely in floral morphology (tubular corolla vs. long, thin spur, respectively), nectar accessibility (6–7-mm corolla tube vs. ≈26-mm spur), and pollinator composition and diversity (dozens of insect species vs. a single hawk moth species; Herrera 1987a, 1993). Despite these differences, they are similar in exhibiting a significant negative relationship across individual plants between maternal pollination success (mean number of pollen tubes per flower, *L. latifolia*) or fruit set (percent flowers setting fruit, *V. cazorlensis*), and within-plant variability in accessibility to nectar reward, as determined by CV_{within} for corolla-tube length and spur length (fig. 9.3). In both species, therefore, increasing within-plant morphological variability of flowers predictably resulted in reduced reproductive performance of individual plants, which is consistent with the prediction derived from consideration of the variance-averse behavior of pollinators. In two previous studies on these species I had looked for effects of *mean* corolla-tube length and *mean* spur length on the reproductive success of individual plants, but found none (Herrera 1993; Herrera, Castellanos, and Medrano 2006). The relationships depicted in figure 9.3 remained unnoticed because I lacked at that time any a priori biological reason to look for them.

POLLINATION QUALITY   There is yet a third class of mechanisms that can set a causal link between within-plant variability in floral features and the reproductive performance of individual plants. In contrast to the two mechanisms examined to this point, which involve variations in the *amount* of pollinator service received by individual flowers or whole

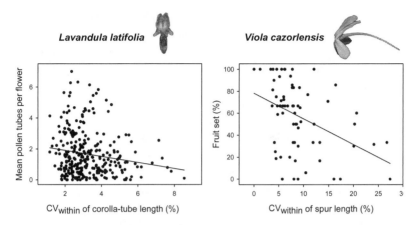

FIG. 9.3 Mean per-flower reproductive success, measured as mean number of pollen tubes per flower or percentage of flowers setting fruit, of *Lavandula latifolia* (*left*) and *Viola cazorlensis* (*right*) plants declines significantly with increasing within-plant variability in accessibility to floral nectar, measured as CV_{within} of corolla-tube length or spur length. Each symbol corresponds to a different plant, and lines are the least-squares fitted regressions ($N = 300$, $R^2 = 0.03$, $P < 0.01$ for *L. latifolia*; $N = 69$, $R^2 = 0.16$, $P < 0.001$ for *V. cazorlensis*). Based on reanalyses of data from Herrera 1993; Herrera, Castellanos, and Medrano 2006.

plants, this third class of direct effects implicates variations in the *quality* of the maternal (pollen receipt) or paternal (pollen dispersal) components of pollination success. Within-plant variation in certain floral traits may influence the probability of geitonogamous pollinations (involving transfer of self pollen between flowers of the same plant; for reviews see de Jong et al. 1993; Snow et al. 1996), as well as the dispersal patterns of pollen grains away from parent plants, including the mean and variance of dispersal distances. Some of these effects are expected to be mediated by variance-sensitive foraging responses of pollinators to within-plant variation, while others may occur independently of pollinator responses to such variation. Among the latter, morphological differences among flowers of the same plant that imply variations in the relative positions of stamens and stigmas may influence the genetic characteristics of the progeny as well as the patterns of self-pollen deposition and carryover.

Within-plant heterogeneity in nectar production per flower may reduce the level of geitonogamy through its effects on the foraging behavior of pollinators (Pleasants 1983; Hodges 1995; Boose 1997). Both theoretical models and empirical data of pollen dispersal and carryover show that, as more flowers are visited consecutively on the same plant by a pollinator, the proportion of self pollen deposited on stigmas increases rapidly (de Jong et al. 1993; Hodges 1995; Harder and Barrett 1996; Snow et al. 1996). Since plants bearing more variable flowers in amount or accessibility of nectar are expected to have, on average, fewer flowers visited consecutively per pollinator approach, they will consequently experience a reduction in the transfer of self pollen among their flowers.

Experimental support has been provided by studies showing that, when everything else is kept constant, pollinators generally probe fewer flowers per visit on artificial inflorescences whose flowers contain variable nectar rewards than on inflorescences where all flowers contain equal amounts of nectar (Pappers et al. 1999; Biernaskie et al. 2002; Hirabayashi et al. 2006). In a particularly thorough study, Hirabayashi et al. (2006) used a factorial design to investigate if the foraging response of bumblebees to within-inflorescence variance in nectar reward (constant, gradual, random) was contingent on concurrent variation in display size (large, small) and inflorescence type (panicle, umbel, raceme). They found that random nectar distribution significantly decreased the number of successive visits and staying time within inflorescences, regardless of the display size and architecture type. Biernaskie et al. (2002) reported similar results from an experiment using artificial inflorescences, where bumblebees and hummingbirds visited fewer flowers on inflorescences with variable nectar distribution than on those with constant distribution.

Geitonogamy has a number of reproductive costs (Snow et al. 1996). In self-incompatible species, geitonogamous pollinations may decrease mating success by clogging stigmas with incompatible pollen or, in less extreme cases, by diluting outcross pollen in stigmatic pollen loads. In self-compatible species, geitonogamy will affect the mating system by increasing the proportion of selfed progeny, and may affect progeny vigor, particularly in species with significant inbreeding depression (Snow et al. 1996). In both self-compatible and self-incompatible species, geitonogamy may also reduce siring success through pollen discounting, that is, a reduction in the number of pollen grains that would otherwise be carried away from a plant and have the potential to reach the stigmas of other conspecifics (Harder and Barrett 1996). Differences among individual plants in frequency of geitonogamous pollinations arising from pollinator responses to floral trait variability will therefore translate into variation in one or more components of the plants' reproductive success.

In addition to generating differences in the frequency of geitonogamy, within-plant variation in nectar reward or morphological floral traits may also influence pollen-dispersal patterns. Ott et al. (1985) found that individuals of *Bombus pennsylvanicus* foraging on an artificial flower system responded to increased variation in nectar reward by increasing both the mean and the variance of interplant flight distances, mainly as a consequence of increasing the frequency of long-distance flights. Assuming that the distribution of pollen-dispersal distances is reasonably congruent with the distribution of pollinator flight distances (an assumption that is not necessarily always true; Gaudeul and Till-Bottraud 2004), Ott et al.'s study has two important implications. (1) Pollinator responses to within-plant floral variation may increase the number of potential mates of each plant and, therefore, the overall genetic diversity of its progeny. (2) Neighboring individuals in plant populations are often more closely related genetically to one another than to distant individuals, and pollen performance increases with distance between mates (Robertson and Ulappa 2004; Glaettli et al. 2006). Increased average distance between mates arising from pollinator responses to within-plant floral variability may therefore enhance plant fecundity through reductions in the degree of relatedness between mates.

Hermaphroditic flowers on the same plant may differ in the length and placement of stamens and pistils, leading to variations in their positions relative to the corolla aperture (e.g., stigma exsertion) as well as in herkogamy, the separation between anthers and stigma of the same flower (Dronamraju 1961; Waser and Price 1984; Seburn et al. 1990; Barrett and Harder 1992). The mating system of individual plants, and hence the genetic constitution of their progeny, may be affected by this form of within-plant variation. Variations in the arrangement of male and female floral parts may influence patterns of pollen pickup and delivery (Barrett 2003; Kudo 2003), and within-plant variation in the placement of stamens and pistils may affect pollen carryover, as shown by Waser and Price's experimental study of hummingbird-pollinated *Ipomopsis aggregata* (1984). These authors manipulated the extent of variability among flowers in anther and stigma placement, and observed the effect on the transfer of pollen analogues (fluorescent dye) by hummingbird pollinators. The slopes of the regressions relating dye deposition to flower position in a visitation sequence steadily declined as variation in style length and anther position increased from treatment to treatment, revealing a positive effect of floral variability on pollen carryover and, thus, dispersal distance. Variation in pollen carryover may influence the genetic diversity of maternal progenies through effects on the number and genetic relatedness of mates, as noted above.

In self-compatible plants, variation among species or conspecific individuals in herkogamy levels tends to be inversely related to the frequency of self-pollination (reviews in Medrano et al. 2005; Parra-Tabla and Bullock 2005). Within-plant variation in herkogamy may also lead to some flowers producing selfed progeny more frequently than others, which will influence the mating system of individual plants. Broad within-plant variation in herkogamy has been reported by Dronamraju (1961) for *Bauhinia acuminata*, and Barrett and Harder (1992) for the tristylous, self-compatible *Eichhornia paniculata*. In individuals of the midstyled morph of this latter species, the position of the short stamens varied widely among flowers, which amounted to broad variation in herkogamy and in the frequency of automatic selfing. It may be predicted that, everything else being equal, the greater the variability in herkogamy among the flowers of the same plant, the more heterogeneous the mixture of selfed and outcrossed seeds produced, and the larger the fraction contributed by the assortment of seeds to overall genetic variation of the plant's progeny. Furthermore, as noted in chapter 5, coexistence of different-sized selfed and outcrossed seeds in a plant's seed crop will contribute to within-plant heterogeneity in seed size, with the potential consequences discussed later in this chapter.

Cascading Consequences: Heterogeneity of Fruits and Seeds

Because of the developmental continuity existing between flowers and fruits, within-plant variation in flower traits may represent a starting point for a cascade of consequences involving variation in fruit and seed traits (e.g., fruit size and seediness, seed size). The simplest, most obvious mechanism whereby within-plant heterogeneity in floral features may cascade into within-plant heterogeneity in fruits and seeds arises directly from the ontogenetic continuity between flowers and fruits, and it was already discussed in chapter 6. Floral traits such as ovary size or ovule number are often closely correlated with fruit size or fruit seediness, respectively, so that intraplant variation in the latter will arise as a consequence of floral variation.

Within-plant heterogeneity in pollinator service resulting from floral variation may enhance intraplant variation in fruit or seed traits. It has been repeatedly shown that the number of pollinator visits received by individual flowers, as well as the taxonomic identity and diversity of the floral visitors involved, may affect the size and seediness of the resulting fruits, as well as the average size of their enclosed seeds (Vander Kloet and Tosh 1984; Jarlan et al. 1997a, 1997b; Dogterom et al. 2000; Morandin et al. 2001; Karron et al. 2006; Roldán Serrano and Guerra-Sanz 2006; Kenta et al. 2007). Therefore, whenever intraplant variation in floral traits leads to heterogeneity among flowers in pollination history (e.g., number of pollinator visits, composition and diversity of visitors), it may also lead to within-plant heterogeneity in fruit and seed traits, which will have the consequences for individual plants discussed in the next two sections.

Continuous Variation: Fruits

Differences among conspecific plants in continuous within-plant variation in fruit traits like size or seediness may lead to among-plant differences in reproductive success through their influence on the spatial characteristics of postdispersal seed shadows, seed-dispersal success, or some combination of these. This holds for wind- and animal-dispersed species alike.

Wind-Dispersed Plants

In wind-dispersed species with multiovulate ovaries, intracrop variation in fruit seediness may influence the spatial characteristics of postdispersal seed shadows. Some tropical wind-dispersed legume trees produce winged legumes with a variable number of seeds (e.g., *Ateleia*, *Lonchocarpus*, *Platypodium*; Janzen 1978, 1982c; Augspurger 1986). In these plants, variation in seed number per fruit is correlated with characteristics expected to affect dispersal distance under field conditions, namely fruit mass, fruit area, wing loading (the mass/area ratio), and rate of descent in still air (Augspurger and Hogan 1983; Augspurger 1986). Consequently, fruits differing in seed number will tend to disperse at variable distances from the parent plant, with those containing more seeds falling nearest to the parent on average, and the fewest-seeded ones traveling farther (Augspurger and Hogan 1983; Augspurger 1986). In *Lonchocarpus penthaphyllus*, for example, 14, 20, and 32% of fruits with one, two, and three seeds, respectively, fall within 5 m of the base of the parent plant. Maximal dispersal distance was 66, 44, and 24 m from the parent for fruits with one, two, and three seeds, respectively (Augspurger and Hogan 1983). Intracrop variation in fruit seediness of anemochorous fruits will therefore affect the postdispersal distribution of seeds around parent plants (Augspurger and

Franson 1987, 1993). The arguments discussed in detail below in the section on the consequences of seed variation also apply here, and individual differences in the extent of within-plant variation in fruit seediness may lead to variation in seed-shadow characteristics that eventually result in differential seedling recruitment.

Animal-Dispersed Plants

In fleshy-fruited plants that rely on frugivorous vertebrates for seed dispersal, individual differences in the magnitude and characteristics of within-plant variation in fruit features may influence the reproductive success of plants through their effects on various components of seed dispersal. The components involved include both "departure-related" (e.g., fruit removal by dispersers, losses to damaging agents) and "arrival-related" (e.g., features of seed shadows; sensu Herrera 2002b).

Departure-related effects include the possible influence of within-plant variation in fruit features on the fraction of fruit crops taken away by frugivores that disperse seeds legitimately, that is, without damaging them in the process. Almost without exception, fruit crops of vertebrate-dispersed plants are not completely depleted by seed dispersers, and individuals differ in the proportion of fruits removed (Howe and Vande Kerckhove 1980; Courtney and Manzur 1985; Jordano 1987; Herrera et al. 1994; Laska and Stiles 1994; Herrera 1995b). If, as ordinarily assumed, differences in fruit-removal rates translate into differential reproductive success, then any possible effect of within-plant variation in fruit traits on the likelihood of fruits being taken away by dispersers will eventually give rise to fitness differences among plants. This may occur, for example, if the variance-sensitive behavior of frugivores leads to discrimination and choice among plants in response to differences in fruit variability.

Although a considerable number of frugivory studies have looked for correlates of individual variation in fruit-removal success, plants have invariably been characterized by their mean fruit traits, and not a single investigation seems to have ever considered the possibility that levels of within-plant variation in fruit traits might partly account for individual differences in dispersal success. I suspect that such effects of fruit variability on dispersal success have frequently gone unnoticed in studies that, while finding significant relationships between fruit-removal success and plant means for some fruit traits, simply did not look for similar relationships with plant variances. That very real possibility is exemplified by one of my own investigations. In a study of seed dispersal by frugivorous passerine birds of the fleshy-fruited Mediterranean tree *Phillyrea latifolia*, a direct relationship between percent fruit removal and mean fruit size of individual plants was reported (Herrera et al. 1994). Nevertheless, as shown in detail in the next chapter, the dispersal success of individual *P. latifolia* plants in our study not only depended on mean fruit size, but also was inversely related to within-plant variance in fruit size, as expected if frugivorous birds were variance-averse and tended to avoid individuals with the most variable fruits. As with the examples of *Lavandula latifolia* and *Viola cazorlensis* reported above, the relationship between variability and dispersal success remained hidden in our data set just because we did not look for it.

Departure-related effects of within-plant variation in fruit traits may also involve mechanisms other than its direct effects on the behavior and preferences of seed dispersers. Predispersal fruit and seed predators may also respond to within-plant variability in fruit traits and, in so doing, may influence plant reproductive success directly through reductions in fecundity or indirectly through impaired dispersal success, since vertebrates tend to avoid fruits that are infected or damaged by invertebrate pulp and seed predators (Sallabanks and Courtney 1992). The first of these mechanisms is illustrated by the response of the bruchid beetle *Amblycerus cistelinus* to individual differences in the levels of within-plant variability in fruit size occurring in a population of its host tree *Guazuma ulmifolia*. This seed predator exhibits within-crop fruit selectivity, with longer fruits in a crop being infected proportionally more often than shorter ones (fig. 8.2). As discussed in chapter 8, the costs associated with such selectivity should give rise to ovipositing beetles exhibiting discrimination and variance-averse behavior toward trees differing in the variability of fruit traits influencing oviposition. This is clearly supported by the inverse relationship between percent fruit predation rate and CV_{within} for fruit length found in a Costa Rican population of *G. ulmifolia*. As depicted in figure 9.4, the greater the within-plant variability in fruit length of a tree, the lower the proportion of its fruit crop that was infected by *A. cistelinus* larvae and, therefore, the greater its relative reproductive success.

A second, indirect mechanism whereby within-crop fruit variability may influence the departure-related component of an individual plant's seed-dispersal success is through its effects on fruit predators. Seed or pulp damage by invertebrates generally renders fruits objectionable to legitimate seed dispersers, either because the pulp becomes relatively

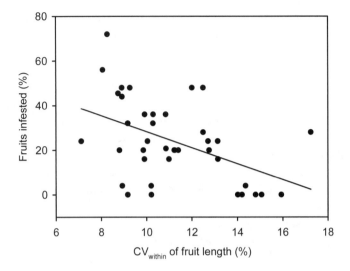

FIG. 9.4 In a Costa Rican population of the vertebrate-dispersed tree *Guazuma ulmifolia*, the proportion of individual fruit crops infested by the bruchid beetle *Amblycerus cistelinus* was inversely related to the magnitude of within-plant variability in fruit length (CV_{within}), a fruit trait that influences within-plant fruit selectivity by these beetles (fig. 8.2). Each symbol corresponds to a different tree, and the line is the least-squares fitted regression ($N = 40$, $R^2 = 0.22$, $P = 0.002$). C. M. Herrera, unpublished data.

unrewarding or because substances produced by the larvae make the fruit unpalatable (Manzur and Courtney 1984; Sallabanks and Courtney 1992; Traveset et al. 1995; García et al. 1999). In these cases, correlations between within-crop fruit variability and crop infestation rate similar to that described above for *G. ulmifolia* will give rise to individual variation in the attractiveness of plants to dispersers. At an English locality, Manzur and Courtney (1984) found that the bird-dispersed fruits of *Crataegus monogyna* containing seeds infested by larvae of the microlepidopteran *Blastodacna hellerella* were actively rejected by blackbirds (*Turdus merula*), the plant's main dispersal agent, possibly because of their bitter taste. This decreased the foraging efficiency of birds in heavily infested plants, where the increase in frequency of fruit rejections slowed down the removal rate of good undamaged fruits and eventually prevented dispersal of many healthy, uninfested fruits away from the parent plant. As a consequence, the proportion of good seeds dispersed per plant declined with increasing crop infestation rate. If observed differences between *C. monogyna* plants in fruit infestation rate by larvae were, as in *G. ulmifolia*, at least partly a consequence of individual differences in fruit variability, then Manzur and Courtney's results would illustrate another mechanism whereby within-plant variation in fruit traits may influence seed-dispersal success through effects on invertebrate fruit damage rates.

Within-plant variation in fruit traits may also affect the reproductive success of plants through its influence on arrival-related aspects of seed dispersal. Several different mechanisms may link within-plant variation in fruit traits with features of postdispersal seed shadows that, although unexplored, might in the long run be even more consequential to plants than variation in fecundity caused by departure-related effects. Different species of seed-dispersing frugivorous birds and mammals tend to select fruits that differ in a variety of features, including those that ordinarily vary widely within individual crops such as size or seediness. Such differential preferences have usually been documented by comparing the consumer spectra of plant species with contrasting fruit features (Herrera 1984c; Pratt and Stiles 1985; Wheelwright 1985; Lambert 1989; Kitamura et al. 2002), but frugivores may also express their preferences when confronted with intraspecific or intracrop variation in fruit traits. Consequently, different fractions of individual fruit crops will be consumed by different species of frugivores characterized by contrasting fruit preferences. This idea was originally suggested by Janzen (1977a, 720) as a possible answer to the question "What is the meaning of variation in fresh ripe fruit weight within a tree's crop?" It has received surprisingly little attention in the voluminous literature on the dispersal ecology of fleshy-fruited plants. The scarce evidence available suggests that consumption of different segments of fruit crops by different frugivores would have frequently been revealed if researchers had actually looked for it.

Bonaccorso (1975) noted that fruits of different weights from the same *Ficus insipida* tree were taken by different species of bats. According to Janzen (1977a), this should generate a quite different seed shadow than if all the figs were of the same weight and thereby taken by only one species of bat. In Israel, Izhaki et al. (1991) found that coexisting species of frugivorous passerines foraged for fruits at different heights in the canopy of fruiting plants of the same species. Since fruits produced at different heights within the crown of the same plant often differ in their characteristics (chapter 4), the results of Izhaki et al. suggest that different portions of local fruit trait distributions were probably being dispersed by different species of frugivores. This possibility is confirmed by other studies of Mediterranean bird-dispersed plants. *Rubus ulmifolius* is characterized by broad variation in seed size, both within and among plants, which is

largely concomitant with variation in size and seediness of fruits. In southern Spain, Jordano (1984) found that the frequency distribution of the size
of *R. ulmifolius* seeds found in bird feces differed significantly among species of seed dispersers. Similar size-based selection of conspecific fruits by
coexisting species of frugivorous birds also occurs in cultivated *Olea europaea*, as shown experimentally by Rey and Gutiérrez (1997). The main
seed dispersers of *Phillyrea latifolia* differ significantly in the size distribution of dispersed seeds, with each species characterized by its own combination of central tendency and relative variability of seed size (fig. 9.5).
Since fruit and seed size are closely correlated in *P. latifolia*, these interspecific differences reveal that different segments of local fruit and seed-
size distributions are taken by slightly different assemblages of avian frugivores. Given that a substantial portion of population-wide variance in
fruit traits usually occurs within the crops of individual plants (chapter 3)
and that conspecific plants generally differ in the extent of within-plant
variability in fruit traits (chapter 7), partitioning of the available spectrum
of seed-dispersing frugivores by fruits of different characteristics may
partly account for among-plant variation in diversity and species compo-

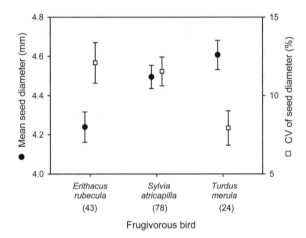

FIG. 9.5 Mean size (average cross diameter, *filled circles*) and relative size variability (CV of
cross diameter, *open squares*) of seeds recovered from feces or regurgitations of the three
main avian seed dispersers of the fleshy-fruited tree *Phillyrea latifolia* at a site in southern
Spain (Herrera et al. 1994). This species produces single-seeded drupes, and fruit and seed
size are closely correlated; hence these data reveal that the three species of frugivores disperse distinct subsets of the local fruit and seed size distributions. Vertical segments denote ±1
SE of estimates. Sample sizes are in parentheses below species names. C. M. Herrera, unpublished data.

sition of disperser assemblages frequently observed in the field (Herrera and Jordano 1981; Englund 1993; Traveset 1994). Such variation, in turn, may have important implications for plants, as shown below.

The larger the variance in fruit traits occurring within a plant, the greater the taxonomic diversity of fruit consumers that can be expected to visit it regularly. Disperser species generally differ in quality-related aspects of their seed-dispersal services; hence diverse features of the post-dispersal seed shadow of each individual plant depend closely on the species composition of its disperser assemblage. Different animals tend to disperse seeds to microhabitats of different quality in terms of survival prospects for the seeds and the resulting seedlings (Herrera and Jordano 1981; Reid 1989; Izhaki et al. 1991; Wenny and Levey 1998). In addition, disperser species differ in the mean and variance of the time elapsed from seed ingestion to defecation or regurgitation, which will generate different distributions of dispersal distances and, consequently, contrasting spatial distributions of seeds (Janzen 1982a; Thomas et al. 1988; Clark et al. 2005; Spiegel and Nathan 2007). Furthermore, dispersers differ in their effects on the length of dormancy of ingested seeds, which leads to species-specific temporal distributions of seedling emergences (Izhaki and Safriel 1990; Traveset et al. 2001). All these effects, acting in concert, will lead to an increase in the spread and evenness of the progeny distribution of a given individual over *both* time and space as more disperser species, each of which treats the seed in its gut differently and transports seeds to different microsites and different distances, ingest its fruits (Izhaki and Safriel 1990). Since spatiotemporal variation in the distribution of seeds may influence plant fitness through spreading the risks encountered during germination, individual differences in the magnitude of fruit trait variation may translate into fitness differences via their effects on the composition and diversity of the animals that disperse their seeds.

Continuous Variation: Seeds

Interpretations, Expectations, and Models

Differences between individual plants in the magnitude and characteristics of continuous within-plant variation in the size, germinating behavior, and dispersal capacity of their seeds may translate into individual differences in reproductive success. At a time when seed size was generally treated as a quintessentially species-specific trait, Janzen (1977b,

1977c, 1978, 1982b) provided overwhelming evidence of extensive variation in seed mass within individual seed crops of some tropical plants (see
also chapter 3). He conferred great importance on those findings, because
"the very large variation in seed weights within the crop is potentially of
great ecological significance to the seedlings" (Janzen 1977b, 349). This
expectation was soon confirmed by Howe and Richter's demonstration
(1982) that variation in seed size occurring within single crops of the tropical tree *Virola surinamensis* was large enough to promote broad differences in the size and vigor of the seedlings produced. Janzen suggested
that, contrary to theoretical models predicting that plants should produce
similar seeds of uniform optimal size (e.g., Smith and Fretwell 1974), there
should be nothing like a species-specific optimal seed size. Janzen (1977a,
719) said, "There is no optimal seed." He proposed instead that it is the
shape of the within-plant distribution of the mass of single seeds that is
actually most important in relation to a parent plant's fitness prospects.
Associated with this suggestion was the idea that individual differences
in the magnitude and features of within-plant variation in seed size would
translate into variation in the number of successfully recruited offspring
per mother plant, via effects on postdispersal seed shadows (e.g., mean
dispersal distance, seed density heterogeneity, maximum reach) and probability of escape from seed predators.

Similar arguments have been advanced in relation to continuous
within-plant variation in other seed traits, such as germinating behavior
(Silvertown 1984) and dispersal capacity (Augspurger and Franson 1993).
If seedlings emerging at different times of year or on different years experience differential success, then within-plant variation in the germination time of seeds can eventually affect the number of successfully established seedlings produced by individual plants and hence their maternal
fitness. Elaborating on this intuitively appealing assumption, the early
theoretical models of Cohen (1966) predicted that, provided there is sufficient uncertainty as regards the establishment success of seeds germi-
nated at different times, plant individuals producing crops of seeds with
diverse germination periods will spread the risk and experience a fitness
advantage over plants producing seeds with identical germination periods (a diversified bet-hedging strategy; Philippi and Seger 1989). Subsequent more-elaborate models showed also that the sign and magnitude
of the fitness effects of variation in germination time depend not only on
the unpredictability of environmental conditions, but also on the shape of
the fitness curve for emergence time in a given environment (Lacey et al.

1983). Regardless of detailed aspects of their assumptions and predictions, what these and other (e.g., Philippi 1993) theoretical models demonstrate is that within-plant variation in seed germination time is bound to have important consequences for the fitness of maternal parent plants.

In wind-dispersed species, seed settling or "terminal" velocity, an inverse surrogate for dispersal ability (slower settling velocities correspond to greater dispersal ability), may vary extensively among seeds of the same crop as a consequence of continuous variation in the proportion between seed mass and the size of the ancillary plume (e.g., pappuses of the Compositae) or wing (e.g., species with single- or multiseeded samaras; Augspurger and Hogan 1983; Andersen 1992; Sipe and Linnerooth 1995). Augspurger and Franson (1993) used a computer simulation to quantify how intracrop variation in wing loading of seeds may affect their postdispersal distribution around parent plants. Increasing within-plant variation in wing loading increased the area and uniformity of density of the postdispersal seed distribution without changing mean dispersal distance. Decreasing mean and variance in wing loading increased the area and uniformity of the seed distribution, as well as the mean dispersal distance. Results of the simulation were similar regardless of whether the differences in wing loading arose by altering seed mass or wing area. A similar relationship linking within-plant variation in seed size with features of postdispersal seed shadows was also postulated by Janzen (1977b) for the floating diaspores of the water-dispersed tropical vine *Mucuna andreana*.

Empirical Evidence

Studies examining continuous intraspecific seed variation have frequently echoed or revisited Janzen's insightful interpretations (1977a, 1977b) of within-plant variability in seed features (e.g., Michaels et al. 1988), but I am not aware of any attempt to test under natural conditions the predicted relationship between within-plant variation in seed traits and individual plant fitness. This dearth of empirical evidence is hardly surprising, given the nearly insurmountable practical difficulties inherent in accurately tracking the postdispersal seed shadows and eventual dispersal success of individual parent plants under natural field conditions (Godoy and Jordano 2001; Herrera 2002b). This greatly limits the possibilities of relating the traits of parent plants, including levels of within-plant variation in seed features, to the spatial characteristics of their seed shadow and their realized dispersal success. The few available field experiments and observational studies, however, provide circumstantial evidence that within-plant variation in seed features influences plant fitness in ways that are consistent with the insights mentioned previously.

Using artificial fruits closely mimicking the winged seeds of the tropical wind-dispersed tree *Tachigalia versicolor*, Augspurger and Franson (1987) showed that variation in wing loading (achieved by manipulating seed mass and wing area) effectively translated into variation in mean seed-dispersal distance, and in the total area and average seed density of the postdispersal seed shadow, all of which might eventually translate into differential numbers of recruited seedlings. Field studies by Biere (1991b) on *Lychnis flos-cuculi* suggest that a high variance in germinating time per se may effectively contribute to increased survivorship of seedling progenies, and that individual differences in within-season variance in germination time could account for survival differences among maternal genotypes (Biere 1991a). Possibly the most direct demonstration that individual differences in variability of seed germination time may translate into variation in realized recruitment is provided by the results of a long-term field experiment involving the sowing of maternal seed progenies of *Lavandula latifolia* in the field, and subsequently following the emergence and survival of seedlings over several years (Herrera 2000). Individual mother plants differed very little in the mean germination date of their seeds, but they differed nearly twofold in the variability of these dates, as described by the CV_{within} of seedling emergence date. The number of offspring of each plant surviving past the first summer (the key demographic bottleneck in this species) was unrelated to the plant's mean germination date, but it was directly related to the plant's variability in emergence date (fig. 9.6). As predicted by theory, mother plants producing the most variable seed crops with respect to emergence time eventually contributed most seedlings to the population.

The preceding examples illustrating that within-plant variability in seed traits may in itself influence a plant's fitness refer to situations in which the causal factors are unrelated to interactions with animals. Diverse mechanisms related to interactions with animals may also be responsible for within-plant variation in seed features eventually translating into differential reproductive success. In some cases animals play an exclusively passive role, while in others they play an active role based on discrimination and choice among plants.

In plants whose seeds are dispersed endozoochorously by vertebrates,

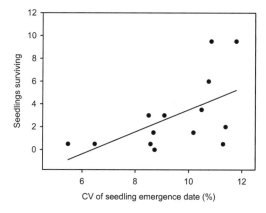

FIG. 9.6 An experiment involving the sowing of seed progenies from individual *Lavandula latifolia* shrubs in the original habitat reveals that the number of seedlings surviving past the first summer after emergence (vertical axis; mean number of seedlings alive per plot one year after emergence, out of 100 seeds sown per plot) increased with increasing within-plant heterogeneity in the germination date of seeds (horizontal axis; mean coefficient of variation of seedling emergence date). Each symbol corresponds to a different plant ($N = 14$). Based on a reanalysis of data from Herrera 2000.

the guts of these animals may play the role of passive seed-size sorters and, in so doing, may connect within-plant variation in seed size with some characteristics of postdispersal seed shadows, a possibility suggested by Janzen (1977a, 1977c). The passage rate through the digestive tract of vertebrates of small hard objects is influenced by their weight and specific gravity. In interspecific contexts, relationships between seed size or specific gravity, on one hand, and mean gut retention time or temporal spread of seed defecations, on the other, have been shown for tropical fleshy-fruited plants fed upon by frugivorous birds (Levey 1986; Holbrook and Smith 2000), and for legume and grass seeds placed directly into the rumens of cattle (Gardener et al. 1993). Analogous relationships are expected to hold in intraspecific contexts, too. Because of their presumably size-dependent gut-passage rates, therefore, the different-sized seeds eaten by a vertebrate in a single meal from an individual seed crop will tend to be passed in separate defecations and at different places. Consequently, variations in seed size within a plant are likely to exert some influence on the spatial pattern of the seed shadows generated by animal dispersal agents, and hence eventually on its reproductive success. This effect should be particularly important in plants with seeds dispersed by large birds and mammals characterized by extended gut-passage times. It should not be restricted, however, to species conventionally classified as endozoochorous. Grazing mammals are important dispersal vectors for species conventionally classified as having "unspecialized" dispersal, a nonnegligible portion of whose seeds survive passage through the digestive tract of these mammals (Mouissie et al. 2005; Bruun and Poschlod 2006).

Within-plant variation in seed traits may lead to variation in the seed shadows also in species exhibiting adaptations for epizoochory. By means of a series of standardized experimental comparisons, Tackenberg et al. (2006) showed that the "retention potential" (the proportion of diaspores still attached to the animal coat after a certain time period) of epizoochorous seeds on the coat of cattle and sheep was correlated across species with variations in the size, length, width, and volume of diaspores. It is reasonable to predict that variation within a plant in diaspore traits influencing adhesion to and persistence on mammalian fur coats will also influence the postdispersal seed shadows.

Seed-eating animals may also play an active role in connecting within-plant variation in seed traits and plant reproductive success. As argued in chapter 8, the active response of individual foragers to plant differences in extent and characteristics of within-plant variation in seed traits may lead to discrimination among plants and differences among the latter in reproductive success. Studies of parthenocarpic species, where individual seed crops contain both filled and empty seeds inside otherwise normal fruits, provide some of the clearest evidence to date that within-plant variability in seed traits is often consequential for the reproductive success of individual plants through its effect on the foraging behavior of animals. From the viewpoint of predispersal seed predators, individual plants of parthenocarpic species represent extreme cases of highly variable food patches, since their crops are made up of variable mixtures of potentially useful and downright useless food items (filled and empty seeds). Because of the foraging and handling costs associated with exploiting this type of food, seed predators should generally prefer less variable crops, with low proportions of parthenocarpic fruits with empty seeds.

Most studies that have so far examined the relationship between seed predation rate and frequency of empty seeds found support for this prediction (but see García et al. 2000). In *Pistacia terebinthus*, a Mediterranean treelet with plants producing variable proportions of parthenocarpic fruit, the proportion of sound seeds escaping damage by chalcidoid wasps increases across individuals as the proportion of empty seeds increases

(Traveset 1993). In the congeneric *Pistacia lentiscus*, the percentage of empty seeds is negatively correlated with seed predation by chalcidoid wasps and granivorous birds (Verdú and García-Fayos 1998, 2001). In *Juniperus osteosperma*, another species producing fruits containing empty seeds, trees attacked less frequently by the seed-eating bird *Parus inornatus* had higher proportions of empty seeds (Fuentes and Schupp 1998). In *Yucca schottii*, a species that produces variable proportions of infertile and fertile seeds in developing fruits, a close inverse correlation was found between the proportion of infertile seeds in fruits and the number of seeds eaten by lepidopteran larvae, which tended to leave fruits more frequently when encountering infertile seeds (Ziv and Bronstein 1996). These studies support the view that within-plant variation in the food value of seeds for seed predators, occurring when empty and sound seeds coexist on the same crop, will generally induce variation among plants in the impact of predispersal seed predators. The greater the within-plant variability in the food value of seeds, the higher the likelihood of filled seeds escaping predators.

Similar relationships between the impact of predispersal seed predators on individual plants and the magnitude of their within-plant variation in seed size should also occur in "ordinary," that is, nonparthenocarpic species exhibiting less extreme, continuous seed-size variation, but no studies seem to have explicitly looked for them so far. For the reasons given in chapter 8, relationships of this sort should be particularly frequent in the case of small, very selective invertebrate predispersal seed predators whose larvae spend their whole lifetime confined within single seeds. This suggestion is supported by data on seed-size variability and predation rate by chalcidoid wasps presented by Chung and Waller (1986) for clones of *Rhus glabra*. A reanalysis of their figures reveals a marginally significant, inverse correlation across clones between CV_{within} for seed length and percent seeds consumed by wasps ($r_s = -0.533$, $P = 0.13$). Differences among clones in seed-size variability may thus effectively translate into differential realized fecundity because of differential responses of seed predators.

Overview and Synthesis

The preceding sections have discussed a variety of mechanisms through which within-plant variability in features of reiterated structures may come to influence individual plant fitness. Whenever one or more of these mechanisms apply, individual differences in variability levels may eventually translate into fitness differences. Depending on the type of mechanism, the organ involved, and the biological particularities of each system, the effect on fitness of within-plant variability may be predominantly positive, predominantly negative, or variable, as summarized in table 9.2. Mechanisms typically leading to positive effects on fitness involve either some improvement in the exploitation of patchy, unpredictable, or heterogeneous environments, either biotic or abiotic, or a reduction of the detrimental effects caused by antagonistic organisms such as herbivores and fruit and seed predators. Conversely, negative effects will generally arise as a consequence of within-plant variability bringing about a reduction in the strength or frequency of interactions with mutualistic counterparts, like pollinators or animal seed-dispersal agents. The fitness consequences of broadening and partitioning the available spectrum of mutualistic counterparts with which each individual plant interacts (pollinators, dispersers) may be either positive or negative, depending on whether such expansion of interacting biotic agents incorporates taxa that are, respec-

TABLE 9.2 **Summary of mechanisms by means of which within-plant variability in characteristics of reiterated structures may have effects on individual plant fitness.**

| Organ | Most likely effect of within-plant variability on individual fitness | | |
	Positive	Negative	Variable
Leaves	Optimal exploitation of spatial and temporal abiotic environmental gradients		
	Reduction of herbivore damage and/or impact of damage		
Flowers	Reproductive assurance (dimorphic cleistogamy)	Impairing plant attractiveness to pollinators	Broadening the spectrum of pollinators
	Reduction of geitonogamous pollinations and enhancing pollen dispersal		Partitioning the spectrum of pollinators
			Heterogeneity in pollination success
Fruits and seeds	Coping with environmental uncertainty in biotic and abiotic factors ("diversified bet hedging")	Impairing plant attractiveness to mutualistic animal dispersers	Partitioning the spectrum of dispersal agents
	Broadening the postdispersal seed shadow in time and space		
	Reduction of damage by fruit and seed predators		

tively, above or below average in the quality of their services from the perspective of the plants. Similarly variable will be the sign of the fitness consequences of within-plant heterogeneity in pollination success, which will depend on how the shape of the within-plant frequency distribution of per-flower pollination service is affected by variability. The effect on fitness may be positive if heterogeneity implies a longer reach of the right tail of the distribution, with a consequent decline in the proportion of flowers failing to reach some minimum pollination service threshold, but negative if heterogeneity lengthens the left tail of the distribution much beyond such a threshold.

The significance of discrete, discontinuous within-plant variability as an improved means of exploiting patchiness in the physical environment has previously been emphasized in relation to different functional aspects and ecological contexts (Ray 1987; Lloyd 1984; Wells and Pigliucci 2000; Imbert 2002). Recent reviews summarized in the preceding sections, however, make it clear that this generally accepted idea is rather imperfectly backed by empirical evidence, as denoted especially by the striking scarcity of studies supporting the mixed-superiority condition (individuals producing a mixture of variants are superior to those producing one of the variants). One comes to the conclusion that the fitness effects of within-plant variation in organ traits is not much better established to date for discrete than for continuous variation. I have shown above that "division of labor" (Ray 1987), exploitation of environmental patchiness, or partitioning of environmental gradients, whichever expression we choose to call it, need not be restricted either to discontinuous variation, as traditionally implied in the literature, or to the partitioning of the physical, abiotic environment. The biotic environment represented by pollinators or seed dispersers is also susceptible to partitioning, or "division of labor," by structures borne on the same plant performing the same function but differing slightly in their phenotypic characteristics, and there are reasons to suspect that these effects may prove to be considerably more common than hitherto recognized. As illustrated by the reanalyses of my own data on *Viola*, *Lavandula*, and *Phillyrea*, the opportunity to detect the phenomenon is missed if one focuses exclusively on the relationships between individual trait means and measurements of reproductive success, as is commonly done. One wonders how many other researchers would come across significant but previously overlooked effects of within-plant variance on measurements of reproductive success if they subjected their data to similar reanalyses.

Having documented in this chapter the existence of mechanisms that may link within-plant variability in organ traits with individual fitness, the next logical step is to consider the frequency of such mechanisms in nature, the quantitative importance of their effects, and the likelihood that they become evolutionarily relevant. These issues are addressed in the next chapter.

Evolutionary Implications of Within-Plant Variability in Organ Traits

Subindividual multiplicity of organs can affect the evolutionary trajectory of organ traits by setting upper limits on responses to selection, opening the possibility of selection by animals on plant-level variability, and conditioning the size of realized phenotypic space at the individual and population levels.

The two preceding chapters have considered the ramifying ecological consequences of within-plant variation in characteristics of reiterated organs. Within-plant variation affects the animals that interact with plant organs for food or breeding sites, and can affect the plants themselves as a result of the foraging responses of the animals. Ecological phenomena associated with within-plant variation, particularly those associated with the discriminating behavior of animals toward individual plants, can also have evolutionary repercussions. For example, discrimination by feeding or ovipositing animals among conspecific plants based on their different levels of within-plant trait variation can engender phenotypic selection on levels of within-plant phenotypic variance. This phenotypic selection on variability levels can in turn have direct evolutionary effects, such as contributing to maintain adaptive levels of subindividual variation. In this chapter I present an overview of these and other evolutionary implications that can be inferred from the existence of within-plant variation in organ traits and the associated ecological phenomena mediated by interactions with animals that have been described in chapters 8 and 9. I do not separately consider the evolutionary implications of within-plant variation that are most directly associated with the influence of abiotic elements of the plant environment (e.g., adaptive value of continuous or discontinuous variation in leaf form and photosynthetic characteristics), as these have been reviewed on a number of occasions (Cook and Johnson 1968; Lloyd 1984; Sultan 1995; Winn 1996a, 1999a; Wells and Pigliucci 2000; Imbert 2002). Nevertheless, the core of the arguments below, although developed specifically in the context of the interaction between animals and subindividually variable plants, can easily be extrapolated to situations where selection on within-plant variation is exerted by components of the abiotic environment (e.g., vertical light gradients, heterogeneity of soil properties).

Evolutionary implications of within-plant variation fall into two main categories concerning the adaptive responses of organ-level traits (e.g., fruit diameter or corolla length) and plant-level features (e.g., level of within-plant variability in fruit size or corolla length). Adaptive responses occurring at the organ and whole-plant levels may be linked or may take place independently of each other, as discussed later. At the organ trait level, on one hand, because of its contribution to environmental phenotypic variance, within-plant variation will generally set limits on adaptive responses of organ traits to selection. On the other hand, within-plant variation and its suite of associated ecological phenomena may affect the evolution of whole-plant features, such as levels of phenotypic plasticity or developmental instability, and the shape of reaction norms. I consider these several implications below.

Limits on the Response to Selection

All kinds of phytophagous animals potentially discriminate among conspecific plants on the basis of phenotypic differences in features of leaves, flowers, fruits, or seeds, as discussed in chapter 8. Consequently, some individual plants experience more frequent, temporally more consistent, stronger, or more consequential interactions with their animal associates than others. If these individual differences are related to variation in average organ features, and if such variation has some genetic basis, then there is no question that selectivity by animals can become a crucial selective mechanism driving the adaptive evolution of organ traits at the population level. This reasoning prevails in recent treatments of the evolutionary ecology of plant-animal interactions (several authors in Herrera and Pellmyr 2002), and is supported by many studies showing that individual variation in average leaf, flower, fruit, or seed traits are predictably related to differences in fitness-related measurements, as shown in the next section. What has been much less frequently acknowledged is that, regardless of the particularities of the plant-animal interaction involved and of the nature and strength of the selective forces at work, within-plant variation in organ traits will generally act as a constraint on responses to selection by animals on organ traits.

The "depressing" effect of within-plant variation on adaptive changes of reiterated organ traits can be deduced from simple quantitative genetics considerations that largely follow from ideas in texts such as Falconer and MacKay (1996) and Lynch and Walsh (1998). Consider some phenotypic trait of a reiterated organ (e.g., leaf length, fruit width). Leaving aside the rare cases in which genetic mosaicism may be involved (chapter 5), the variance among repeated measures of the trait on the same individual is environmental in origin, representing variation arising from localized circumstances operating during development (chapter 6). It thus follows that an upper-bound estimate of the genetic variance of an organ trait is given by

$$Var(G)_{max} = Var(z) - Var(e_w), \quad 10.1$$

where $Var(z)$ is an estimate of the total phenotypic variance for the trait in the population, and $Var(e_w)$ is an estimate of the within-individual component of variance. Measurement error will inflate estimates of $Var(e_w)$ relative to its true value, but since it also contributes to $Var(z)$, measurement error cancels out in this equation. Despite this, however, measurement error may still be a problem in estimating the fraction of the true phenotypic variance that is accounted for by $Var(G)_{max}$; hence it is desirable to have estimates of $Var(z)$ free of measurement error. Letting $Var(e_m)$ denote the variance associated with measurement error, the repeatability r of the trait under consideration becomes

$$r = \frac{Var(z) - Var(e_w)}{Var(z) - Var(e_m)}. \quad 10.2$$

If measurement error is sufficiently small in relation to $Var(z)$, then it is possible to rewrite equation 10.2 as follows, using the notation introduced in chapter 3:

$$r \approx 1 - \frac{Var_{within}}{Var_{total}}; \quad 10.3$$

its bias will depend on the actual $Var(e_m)$ value.

Using equation 10.3, I computed repeatability values for 228 data sets listed in tables 3.1 to 3.4, which comprise a variety of continuously varying leaf, flower, fruit, and seed traits from many species. For several of the data sets involving flower and fruit traits, I was able to estimate $Var(e_m)$ from repeated, independent measurements taken on the same structures. The $Var(e_m)/Var(z)$ ratios obtained ranged between 0.01 and 0.03, which suggests that the upward bias of equation 10.3 is probably unimportant in most of these data sets. Figure 10.1 shows that leaf traits tend to exhibit the smallest repeatabilities and flower traits the largest, while seeds and fruits are intermediate. Despite these differences, there is considerable overlap among organ types, and repeatabilities are substantially smaller than unity in the majority of cases. For the number of petals in flowers of *Nyctanthes arbor-tristis* (fig. 2.3, table 3.2), a trait that can be safely assumed to be measured without error, repeatability is as low as 0.10.

The repeatability of a trait provides an upper-bound estimate of its broad-sense heritability (H^2), or the fraction of the total phenotypic vari-

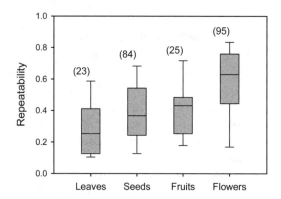

FIG. 10.1 A summary of repeatability estimates for continuously varying leaf, flower, fruit, and seed features, obtained by applying equation 10.3 (which assumes a negligible measurement error) to the data sets of tables 3.1 through 3.4. Figures in parentheses beside boxes denote the number of distinct data sets included in each case. Box plots show the 10%, 25%, 50%, 75%, and 90% percentiles of the distributions.

ance that has a genetic basis, both additive and nonadditive. "A large value of r offers the possibility that a considerable amount of the character variance is genetic, while a small value of r informs us that environmental variance dominates" (Lynch and Walsh 1998, 122). Broad-sense heritability, in turn, sets an upper limit to narrow-sense heritability (h^2; the proportion of phenotypic variance due to additive genetic variance alone), a central element in the familiar breeder's equation

$$\Delta\mu = h^2 \, S, \quad 10.4$$

where $\Delta\mu$ is the response to selection across generations (i.e., the change in mean organ phenotype caused by selection), and S is the selection differential. Equation 10.4 shows that, by setting an upper limit on h^2, repeatability of a trait will thus set an upper limit to the response to selection across generations. It is clear from figure 10.1 that, particularly in the case of leaf, fruit, and seed traits, repeatabilities are often low enough as to confidently predict low heritabilities and therefore small responses to selection. Dohm's caveats (2002) on the use of repeatability to infer heritability's upper bounds will quite rarely apply to situations such as those included in figure 10.1, in which repeated phenotypic measurements are taken on truly homologous organs produced by the same individual (Michael Dohm, personal communication). It is also worth noting that most repeatability estimates summarized in figure 10.1 are probably overestimates, since most phenotypic variance partitions shown in tables 3.1 to 3.4 refer to wild-grown plants. In these cases, estimates of among-individual phenotypic variance components are inflated to an undetermined degree by environmental phenotypic variance arising from plastic responses of individuals to variation in features of their local growing environments.

The constraining role of within-plant variation on adaptive changes of organ traits in response to selection has been rarely suggested for wild plants (Obeso and Herrera 1994). In contrast, plant breeders have long been aware that within-plant variation may become a serious hindrance limiting the responses to artificial selection on organ-level traits, thus threatening the success of crop improvement programs (Brim et al. 1967; Jellum 1967; Kondra and Downey 1970; Fick and Zimmerman 1973; Bramble et al. 2002; Calderini and Ortiz-Monasterio 2003). Given that the strength of directional selection gradients experienced by wild plants under natural conditions are generally weaker than those applied in crop improvement programs, one would therefore expect extensive within-plant variation in organ traits to represent also an important factor limiting the responses to selection exerted by animals on wild plants. Among other consequences, this may slow down the realization of adaptive phenotypic changes in response to alterations in the selective regimes imposed by animals on organ traits, such as those resulting from extinctions or replacements of pollinators, seed predators, or frugivores. All else being equal, the greater the relative importance of the within-plant component of phenotypic variance for a given organ trait in a population, the slower the expected adaptive changes in response to a changing selective environment, and the stronger and more consistent the selective pressures needed to efficaciously induce such changes.

Levels of within-plant phenotypic variance vary widely among traits and organs and, for a given trait, among species and populations within species (chapter 3). The potential limiting role played by $Var(e_w)$ on the responses to selection on organ traits will thus vary accordingly. It is therefore important to understand what maintains the existing levels of environmental variance, and particularly whether such levels are maintained by selective processes. Although standard genetic models have traditionally assumed that the distribution of $Var(e_w)$ for a given trait among the individuals of a population is unrelated to their genetic differences, recent theoretical models have increasingly shown that levels of environmental variance can be maintained by selection under a broad range of environmental and selective conditions (Bull 1987; Gavrilets and Hastings 1994; Wagner et al. 1997; Zhang 2005; Zhang and Hill 2005). As shown in the next section, animals may become effective agents of selection on the environmental component of phenotypic variance of organ traits via selection on within-individual variation. This reveals an important ecological mechanism for the maintenance of adaptive levels of environmental variance in these traits, to be added to those considered so far by quantitative genetics models.

Selection on Variability

The most direct and important evolutionary consequence of within-plant variation in organ traits is opening the possibility for animals to exert selection on levels of within-plant variance. The preceding chapters have shown that animals can discriminate among organs of the same plant that differ in phenotypic characteristics; that within-plant variation in organ traits impose direct and indirect costs on foraging animals; that within-plant variability is an individual trait that may have a genetic basis; that individual differences in levels of within-plant variability may give rise to among-plant discrimination by animals; and that individual variation in variability may have consequences for fitness-related traits such as pollination or seed-dispersal success, through a variety of mechanisms. Taken together, all these facts indicate that better insight on the evolutionary consequences of plant-animal interactions would be gained if the possibility of selection on variability were incorporated into studies of selection by animals on plant traits. Below I provide the background and justification for a proposed extension of Lande and Arnold's regression-based phenotypic selection models (1983), which I then formalize. The proposed model of phenotypic selection incorporates subindividual variabilities as just another set of descriptors of a plant's phenotype in addition to customarily used means. Then I present some examples showing that the application of "variability-aware," expanded phenotypic selection models that incorporate within-plant variances as phenotypic descriptors can at times lead to conclusions somewhat different from those reached using the conventional procedure of characterizing individual plants by trait means alone. But more importantly, incorporating selection on within-plant variance into selection models provides a tool to broaden our views of the evolutionary scenarios and selective pathways associated with the evolution of phenotypic features of reiterated organs.

Expanding Phenotypic Selection Models: Background and Justification

Phenotypic selection occurs when individuals possessing different characteristics (i.e., phenotypes) differ in some fitness-related measurement, such as fecundity or survival. Studies of phenotypic selection on animals and plants in the wild have proliferated in the past 25 years, following the development of methods for estimating the strength of selection on quantitative traits (Lande and Arnold 1983; Arnold and Wade 1984a, 1984b), and following Endler's influential book *Natural Selection in the Wild* (1986; for recent reviews see Kingsolver et al. 2001; Kingsolver and Pfennig 2007). In a review of selection studies published between 1984 and 1997, Kingsolver et al. (2001) identified 63 studies yielding estimates of directional selection for a wide range of taxa and types of traits. Nearly 75% of studies focused on animals, which points to a zoological bias of phenotypic selection studies (see also table 5.1 in Endler 1986). This slant was already visible in the early works by Lande and Arnold (1983) and Arnold and Wade (1984b), which sparked much of the subsequent interest on phenotypic selection, because the examples chosen to illustrate the application of their novel methods involved exclusively animals. Despite this initial bias toward animals, the phenotypic selection approach was soon adopted by, and its attendant statistical methods "adapted" to, plant selection studies (Kalisz 1986; Campbell 1989; Schemske and Horvitz 1989; Campbell et al. 1991; Herrera 1993). By "adapted" I mean that some tinkering with the original model was unavoidable in order to accommodate the reality that, in plants, many phenotypic traits of interest to evolutionary ecologists actually refer to reiterated organs rather than to the whole plant itself. In other words, plant modularity and reiteration of structures had to find its way into the framework of selection models that had been originally devised for nonmodular animals whose phenotypic traits could be completely characterized by a single figure each. Lande and Arnold (1983) and Arnold and Wade (1984b) used their methods for the first time to examine selection on body size or linear dimensions of body parts of vertebrate and invertebrate animals, each of which require only one or two (in the case of paired structures) values per individual. In contrast, phenotypic selection studies of plants have typically included traits that require a multiplicity of phenotypic values per individual (table 10.1). The solution routinely adopted to date to incorporate this multiplicity of "organ phenotypes" into models originally devised for single-figure, phenotypic characters of individuals has been to use plant means of organ-level phenotypic traits as descriptors of individual phenotypes. This decision to represent individual plant phenotypes by means alone would be justified if subindividual phenotypic variation in organ traits did not exist; or if, although existing, its magnitude were identical in all individuals; or if, although it did exist and differed among individuals, the differences were inconsequential to fitness. As earlier chapters have clearly shown that there is little support for any of these three premises, the prevailing habit of representing individual phenotypes in plant selection studies exclusively by their mean organ traits should be revised accordingly.

Plant features influencing fitness via their effects on interactions with animals fall into two distinct classes that should be treated differently, namely those describing features of the plant as a unitary whole and those describing properties of individual reiterated organs. Phenotypic selection studies of plants have traditionally examined the influence on fit-

TABLE 10.1 **Examples of continuously-varying traits of reiterated organs, or "multiple-value traits," included in phenotypic selection studies of plants.**

Trait	Reference
Leaf	
Area	
Length and width	Irwin 2000
Stomatal conductance	Heschel and Riginos 2005
Succulence, nitrogen content, and water-use efficiency	Donovan et al. 2007
Weight	Totland et al. 1998; Totland 1999
Flower	
Anther or stigma exsertion	
Corolla length and width	
Flower shape	Herrera 1993; Gómez et al. 2006
Mass of flower parts	Parra-Tabla and Bullock 2000
Nectary-stigma distance	
Number of ovules per flower	
Petal length and width	Herrera 1993; Irwin 2000
Pollen production per flower	
Spur or corolla-tube length	Herrera 1993; Totland et al. 1998; Maad 2000
Stigma-anther separation	
Fruit	
Carpel length	Irwin 2000
Fruit cross diameter	Herrera et al. 1994; Jordano 1995b
Seed	
Germination date	
Seed size	

Source: Taken from Kingsolver et al.'s phenotypic selection database (2001), except where otherwise indicated.

ness of these two types of features without making any explicit distinction between them. Recognizing the differences is a first step toward improving current models of phenotypic selection by animals on plants.

The first class of plant features that may influence fitness through effects on animal selectivity include whole-plant traits that, at a given time, can be unambiguously expressed using a single figure per plant, such as height, crown diameter, trunk diameter, number of flowers or inflorescences, or fruit crop size. It is well known that individual differences in one or more of these "single-value traits" may decisively influence plant fitness via effects on the nature and strength of interactions with frugivores (Howe 1980; Jordano 1987; Herrera et al. 1994), seed predators (De Steven 1983; Christensen et al. 1991; Rabasa et al. 2005), fruit predators (Prokopy and Owens 1983; Jordano 1987; Raghu et al. 2004), foliage-feeding herbivores (Karban and Courtney 1987; Forsberg 1987; Alonso and Herrera 1996), or insect pollinators (Dudash 1991; Mitchell et al. 2004; Grindeland et al. 2005).

The second class of plant traits that may likewise influence plant fitness via effects on interactions with animals include features of reiterated
structures. This class of phenotypic features could be dubbed "multiple-
value," or reiterated, traits. As noted above, studies examining phenotypic
selection by animals on organ traits (e.g., leaf size, corolla-tube length,
seed mass, fruit diameter) have routinely related individual differences in
fitness-related measurements to variation in the *central tendency* of such
multiple-value traits as expressed by variation of individual means. This
is, in principle, a sound procedure justified by the frequent observation
that phytophagous animals respond selectively to variation among plants
in mean values of organ traits. Herbivores discriminate among plants on
the basis of variation in mean concentration of nutrients and secondary
metabolites (Lightfoot and Whitford 1989; Vourc'h et al. 2001; Moore and
Foley 2005). Frugivores respond to individual variation in mean fruit size,
pulp/seed mass ratio, and pulp composition (Howe and Vande Kerckhove
1980; Herrera 1988; Foster 1990; Wheelwright 1993; Herrera et al. 1994).
Pollinators discriminate among plants on the basis of differences in mean
corolla dimensions, nectar production, and nectar content (Herrera 1993;
Mitchell 1994; Pappers et al. 1999; Maad 2000; Caruso 2001). Predispersal fruit and seed predators discriminate among conspecific host trees on
the basis of differences in mean fruit size, seed mass, seed/fruit wall mass
ratio, and nutrient and secondary metabolite content of seeds (Courtney
and Manzur 1985; Greig-Smith and Wilson 1985; Christensen et al. 1991;
Crowley and Garnett 2001).

Nevertheless, whenever multiple-value traits are involved in phenotypic selection studies, there are at least two important reasons why the
usual practice of characterizing individual plant phenotypes by character
means alone can lead to incomplete or flawed representations of selection
patterns by animals on plants: (1) Leaving phenotypic characters affecting fitness out of phenotypic selection models may complicate inferences
regarding selection and bias estimates of selection strength (Mitchell-
Olds and Shaw 1987). Because the level of within-plant variability in a
given organ trait is an individual attribute whose variation may in itself
give rise to differences in components of fitness (chapters 7 and 9), important aspects of the selective regime on plant phenotypes may be missed
if the contribution to fitness of differences in variability is neglected. (2)
Plant means and variances of subindividually variable traits tend to be
correlated across individual plants (fig. 3.1). This raises the possibility of
obtaining spurious, distorted, or biased selection estimates if the effects of

within-plant variances on fitness are not explicitly accounted for in phenotypic selection models along with those of the means. Leaving the effect of subindividual variability out of selection models may thus lead to estimates of selection gradients on means that would be "contaminated" to an unknown degree by concurrent selection on variability. Properly dissecting the roles of selection on central tendencies and selection on variability will require analyzing simultaneously the effects on fitness of individual means and variances of traits. One way to accomplish this is by a simple expansion of regression-based phenotypic selection models as ordinarily applied to plants.

A "Variance-Aware" Phenotypic Selection Model

It follows naturally from the verification of the Haldane-Roy conjecture (chapter 7) that it makes biological sense to include the set of within-plant variances of reiterated organ traits as yet another set of descriptors of a plant's phenotype in addition to customary trait means. Since individual plants commonly differ in their within-plant variabilities, and interacting animals may discriminate among plants on the basis of this trait, the incorporation of variabilities into phenotypic selection models is not only justified, but also captures natural patterns of selection by animals on organ traits more realistically.

Different methods have been proposed to estimate the contribution to fitness of many traits by means of multiple regression (Endler 1986). Among these, the Lande-Arnold model (Lande and Arnold 1983; Arnold and Wade 1984a, 1984b), building on earlier work by Pearson (1903), has been used most frequently in studies of phenotypic selection on wild plants. Like other regression-based models, the Lande-Arnold method explicitly takes covariance among traits into consideration, a feature that makes it particularly well suited to deal with phenotypic traits that are a priori correlated between themselves, such as plant means and within-plant variances of reiterated organ characters. In the Lande-Arnold model, an estimate is made of the function relating the fitness of an individual (W) to the values X_i of each trait i:

$$W = a + \sum b_i X_i, \quad 10.5$$

where a and the coefficients b_i are constants relating W to each X_i holding the other X_i constant. Absolute fitnesses (W) are divided by mean fitness

($\overline{W}$) to yield relative fitnesses w with $\overline{w} = 1$. The b_i coefficients ("gradients") estimate the contribution to fitness of trait i, holding the effects of other traits constant. Standardized coefficients (β_i) estimate selection on trait i in terms of the effects on relative fitness in units of phenotypic standard deviations of the trait, which allow direct comparisons among traits, fitness components, and study systems (Kingsolver et al. 2001).

As noted above, application of the Lande-Arnold model to plant selection studies has routinely used plant means of reiterated organ traits as the X_i, and consequently the selection gradients obtained are informative on the strength of selection on the central trends of such multiple-value characters within individual plants. Given that the within-plant variances of multiple-value characters are also individual phenotypic traits potentially subject to selection, they can also be included as predictors in the model, and equation 10.5 can then be rewritten as follows:

$$W = a + \sum b_i X_i + \sum b_j M_j + \sum c_j V_j, \quad 10.6$$

where subscripts i and j stand for single-value (e.g., height, number of inflorescences, fruit crop size) and multiple-value (e.g., corolla length, fruit width, leaf area) phenotypic traits, respectively; X_i are individual values for single-value trait i; and M_j and V_j are individual means and variances for multiple-value trait j, respectively. This equation makes explicit the distinction between the three classes of phenotypic traits that should be considered in plant selection studies, namely single-value traits, individual means of multiple-value traits, and individual variances of the same multiple-value traits. Estimates of linear selection gradients on X_i, M_j, and V_j may be obtained using multiple regression, in the same way as phenotypic selection studies that follow the Lande-Arnold model (Endler 1986). Advantages of equation 10.6 over the usual procedure, considering only the means of multiple-value traits and ignoring their within-plant variances, include allowing dissection of selection on individual means and variances, as well as removing possible spurious effects of the mean-variance correlations on estimates of linear selection on plant means.

Equation 10.6 expands conventional phenotypic selection models by incorporating only one of the possible additional measurements characterizing within-plant variation in organ traits. It was shown in chapter 7 that, in addition to the amount of subindividual variation described by the variance, other moments of the within-plant distribution of trait values such as skewness and kurtosis, as well as descriptors of the spatial organization of variation, can also be treated as individual traits with a genetic basis and potentially subject to selection by animals. Further extensions of phenotypic selection models could easily be achieved by incorporating one or more of these parameters in equation 10.6. In the examples presented in the next section, however, only within-plant variances are included in the models, for simplicity.

The above expansion of the Lande-Arnold model is only one possible way to handle reiteration of functional units within an individual plant. Other ways may be called for in certain situations. For instance, it would be possible to treat the individual as a "group" of modules, with the modules as units of statistical replication in the process of phenotypic selection. Suppose one is studying a plant in which one flower per day opens and experiences an episode of selectivity by pollinators. It might be best to do the regressions of a flower's relative fitness on its characters, and after the regression calculate the part of selection that is on the central tendency and the part of selection that is on the dispersion within individuals. Having adopted this multilevel approach, one could take advantage of methods for analyzing selection in hierarchically structured populations (Heisler and Damuth 1987; Damuth and Heisler 1988). I shall not discuss further extensions on the study of phenotypic selection.

Phenotypic Selection on Variability: Examples

I estimated standardized directional selection gradients on plant means and within-plant variances of leaf, flower, and fruit traits using several data sets for which I had access to the raw data (table 10.2). These examples illustrate only a subset of the possible mechanisms whereby subindividual variability may influence plant fitness (table 9.2). They suffice, however, to illustrate the point that phenotypic selection on within-plant variance in organ traits would surface as a frequent phenomenon if it were sought more often. One analysis involves selection on leaf size of *Prunus mahaleb* by larvae of the monophagous lepidopteran *Yponomeuta mahalebella*. Two analyses involve selection on fruit size, by the seed predator *Amblycerus cistelinus* on *Guazuma ulmifolia* and by seed-dispersing frugivorous birds on *Phillyrea latifolia* (see chapter 9 for more on these systems). The remaining seven analyses examine selection on floral traits presumably exerted by insect pollinators on *Helleborus foetidus*, *Lavandula latifolia*, and *Viola cazorlensis*. Two models were fitted in each case, a conventional one in which plant means alone are included among predictors of

TABLE 10.2 **Examples of conventional and expanded phenotypic selection analyses based on regressing fitness-related measures on traits of fruits and flowers.**

			Phenotypic selection model		
				Expanded	
			Conventional		
Species	Fitness-related variable	Phenotypic trait	β_{mean}	β_{mean}	β_{var}
Guazuma ulmifolia	Percent fruit crop infested[a]	Fruit length	**−0.32**	**−0.41**	**+0.38**
Phillyrea latifolia	Percent fruit crop dispersed	Fruit width	**+0.31**	**+0.39**	**−1.11**
Prunus mahaleb	Herbivore incidence[a]	Leaf area	−0.17	**−0.75**	**+0.70**
Helleborus foetidus[b]	Per-plant follicle production	Flower size	+0.06	+0.09	**−0.24**
	Per-plant follicle production	Flower size	**+0.41**	**+0.46**	**−0.25**
	Per-plant follicle production	Flower size	**−0.28**	**−0.27**	+0.09
Lavandula latifolia	Pollen tubes per flower	Corolla-tube length	**−0.14**	−0.08	**−0.12**
	Pollen tubes per flower	Corolla-lip length	+0.08	+0.04	**−0.17**
Viola cazorlensis	Percent fruit set	Flower-spur length	+0.05	+0.04	**−0.26**
	Percent fruit set	Floral-pedicel length	+0.17	+0.18	**−0.32**

Sources: Based on data from Herrera 1993; Herrera et al. 1994; Alonso 1997b; Rey et al. 2006; Herrera, Castellanos, and Medrano 2006; C. M. Herrera unpublished data.
Notes: In conventional analyses only plant means were included in the regression equations, while both plant means and within-plant variances were included in the expanded models. β_{mean} and β_{var} represent the standardized linear selection coefficients on plant means and variances, respectively. Significant and marginally significant ($P < 0.07$) selection coefficients are in bold type. Fitness-related measurements were in each case divided by the corresponding population mean to obtain relative measures with mean unity.
[a]Signs of selection coefficients were reversed, on the assumption that the fitness-related variable used and fitness itself were inversely related.
[b]The three analyses for *Helleborus foetidus* correspond to different populations.

fitness-related measurements, and an expanded one that includes both the means and variances of traits, as in equation 10.6. As these analyses are intended only for illustrative purposes, only linear selection coefficients and multiple-value traits were included in the models. It would be feasible to incorporate quadratic selection coefficients for trait variance into expanded models if one were interested in testing the possibility of stabilizing or disruptive selection on levels of within-plant variability.

The expanded models reveal the existence of significant phenotypic selection on within-plant variance in all but one of the analyses (table 10.2). In fact, the limited examples presented suggest that selection on within-plant trait variance may be more prevalent than selection on plant trait means (nine vs. five significant selection coefficients, respectively). Keeping constant the effect on fitness-related measurements of individual variation in organ trait means, there is positive directional selection on the within-plant variance of *P. mahaleb* leaf size and *G. ulmifolia* fruit size, and negative directional selection on the within-plant variance of fruit size in *P. latifolia* and all the floral dimensions considered for *H. foetidus*, *V. cazorlensis*, and *L. latifolia*. In the four cases where selection gradients on plant means and variances are simultaneously significant, the two gradients have opposite signs. Selection simultaneously favors individuals with smaller and more variable leaves in *P. mahaleb*; smaller and more variable fruits in *G. ulmifolia*; individuals with larger and less variable fruits in *P. latifolia*; and individuals with larger and less variable flowers in *H. foetidus*. Coupled selection gradients on individual means and variances, particularly when they are of opposite signs, may condition adaptive levels of within-plant variance in organ traits, as discussed in the next section.

The comparison of results of conventional and expanded phenotypic selection models reveals some additional aspects that are worth noting. In eight out of ten analyses in table 10.2, the sign and magnitude of selection coefficients on means remain unaltered after including within-plant variances as an additional set of phenotypic descriptors. In the remaining two analyses (selection on leaf size in *P. mahaleb* and corolla-tube length in *L. latifolia*), the expanded model leads to different conclusions from that of the conventional one: the nonsignificant selection on leaf size becomes significant, and the negative selection on mean corolla-tube length vanishes, when the effect of within-plant variance is simultaneously accounted for. The examples shown in table 10.2 are thus somewhat reassuring in that conclusions of most earlier phenotypic selection studies of organ traits are expected to be robust to the incorporation of within-plant variances. But there is also some hint that a certain number of patterns of selection on organ traits obtained from conventional models might have to be revised following the incorporation of within-plant variances as additional phenotypic descriptors.

Warnings have recently been issued on the interpretation of results of regression-based phenotypic selection models, as it has been suggested that they may be biased by environmentally induced covariances between phenotypes and fitness. If environmental conditions affect fitness-related measurements, and individuals with different phenotypic traits experience different environmental conditions, this can alter the measured relationship between traits and fitness, and hence selection coefficient estimates (Scheiner et al. 2002; Stinchcombe et al. 2002; Winn 2004). Phenotypic selection studies considering the means and variances of reiterated organ traits may also be subject to this complication. As for the rest of phenotypic studies of wild populations, there is not yet sufficient information to assess the frequency and magnitude of biases on selection estimates caused by environmental effects in these models. Keeping these caveats in mind, the results of expanded phenotypic selection models shown in table 10.2 serve at least to suggest that phenotypic selection on within-plant variability is sufficiently frequent as to warrant consideration on its own.

Implications of Selection on Variability: Theoretical Scenarios

The most direct and important consequence of selection on subindividual variability in organ traits is that it furnishes a mechanism that opens the way for adaptive responses. The fact that within-plant variation is environmental or developmental in origin does not mean that it cannot be selected for and become adaptive. The amount of variation or the capacity to be variable in the right circumstances can have a genetic basis (chapter 7). In this way, the capacity of individual genotypes to produce an array of alternative phenotypes can become both a target of selection and an important factor in evolution (Waddington 1942; Schmalhausen 1949; Schlichting and Pigliucci 1998; West-Eberhard 2003). As expressed by Schmalhausen (1949, 4), "The statement that [phenotypic] modifications [of the same genotype] are not heritable is not precise. The ability to undergo modification is strictly hereditary." The same idea was put forward by Bull (1987, 303) in more precise terms: "Although the environmental component of a phenotype is itself not heritable, the susceptibility of an individual to environmental effects is a property that often has a genetic basis ... and thus can evolve under natural selection." Empirical support for the genetic basis of environmental effects has been provided for a number of traits and organisms (for animals see Waddington 1959; Roff and Réale 2004; Ros et al. 2004; for plants see references in chapter 7). Quantitative genetic models have also provided theoretical support for the evolvability of environmental effects, as shown below.

Some of these models were built without explicit reference to the particular mechanism(s) causing the environmental component of phenotypic variance, while others were specifically designed for certain types of environmental variance, like that caused by developmental instability or phenotypic plasticity. In this section I briefly summarize conclusions of both classes of genetic models that have theoretically explored the evolution of environmental phenotypic variance. My treatment is illustrative rather than exhaustive, mainly aimed at highlighting conclusions from these models that support or are compatible with the view that selection by animals on levels of subindividual variability is evolutionarily relevant because of its possible contribution to the maintenance of adaptive levels of within-plant variation. The models I consider generally do not refer explicitly to within-plant variation, or even to plants, and they are mainly concerned with other types of environmental phenotypic variation. I know of no reason, however, why such models could not be applicable in the present context, where a substantial proportion of the environmental component of phenotypic variance arises from the multiplicity of homologous organs produced by the same genotype.

General Adaptive Scenarios

Any form of phenotypic selection on the within-plant variance of some organ trait, if combined with a genetic basis of such variance, is expected to lead to the evolution of adaptive levels of within-plant variance, regardless of the organismal mechanism(s) accounting for such variation. This is clearly supported by predictions from genetic models that have explored the adaptive consequences of selection on environmental components of phenotypic variance without emphasizing any particular causal mechanisms (Bull 1987; Zhang 2005). Such models have mainly focused on the adaptive value of the environmental component of phenotypic variance in relation to spatial and temporal fluctuations of the environment. Their most important conclusion, namely that the environmental component of phenotypic variance may be adaptive and shaped by selection, does not depend on detailed aspects of the selective environment, however. More specifically, Bull (1987) showed that the two components of phenotypic variance in a trait (i.e., genetic and environmental) may be selectively maintained in a population independently of each other. Selection may favor the maintenance of only the environmental components, or only the genetic component, or be indifferent to the composition of the variance.

It should be stressed that the possibility of selection on the environmental (i.e., within-plant) component of phenotypic variance, and the evolutionary independence of the genetic and environmental components, are both incorporated into the expanded phenotypic selection model proposed in equation 10.6. The results of fitting expanded models to empirical data shown in table 10.2 are nicely congruent with the theoretical expectations of Bull's model (1987), because linear selection on means and variances covaries independently. That limited selection of examples shows selection on means alone (one instance), on the variances alone (five instances), and on both the means and the variances (three instances). Within the latter group, selection on mean and selection on variance have opposite signs in all cases.

Mechanism-Dependent Scenarios: Developmental Instability

In most real-world situations, observed levels of within-plant variability in organ traits will be the composite outcome of an entangled combination of causes and mechanisms that may operate simultaneously or at different times in organ development. These include developmental instability, organ-level reaction norms, and direct and indirect architectural effects (chapters 5 and 6). Some of these, such as direct architectural effects, are very unlikely to be susceptible to adaptive modifications in response to selection by animals on variability, while others are expected to respond much more easily. Under a regime of selection on within-plant variability such as those exemplified in table 10.2, organ-level reaction norms and levels of developmental instability may ultimately become the targets of selection, with a likelihood that will depend on their respective importance as causal agents of variability. As shown by some quantitative genetic models, components of environmental phenotypic variance that stem from developmental instability and reaction norms may be adaptive, in the sense that they can be maintained by selection. I consider here some adaptive scenarios described by genetic models in which developmental stability is implicated as the main source of environmental variance, leaving for the next section those that implicate reaction norms.

The possibility that developmental instability in a character can be maintained by the direct action of selection, or indirectly by selection on some correlated character, was suggested by Simons and Johnston (1997). In their view, developmental noise would be advantageous to individual plants if it had the effect of increasing the phenotypic variance of a bet-hedging trait. In this case, developmental instability may represent an ability to destabilize development up to a certain optimal, adaptive level, rather than an inability to stabilize it. A closely related idea was examined by Wagner et al. (1997) more formally in the broader framework of a general population genetic theory of canalization. These authors examined the conditions conducive to environmental canalization, or the suppression of phenotypic variation of quantitative characters at the micro-environmental level, a kind of environmental phenotypic variance that would reflect developmental instability. Their model predicts a straight-forward selection mechanism whereby stabilizing selection on a trait, as it frequently occurs in nature (Kingsolver and Pfennig 2007), will favor genes that decrease environmental variance (see also Gavrilets and Hastings 1994). This prediction, however, is clearly at odds with the abundant empirical observations revealing extensive environmental phenotypic variance at the microenvironmental level (chapter 4). Several possibilities were advanced by Wagner et al. (1997) to explain the persistence of developmental sensitivity to microenvironmental variation. One of these is that if the genes involved in environmental canalization of a character also have direct effects on the character itself, then the amount of environmental canalization will not depend on the strength of stabilizing selection, but it would rather represent a compromise between selection forces in favor of environmental canalization and selection against direct effects, and a "canalization limit" would be reached. This implication of Wagner et al.'s model is directly relevant to the issue of the adaptive maintenance of within-plant variation in organ traits. If the direct relationship frequently linking within-plant means and variances of organ traits (fig. 3.1) is taken as a statistical expression of a pleiotropic connection between canalization of trait values and trait values themselves, then the occurrence of selection on *both* within-plant means and within-plant variances, of the sort illustrated by some examples in table 10.2, may maintain adaptive levels of within-plant variability in organ traits. A similar inference can be drawn from Gavrilets and Hastings's model (1994) for selection on developmental noise. Under linear directional selection, there is a steep increase of the environmental component of phenotypic variance, and a weak decline in the genetic component, with increasing correlation between trait values and microenvironmental sensitivity (developmental noise). The prediction may thus be tentatively advanced that, if within-plant variance is mainly due to developmental instability, and within-plant means and variances are correlated across individuals, then simultaneous selection on means and variances that are of opposite signs will contribute to the maintenance of adaptive within-plant variation.

It may also be intuited from Wagner et al.'s canalization-limit interpretation that adaptive levels of within-plant variance in organ traits will depend critically on the fitness consequences of individual differences in levels of subindividual variation or, in other words, on the net balance of fitness advantages and disadvantages accrued to genotypes that differ in the amount of environmental variance. This possibility was explored by

Zhang and Hill (2005; see also Zhang 2005). They considered the selective forces that may contribute to the maintenance of environmental phenotypic variance under the assumption that the genes that affect the phenotypic mean also directly control the phenotype variance. These authors proposed that there is an intrinsic cost to the organism in reducing the variability due to developmental noise (homogeneity cost) that would be akin to an engineering cost. Their model reveals that the conflict between the opposing pressure from selection that favors individuals close to the optimal phenotype, and from the homogeneity cost that favors variable individuals, maintains environmental variance. Even a small homogeneity cost may be sufficient to maintain nonnegligible levels of environmental variance. Translating these theoretical results to the particular context of animal-mediated selection on levels of subindividual variability, it can be suggested that even small selection intensities from animals in the direction of favoring subindividual variability may help maintain substantial levels of adaptive within-plant variation.

Mechanism-Dependent Scenarios: Reaction Norms

It is expected that within-plant variation in organ traits will reflect to a variable degree the plastic, programmed phenotypic responses by developing organs to the small-scale, localized variation in their internal and external microenvironments that takes place during certain stages or temporal windows in their developmental process (chapters 5 and 6). In cases where phenotypic plasticity is the main cause of the within-plant component of environmental variance, the within-plant phenotypic frequency distribution of a given organ trait exhibited by an individual plant will be the outcome of the concerted action of the plant's specific organ-level reaction norm and the perceived frequency distribution of the influential environmental variable (fig. 10.2a). The shape and slope of the organ-level reaction norms that govern the plastic responses of organs to environmental variation can thus become in these cases the target of the selection exerted by animals on levels of within-plant variability.

The notion that reaction norms may be shaped by selection was articulated into a coherent evolutionary theory by Schmalhausen (1949), and it has subsequently received extensive support from theoretical models, particularly over the last two decades (Via and Lande 1985; de Jong 1990; Gavrilets and Scheiner 1993a, 1993b; Schlichting and Pigliucci 1998; DeWitt and Scheiner 2004a; Zhang 2005). While a general consensus has

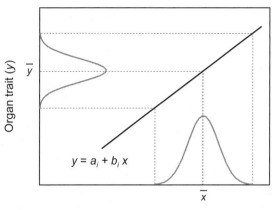

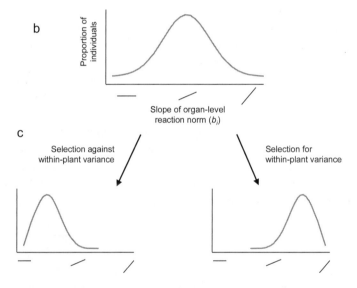

FIG. 10.2 Within-plant variation in organ traits may largely reflect programmed developmental responses of organs to localized environmental variation, either internal or external. In these cases, (*a*) the within-plant phenotypic frequency distribution of a given organ trait exhibited by an individual plant i (vertical axis) will be the outcome of the concerted action of the plant's specific organ-level reaction norm, represented by the equation $y = a_i + b_i x$, and the perceived frequency distribution of the influential environmental variable (horizontal axis). In the example shown, for a given distribution of x, steeper reaction norms will lead to subindividually more variable plants than flatter ones. If individual organ-level reaction norms have a genetic basis, then in a population composed of a mixture of individuals that differ in the slopes b_i of their respective organ-level reaction norms (*b*), directional selection against or for within-plant variance can modify the composition of the population with regard to the distribution of b_i, shifting it toward a greater representation of individuals with either flatter (*left*) or steeper (*right*) organ-level reaction norms (*c*).

emerged that reaction norms can be shaped by selection, there has been a long-standing controversy on the genetic mechanisms underlying adaptive reaction norms. Specifically, the disputed issue is whether adaptive reaction norms arise as direct targets of selection or as by-products of selection acting on the traits themselves, or in other words, whether there are specific "plasticity genes" that through either their expression or their regulatory role modify a trait's expression (for reviews see Via et al. 1995; Sarkar 2004). This dispute, which may be partly semantic (Sarkar and Fuller 2003; Pigliucci 2005), remains unresolved to date. The two important aspects that I wish to emphasize here, however, are not contingent on the resolution of the controversy on the details of genetic mechanisms underlying phenotypic plasticity and the shape of reaction norms. These aspects are that there is ample evidence for a genetic basis of phenotypic plasticity itself, and selection on phenotypic plasticity may lead to adaptive reaction norms (Schlichting and Pigliucci 1998; DeWitt and Scheiner 2004a; Pigliucci 2005; Zhang 2005).

In this scenario, directional selection by animals on levels of within-plant variability may drive the adaptive evolution of organ-level reaction norms in plant populations. A simple example of a possible selective mechanism is depicted in figure 10.2, assuming linear organ-level reaction norms and selection on the within-plant variances of organ traits but not on the within-plant means. It is also assumed that individual organ-level reaction norms have a genetic basis. In a population composed of a mixture of individuals that differ in the slopes of their respective organ-level reaction norms (fig. 10.2b), directional selection against or for within-plant variance can modify the composition of the population with regard to the distribution of such slopes. Selection against within-plant variance will shift the distribution of slopes toward a greater representation of individuals with flatter (i.e., subindividually less variable) organ-level reaction norms, while selection favoring within-plant variance will produce the opposite effect, shifting the distribution toward a predominance of plants with steeper (i.e., subindividually more variable) organ-level reaction norms (fig. 10.2c).

Methodological Remarks

The preceding sections have illustrated a variety of adaptive pathways that, stemming from selection on within-plant variability in organ traits, may eventually contribute to maintain adaptive levels of within-plant variation. Some of these pathways are independent of the mechanisms involved in generating within-plant variation or their respective importances, and thus are rather easily amenable to study by application of expanded phenotypic selection approaches that incorporate variances in addition to the customary means as phenotypic descriptors. Simply obtaining information on whether there is selection on the mean, the variance, or both of a character, and on the signs of the associated selection gradients, will represent a significant improvement of our insight into the possible adaptive scenarios available to plants in their interaction with animals. On a less positive note, a drawback of this mechanism-independent approach is that it will not allow identification of the direct targets of selection on variability, thus greatly reducing the possibility of making inferences on the adaptive mechanisms involved. Detailed hypotheses and predictions on the adaptive consequences of selection on variability would require the identification in each case of the major causes and mechanisms responsible for such variation, since these would presumably be the targets of selection. In practice, this dissection of causal factors will generally have to confront the considerable difficulties noted in chapter 6, and would require a combination of observation and careful experimental work.

It has been frequently highlighted that one of the main challenges of studying phenotypic plasticity stems from the technical difficulties associated with setting up appropriate experiments (Scheiner and DeWitt 2004; Pigliucci 2001, 2005). Plasticity is the property of a genotype, not the individual; hence methods for measuring plasticity traditionally involve raising separate ramets of identical genotypes in different environments. The number of genetically identical individuals needed for such studies increases multiplicatively with the number of environments assayed. Reiterated organs produced, often profusely, by single individuals provide a unique opportunity for approaching the study of phenotypic plasticity from a different angle. In situations where within-plant variability in organ traits reflects, at least partly, the programmed plastic responses of organs to external or internal microenvironmental variation, phenotypic plasticity in organ traits could be measured, and the shape of reaction norms could be determined, by experimentally subjecting developing organs on different modules of the same individual (e.g., flowers on different inflorescences, fruits on different shoots) to different external environmental conditions (e.g., light level, temperature). Although the results obtained using this approach would only complement those obtained with the usual protocol of exposing different individuals of identical genotypes to contrasting environments, some of the technical limitations of the latter method could be overcome because of the large number of genetically identical, homologous copies of the same organ usually produced by the same individual plant. Increased availability of experimental units would allow tackling, for example, the issue of the possible interaction effects of separate environmental factors on the shape of reaction norms.

Adaptive Levels of Subindividual Variability: Empirical Scenarios

Findings suggesting an adaptive value for within-plant variation in organ traits can be found sporadically in the ecological literature (Whitham 1981; Biere 1991a; Biernaskie and Cartar 2004), but there is little rigorous empirical support for these interpretations, and the circumstantial evidence available is almost totally confined to instances of discrete variation (chapter 9). Demonstrating that a certain level of subindividual variation is immediately advantageous to plants, as in the examples discussed in chapter 9 or shown in table 10.2, is only a weak argument for the adaptive nature of such variation. Much stronger evidence would be provided, for example, by the simultaneous demonstration that levels of subindividual variability vary among populations of the same species, that such differences are genetically based, and that among-population variation can be interpreted as reflecting local adaptation in response to the concomitant variation in some biotic or abiotic environmental factors that exert geographically variable selective pressures on such variability. Diverse modalities of this general approach have provided the relatively few demonstrations so far of adaptive phenotypic plasticity (Donohue et al. 2001; Berg et al. 2005; van Kleunen and Fischer 2005; Roiloa et al. 2007) or adaptive within-plant variation of organ traits in heterophyllous, heterocarpous, and cleistogamous species. For example, Cook and Johnson (1968) studied intra- and interpopulation variability in heterophylly in *Ranunculus flammula* by growing plants from different populations in terrestrial and aquatic conditions, and performing reciprocal transplant experiments. The most-heterophyllous individuals were associated with the most unpredictable environments, and such association was the adaptive outcome of disruptive selection acting on populations to produce individuals specially adapted to either persistently terrestrial or persistently aquatic conditions. In species of heterocarpous Asteaceae that produce dimorphic achenes with contrasting dispersal ability, the proportion of widely dispersible achenes often declines along gradients of increasing aridity, which has been interpreted as reflecting variable local adaptations to changing levels of environmental uncertainty (Venable et al. 1987; Kigel 1992). In 40 populations of the cleistogamous grass *Danthonia spicata* studied by Clay (1983b), the mean percentage of cleistogamous flowers produced by individual plants varied widely among populations (5–43%), increasing from those growing in undisturbed forest sites to those occupying more open and disturbed habitats. Across populations, the proportion of cleistogamous flowers was directly related to the degree of cleistogamous seedling establishment, which provides circumstantial evidence of adaptive levels of cleistogamous flower production (Clay 1983a, 1983b).

These and other examples show that adaptive levels of within-plant variability can be maintained in species with discontinuous subindividual variation in organ traits (see, e.g., Wells and Pigliucci 2000 for heterophyllous plants). Similar information is considerably scarcer for continuously varying organ traits, which is not surprising given the usual neglect of this type of variation. Nevertheless, levels of within-plant variability in continuously varying traits may also differ among populations of the same species, which will provide opportunities for formulating and testing hypotheses on the possible adaptive origin of such variation. An example was provided in figure 7.2, showing that average levels of within-plant variability in corolla length differ widely among populations of *Lavandula latifolia*. In the evergreen oak *Quercus coccifera*, the extent of within-crown variation in several structural and functional foliar traits varies among populations, and such variation is correlated with environmental variation and may have a genetic basis (Balaguer et al. 2001; Rubio de Casas et al. 2007).

The study of geographical variation has traditionally played an important role in elucidating the adaptive significance of phenotypic traits (Clausen et al. 1941; Hiesey et al. 1942; Gould and Johnston 1972). Analyses of geographical variation in the extent and characteristics of within-plant variation in organ traits may likewise provide insights into the degree to which they are adaptive and maintained by selection exerted by animals. Examinations of the adaptive value of within-plant variation based on the study of differences among populations could profitably adopt the five-step protocol suggested by Herrera, Castellanos, and Medrano (2006) for studying floral adaptations to animal pollinators: (1) document geographical variation in the composition or abundance of

animal counterparts that putatively exert selection on within-plant variation in the organ trait of interest (e.g., folivores, pollinators, seed dispersers); (2) test whether they actually exert phenotypic selection on levels
of within-plant variation in that trait, for example, as in table 10.2; (3)
examine whether the selection gradient on within-plant variability varies
among populations and whether such variation is related to differences
in the composition or abundance of animals; (4) assess the concordance
between variable selection gradients and variable levels of within-plant
variability; and (5) determine whether population differences in levels of
within-plant variation have a genetic basis.

It is obvious that considerable amounts of empirical information on
patterns of intraspecific variation in magnitude and characteristics of
subindividual variation in organ traits are still needed before tests of its
adaptive value that even minimally satisfy this five-step protocol can be
attempted. For the time being, however, a few general predictions can
be advanced on the basis of the general ecological and behavioral mechanisms discussed in chapters 8 and 9, the expectations summarized in table
9.2, and the results demonstrating phenotypic selection on within-plant
variance shown in table 10.2. Verifications of these predictions would
lend support to the notion that observed levels of within-plant variation in organ traits are, at least in part, the adaptive consequences of
animal-mediated selection.

1. In a given species, those organ traits that are mainly involved in enhancing a
   plant's positive relationship with mutualistic animals that contribute to its fitness (e.g., pollinators, seed dispersers) will often exhibit lower levels of subindividual phenotypic variability than those organ traits that intensify the detrimental interactions between the plant and its antagonists (e.g., folivores, seed
   predators). This prediction may be reversed in the case of pollination-related
   traits of self-compatible species where selfing results in inbreeding depression,
   as floral constancy may reduce plant fitness, which would potentially select for
   increased flower-to-flower variation so as to discourage long visits to individual
   plants.
2. Different traits of the same organ that are predominantly involved in enhancing interactions with antagonists and mutualists will have contrasting levels
   of subindividual variability, being greater for traits that enhance detrimental
   interactions with antagonists than for those enhancing positive interactions
   with mutualists. Trade-offs are expected in the case of organ traits that simultaneously influence the interaction with antagonists and mutualists.

3. Within a species, variation among populations in the relative importance of mutualists and antagonists as selective agents on a given organ trait (e.g., Rey et al. 2006) should result in concomitant among-population variation in within-plant variability of that trait, being highest where antagonists play the most prominent selective role, and lowest where mutualists are the most influential selective agents.

4. Levels of within-plant variation in organ traits that influence host selection by specialized antagonists (e.g., endophagous folivores, small seed predators) will be directly correlated across populations with the local abundance/incidence of antagonists.

5. In species that interact with taxonomically diverse assemblages of mutualists that differ in their preferences for a given organ trait (e.g., fruit diameter, corolla size), subindividual variability should be greatest in populations where mutualists are most heterogeneous in their preferences, and lowest where they are most alike.

6. In animal-dispersed species that inhabit spatiotemporally unpredictable environments, within-plant variation in dispersal-related organ traits (e.g., fruit size, pulp-seed ratio, pulp nutritional value) should increase with increasing unpredictability of the favorable microhabitats for the species.

Concluding Remarks

Two conclusions arising from this chapter are (1) adaptive levels of environmental phenotypic variance in organ traits can be maintained by selection, regardless of the mechanisms producing it, and (2) given that subindividual variability often accounts for nontrivial proportions of total environmental phenotypic variance of organ traits, selection by animals on variability has the potential to modify the magnitude of environmental variance and, in so doing, to shift the balance between the genetic and environmental components. Environmental and genetic factors may be envisaged as "competing" to produce a given level of phenotypic variance, as suggested by Bull (1987). The spatial and temporal dynamics of such competition has manifold evolutionary implications, and animals can play a driving role by shifting the balance toward one side or the other. The examples shown in table 10.2, albeit limited, suggest that a variety of scenarios are possible: selection on means alone, selection on variances alone, and simultaneous selection on means and variances. More data are needed before specific predictions can be advanced for each of these

338

outcomes, but at least the following general expectation can be formulated: phenotypic selection by animals on levels of subindividual variability, by virtue of its potential to shift the balance between genetic and environmental components of phenotypic variance, can have implications that extend well beyond those related to shaping organ-level reaction norms or maintaining adaptive levels of developmental instability.

In cases where within-plant variation mainly reflects the operation of organ-level phenotypic plasticity, there are two possible avenues whereby selection by animals can influence the evolutionary trajectories of organ traits. First, it can contribute to either reduce or increase the costs of plasticity and, in so doing, to alter the evolutionary prospects of plasticity itself. The study of the limits and costs of phenotypic plasticity has received considerable attention in recent years (Poulton and Winn 2002; van Kleunen and Fisher 2005), and the topic has been singled out as a pending issue in our understanding of the evolution and maintenance of phenotypic plasticity (Pigliucci 2005). Acknowledging the possible selective role of animals on levels of within-plant variation can also contribute some insights to this topic. Selection against within-individual variability, such as that exemplified in table 10.2 by insect pollinators or avian frugivores, would tend to amplify the costs of plasticity. In contrast, selection favoring subindividual variability, such as that exemplified by fruit predators, could reduce such costs. Second, by influencing the shape of reaction norms, and particularly their slope (fig. 10.2), selection by animals can contribute to either enlarge or contract the size of the realized phenotypic space at the genotype and population levels. These effects would not be too unrelated to current theories conferring a central role to phenotypic plasticity in differentiation of adaptive strategies, and ultimately macroevolutionary diversification, through phenotypic accommodation (West-Eberhard 1989, 2003, 2005). It seems, however, somewhat speculative at present to pursue this line of reasoning further.

Epilogue

The intellectual eye must work in constant and spirited harmony with the bodily eye, for otherwise the scholar might run the risk of looking and yet overlooking.
—Johann Wolfgang von Goethe, "My Discovery of a Worthy Forerunner"

Because of their modular construction and reiteration of structures, plants inevitably display a certain level of subindividual phenotypic variation in organ traits. Depending on trait and species, nontrivial proportions of the total phenotypic variance exhibited by a population can be found at the relatively restricted spatial scale of single individuals. Quantitative measures can be used to confirm the subjective impression of variability left on us by even casual glances at individual plants, no matter if they are small herbs or big trees. Despite its obviousness, however, subindividual variation in plants seems to have succumbed so far to the "risk of looking and yet overlooking." We have traditionally overlooked one of the main sources of phenotypic variation occurring in plant populations, largely on the misguided understanding that such variation, being the product of the same genotype, is invisible to selection and thus irrelevant from an evolutionary perspective.

It has been my main purpose in this book to convey the message that, when one looks at subindividual variability with open eyes, a feature that was either unnoticed or taken as a nuisance turns into an opportunity for framing new questions, identifying novel biological mechanisms linking sessile plants and mobile choosy animals, and deepening our understanding of the ecological and evolutionary factors involved in plant-animal interactions. Far from being just noise or an annoyance, subindividual variability is part of the essence of being a plant, and I contend that its neglect has hidden from our view significant ecological and evolutionary facets of plant-animal interactions, and autecology more generally. In addition, explicit recognition of subindividual plant variability as a phenomenon worthy of consideration on its own may eventually shed some light on the factors impinging on the entangled relationship between genotypes and phenotypes, and improve our understanding of the evolution of the genotype-phenotype mapping (sensu Wagner and Altenberg 1996), two topics that still figure prominently among the pending issues of the research agenda of evolutionary biology. The modular reiteration that characterizes each individual in a plant population provides us with ample genetically identical copies of homologous organs that, in a very real sense, may be taken as developmental repetitions, or phenotypic reruns by the same genotype in the course of its lifetime under different internal and external environmental conditions, and subject to a variable constellation of constraints. From a pragmatic viewpoint, reiterated organs of the same class produced by a plant can thus be treated as replicated experimental units for which we have a reasonable presumption of genetic identity. Having at our disposal a set of phenotypic replicates of a given organ produced by the same individual provides an opportunity to evaluate the realized phenotypic space of a single genotype and, if such space could be broadened or narrowed experimentally (e.g., by spatially or temporally localized application of hormones or growth substances), to assess the magnitude and regulating factors of hidden, unrealized phenotypic possibilities available to a single plant genotype. In this respect, subindividually variable plants have distinct advantages over animals as study subjects for investigating the ecological and evolutionary significance of developmental plasticity.

Within-plant variability in organ traits may influence individual fitness through a variety of mechanisms, which include coping with spatial and temporal environmental uncertainty through diversified bet hedging, and enhanced exploitation of biotically and abiotically heterogeneous environments through division of labor. It is most remarkable that exactly the same mechanisms have also been implicated in the evolution of within-genotype phenotypic differences in some animals that, like plants, produce a simultaneous multiplicity of homologous but developmentally separate phenotypes. Aphid clones that are produced parthenogenetically from a single stem mother are often phenotypically heterogeneous, and the extent and nature of such intraclonal variation may have adaptive value (Andrade and Roitberg 1995). A bet-hedging strategy similar to that commonly attributed to plants that produce a variety of seed types is exhibited by some fishes and frogs, in which within-female egg-size variability can be interpreted as an evolved response to environmental unpredictability (Koops et al. 2003; Einum and Fleming 2004; Dziminski and Alford 2005; Dziminski and Roberts 2006). Division of labor has played a central role in the evolution of within-genotype phenotypic variance in ant, bee, wasp, and termite societies, each of whose colonies may be seen as sort of modular, albeit physically disconnected, organism. Phenotypically distinct but genetically identical colony members play different roles according to a discernible scheme of division of labor (Wilson 1971; Goulson 2003). As in plants, the phenotypic variation occurring within these insect societies may be either discontinuous or continuous, and within-colony variance in individual-level traits is related to the colony's capacity to exploit resources. There is much more than appealing symmetry and incidental parallelism in the similarity between a plant individual that exploits different pollinators or seed dispersers by producing a range of slightly different flowers or fruits, and a bumblebee colony that is able to exploit a broader range of flower types because of its ability to produce different-sized workers (Morse 1978; Plowright and Plowright 1997; Peat et al. 2005). Or between a plant that exploits a heterogeneous environment by producing diverse seed types, and the colony of granivorous ants that exploits a broad spectrum of seed sizes because of the broad within-colony diversity of worker sizes (Waser 1998). The common theme underlying these examples is that the capacity of a single genotype to produce a multiplicity of slightly different phenotypes has probably represented a significant, convergent evolutionary breakthrough for sessile modular or modular-like organisms, no matter whether they are plants or animals. The odd thing, however, is that although plants are widely recognized as the most vivid example of the capacity of single genotypes to produce a broad range of phenotypes, the evolutionary etiology of phenotypic multiplicity in genetic unity has been studied much more often, and is currently better understood, in the relatively few animals that present it than in higher plants, where the phenomenon is widespread and represents a distinctive feature of the whole lineage. Furthermore, developmental instability and phenotypic plasticity, two of the mechanisms that may contribute to decoupling ("loosening") the mapping of genotypes into phenotypes, have also been studied much more frequently in animals than in plants.

In most biological disciplines dealing with quantitative data, analyses have traditionally focused on mean values alone, on the usually unstated assumption that variation around the mean has little or no biological significance. Attention has been paid to variability only by virtue of its practical value in assessing the statistical significance of differences between means. By highlighting the ecological and evolutionary significance of phenotypic variability occurring around individual means, this book joins the increasing number of voices that have argued in recent years that variability around means of biological phenomena, functionalities, and phenotypes are intrinsically interesting and have manifold ecological, evolutionary, and physiological implications (Suomela and Ayres 1994; Ruel and Ayres 1999; Shelton 2000, 2004, 2005; Amzallag 2001; Orians and Jones 2001; Benedetti-Cecchi 2003; Biernaskie and Cartar 2004; Roslin et al. 2006). At the same time, I have also revisited an old theme, one that was first examined by Karl Pearson and early biometricians a century ago, and again by Thomas Whitham and associates some 20 years ago. The interest in subindividual plant variation of these two forerunners was committed to, respectively, a particular theory of inheritance and a particular mechanism of generation of subindividual variability. In both instances, the interest in subindividual variability quickly faded away following the failure of the specific theory or mechanism with which it had been incidentally associated. Learning from past culs-de-sac, but also because I am convinced that subindividual variability is interesting in itself, I have deliberately kept this book from pivoting on any particular hypothesis or expectation. In particular, I have refrained from reducing the rich, complex phenomenon of subindividual variation to answering the kind of sketchy questions under which it has been sometimes examined. In particular, I have avoided formulating the simplistic, self-defeating question of whether subindividual variation is an adaptation *or* reflects *just* epigenetic noise (e.g., Whitham 1981). It should be clear from the preceding chapters that, although the answer will be highly species- and trait-specific, in most instances it will be, "Possibly both." More and better information is needed on interspecific patterns of subindividual variability in compositional and functional organ traits (e.g., chemistry of leaves, fruits, and seeds), relative importance of different causal mechanisms (e.g., direct architectural effects, organ-level developmental plasticity, developmental instability), and ecological correlates of subindividual variability (e.g., relationships with diverse animal interactors), before all-encompassing questions can be answered. Regardless of the complexities involved, and independent of any particular theory about its adaptive value, there is little doubt that within-plant variation in organ traits is a biological phenomenon whose multifarious implications render it worthy of study in itself. Only time and tests of the general hypotheses and implications I have advanced in this book will eventually tell if, in addition to being interesting, the phenomenon is also important.

Literature Cited

Abbe, E. C., L. F. Randolph, and J. Einset. 1941. The developmental relationship between shoot apex and growth pattern of leaf blade in diploid maize. American Journal of Botany 28: 778–784.

Abrahamson, W. G., and J. N. Layne. 2002. Relation of ramet size to acorn production in five oak species of xeric upland habitats in south-central Florida. American Journal of Botany 89: 124–131.

Ackerly, D. D., and R. K. Monson. 2003. Waking the sleeping giant: the evolutionary foundations of plant function. International Journal of Plant Sciences 164, suppl.: S1–S6.

Acosta, F. J., J. M. Serrano, C. Pastor, and F. López. 1993. Significant potential levels of hierarchical phenotypic selection in a woody perennial plant, *Cistus ladanifer*. Oikos 68: 267–272.

Agrawal, A. A. 2001. Phenotypic plasticity in the interactions and evolution of species. Science 294: 321–326.

Aigner, P. A. 2005. Variation in pollination performance gradients in a *Dudleya* species complex: can generalization promote floral divergence? Functional Ecology 19: 681–689.

Aizen, M. A., and E. Raffaele. 1996. Nectar production and pollination in *Alstroemeria aurea*: responses to level and pattern of flowering shoot defoliation. Oikos 76: 312–322.

Alados, C. L., M. L. Giner, L. Dehesa, J. Escós, F. G. Barroso, J. M. Emlen, and D. C. Freeman. 2002. Developmental instability and fitness in *Periploca laevigata* experiencing grazing disturbance. International Journal of Plant Sciences 163: 969–978.

Alados, C. L., T. Navarro, and B. Cabezudo. 1999. Tolerance assessment of *Cistus ladanifer* to serpentine soils by developmental stability analysis. Plant Ecology 143: 51–66.

Alados, C. L., T. Navarro, J. Escós, B. Cabezudo, and J. M. Emlen. 2001. Translational and fluctuating asymmetry as tools to detect stress in stress-adapted and nonadapted plants. International Journal of Plant Sciences 162: 607–616.

Albani, M. C., and M. J. Wilkinson. 1998. Inter simple sequence repeat polymerase chain reaction for the detection of somaclonal variation. Plant Breeding 117: 573–575.

Ali, M. S., and K. Kikuzawa. 2005. Anisophylly in *Aucuba japonica* (Cornaceae): an outcome of spatial crowding in the bud. Canadian Journal of Botany 83: 143–154.

Alliende, M. C. 1989. Demographic studies of a dioecious tree: 2, The distribution of leaf predation within and between trees. Journal of Ecology 77: 1048–1058.

Aloni, B., E. Pressman, and L. Karni. 1999. The effect of fruit load, defoliation and night temperature on the morphology of pepper flowers and on fruit shape. Annals of Botany 83: 529–534.

Alonso, C. 1997a. Choosing a place to grow: importance of within-plant abiotic microenvironment for *Yponomeuta mahalebella*. Entomologia Experimentalis et Applicata 83: 171–180.

Alonso, C. 1997b. Variaciones en las relaciones planta-insectos fitófagos: efectos de factores bióticos y abióticos. Ph.D. thesis, Universidad de Sevilla.

Alonso, C., and C. M. Herrera. 1996. Variation in herbivory within and among plants of *Daphne laureola* (Thymelaeaceae): correlation with plant size and architecture. Journal of Ecology 84: 495–502.

Alonso-Blanco, C., H. Blankestijn–de Vries, C. J. Hanhart, and M. Koornneef. 1999. Natural allelic variation at seed size loci in relation to other life history traits of *Arabidopsis thaliana*. Proceedings of the National Academy of Sciences of the USA 96: 4710–4717.

Alpert, P. 1999. Clonal integration in *Fragaria chiloensis* differs between populations: ramets from grassland are selfish. Oecologia 120: 69–76.

Alves-Costa, C. P., and C. Knogge. 2005. Larval competition in weevils *Revena rubiginosa* (Coleoptera: Curculionidae) preying on seeds of the palm *Syagrus romanzoffiana* (Arecaceae). Naturwissenschaften 92: 265–268.

Al-Zahim, M. A., B. V. Ford-Lloyd, and H. J. Newbury. 1999. Detection of somaclonal variation in garlic (*Allium sativum* L.) using RAPD and cytological analysis. Plant Cell Reports 18: 473–477.

Amzallag, G. N. 2001. Data analysis in plant physiology: are we missing the reality? Plant Cell and Environment 24: 881–890.

Andersen, M. C. 1992. An analysis of variability in seed settling velocities of several wind-dispersed Asteraceae. American Journal of Botany 79: 1087–1091.

Anderson, P., and J. Agrell. 2005. Within-plant variation in induced defence in developing leaves of cotton plants. Oecologia 144: 427–434.

Andrade, M. C. B., and B. D. Roitberg. 1995. Rapid response to intraclonal selection in the pea aphid (*Acyrthosiphon pisum*). Evolutionary Ecology 9: 397–410.

Anne, P., F. Mawri, S. Gladstone, and D. C. Freeman. 1998. Is fluctuating asymmetry a reliable biomonitor of stress? A test using life history parameters in soybean. International Journal of Plant Sciences 159: 559–565.

Anstey, L. J., D. T. Quiring, and D. P. Ostaff. 2002. Seasonal changes in intra-tree distribution of immature balsam fir sawfly (Hymenoptera: Diprionidae). Canadian Entomologist 134: 529–538.

Antlfinger, A. E. 1986. Field germination and seedling growth of CH and CL progeny of *Impatiens capensis* (Balsaminaceae). American Journal of Botany 73: 1267–1273.

Antolin, M. F., and C. Strobeck. 1985. The population genetics of somatic mutation in plants. American Naturalist 126: 52–62.

Antonovics, J., and J. Schmitt. 1986. Paternal and maternal effects on propagule size in *Anthoxanthum odoratum*. Oecologia 69: 277–282.

Aranda, I., F. Pardo, L. Gil, and J. A. Pardos. 2004. Anatomical basis of the change in leaf mass per area and nitrogen investment with relative irradiance within the canopy of eight temperate tree species. Acta Oecologica 25: 187–195.

Arber, A. 1919. On heterophylly in water plants. American Naturalist 53: 272–278.

Arber, A. 1931. Studies in floral morphology: 2, On some normal and abnormal crucifers, with a discussion on teratology and atavism. New Phytologist 30: 172–203.

Arencibia, A. D., E. R. Carmona, M. T. Cornide, S. Castiglione, J. O'Relly, A. Chinea, P. Oramas, and F. Sala. 1999. Somaclonal variation in insect-resistant transgenic sugarcane (*Saccharum* hybrid) plants produced by cell electroporation. Transgenic Research 8: 349–360.

Armbruster, W. S., V. S. di Stilio, J. D. Tuxill, T. C. Flores, and J. L. Velásquez Runk. 1999. Covariance and decoupling of floral and vegetative traits in nine Neotropical plants: a re-evaluation of Berg's correlation-pleiades concept. American Journal of Botany 86: 39–55.

Armesto, J. J., G. P. Cheplick, and M. J. McDonnell. 1983. Observations on the reproductive biology of *Phytolacca americana* (Phytolaccaceae). Bulletin of the Torrey Botanical Club 110: 380–383.

Arnold, S. J., and M. J. Wade. 1984a. On the measurement of natural and sexual selection: theory. Evolution 38: 709–719.

Arnold, S. J., and M. J. Wade. 1984b. On the measurement of natural and sexual selection: applications. Evolution 38: 720–734.

Arp, P. A., and H. H. Krause. 1984. The forest floor: lateral variability as revealed by systematic sampling. Canadian Journal of Soil Science 64: 423–437.

Aschan, G., and H. Pfanz. 2003. Non-foliar photosynthesis: a strategy of additional carbon acquisition. Flora 198: 81–97.

Aschan, G., H. Pfanz, D. Vodnik, and F. Batic. 2005. Photosynthetic performance of vegetative and reproductive structures of green hellebore (*Helleborus viridis* L. agg.). Photosynthetica 43: 55–64.

Ashby, E. 1948. Studies in the morphogenesis of leaves: 1, An essay on leaf shape. New Phytologist 47: 153–176.

Ashman, T. L. 1992. Indirect costs of seed production within and between seasons in a gynodioecious species. Oecologia 92: 266–272.

Ashman, T. L., and M. S. Hitchens. 2000. Dissecting the causes of variation in intra-inflorescence allocation in a sexually polymorphic species, *Fragaria virginiana* (Rosaceae). American Journal of Botany 87: 197–204.

Ashman, T. L., and D. J. Schoen. 1996. Floral longevity: fitness consequences and resource costs. In D. G. Lloyd and S. C. H. Barrett, eds., Floral biology, pp. 112–139. Chapman and Hall, New York.

Åström, M., P. Lundberg, and K. Danell. 1990. Partial prey consumption by browsers: trees as patches. Journal of Animal Ecology 59: 287–300.

Audergon, J. M., P. Monestiez, and R. Habib. 1993. Spatial dependences and sampling in a fruit tree: a new concept for spatial prediction in fruit studies. Journal of Horticultural Science 68: 99–112.

Augspurger, C. K. 1986. Double- and single-seeded indehiscent legumes of *Platypodium elegans*: consequences for wind dispersal and seedling growth and survival. Biotropica 18: 45–50.

Augspurger, C. K., and S. E. Franson. 1987. Wind dispersal of artificial fruits varying in mass, area, and morphology. Ecology 68: 27–42.

Augspurger, C. K., and S. E. Franson. 1993. Consequences for seed distributions of intra-crop variation in wing-loading of wind-dispersed species. Vegetatio 107/108: 121–132.

Augspurger, C. K., and K. P. Hogan. 1983. Wind dispersal of fruits with variable seed number in a tropical tree (*Lonchocarpus pentaphyllus*: Leguminosae). American Journal of Botany 70: 1031–1037.

Avery, M. L., D. G. Decker, J. S. Humphrey, A. A. Hayes, and C. C. Laukert. 1995. Color, size, and location of artificial fruits affect sucrose avoidance by cedar waxwings and European starlings. Auk 112: 436–444.

Avery, M. L., K. J. Goocher, and M. A. Cone. 1993. Handling efficiency and berry size preferences of cedar waxwings. Wilson Bulletin 105: 604–611.

Avery, M. L., C. L. Schreiber, and D. G. Decker. 1999. Fruit sugar preferences of house finches. Wilson Bulletin 111: 84–88.

Avila-Sakar, G., L. L. Leist, and A. G. Stephenson. 2003. Effects of the spatial pattern of leaf damage on growth and reproduction: nodes and branches. Journal of Ecology 91: 867–879.

Avila-Sakar, G., and A. G. Stephenson. 2006. Effects of the spatial pattern of leaf damage on growth and reproduction: whole plants. International Journal of Plant Sciences 167: 1021–1028.

Bagchi, S. K., V. P. Sharma, and P. K. Gupta. 1989. Developmental instability in leaves of *Tectona grandis*. Silvae Genetica 38: 1–6.

Bairlein, F. 1996. Fruit-eating in birds and its nutritional consequences. Comparative Biochemistry and Physiology 113A: 215–224.

Baker, H. G., and I. Baker. 1982. Chemical constituents of nectar in relation to pollination mechanisms and phylogeny. In M. H. Nitecki, ed., Biochemical aspects of evolutionary biology, pp. 131–171. University of Chicago Press, Chicago.

Baker, H. G., I. Baker, and S. A. Hodges. 1998. Sugar composition of nectars and fruits consumed by birds and bats in the tropics and subtropics. Biotropica 30: 559–586.

Baker, R. R. 1969. The evolution of the migratory habit in butterflies. Journal of Animal Ecology 38: 703–746.

Balaguer, L., E. Martínez-Ferri, F. Valladares, M. E. Pérez-Corona, F. J. Baquedano, F. J. Castillo, and E. Manrique. 2001. Population divergence in the plasticity of the response of *Quercus coccifera* to the light environment. Functional Ecology 15: 124–135.

Bangerth, F., and L. C. Ho. 1984. Fruit position and fruit set sequence in a truss as factors determining final size of tomato fruits. Annals of Botany 53: 315–319.

Bañuelos, M. J., and J. R. Obeso. 2003. Maternal provisioning, sibling rivalry and seed mass variability in the dioecious shrub *Rhamnus alpinus*. Evolutionary Ecology 17: 19–31.

Barrett, B. A. 1994. Within-tree distribution of *Phyllonorycter blancardella* (F.) and *P. crataegella* (Clemens) (Lepidoptera: Gracillariidae) and associated levels of parasitism in commercial apple orchards. Biological Control 4: 74–79.

Barrett, S. C. H. 1998. The evolution of mating strategies in flowering plants. Trends in Plant Science 3: 335–341.

Barrett, S. C. H. 2003. Mating strategies in flowering plants: the outcrossing-selfing paradigm and beyond. Philosophical Transactions of the Royal Society of London B 358: 991–1004.

Barrett, S. C. H., and L. D. Harder. 1992. Floral variation in *Eichhornia paniculata* (Spreng.) Solms (Pontederiaceae): 2, Effects of development and environment on the formation of selfing flowers. Journal of Evolutionary Biology 5: 83–107.

Barrett, S. C. H., L. K. Jesson, and A. M. Baker. 2000. The evolution and function of stylar polymorphisms in flowering plants. Annals of Botany, suppl. A, 85: 253–265.

Barritt, B. H., C. R. Rom, K. R. Guelich, S. R. Drake, and M. A. Dilley. 1987. Canopy position and light effects on spur, leaf, and fruit characteristics of 'Delicious' apple. HortScience 22: 402–405.

Basile, D. V., and M. R. Basile. 1993. The role and control of the place dependent suppression of cell division in plant morphogenesis and phylogeny. Memoirs of the Torrey Botanical Club 25: 63–84.

Baskin, C. C., and J. M. Baskin. 1998. Seeds: ecology, biogeography, and evolution of dormancy and germination. Academic Press, London.

Bassow, S. L., and F. A. Bazzaz. 1997. Intra- and inter-specific variation in canopy photosynthesis in a mixed deciduous forest. Oecologia 109: 507–515.

Baten, W. D. 1935. Constancy in the number of ligulate flowers of *Chrysanthemum leucanthemum*, variety *pinnatifidum*, during the flowering season. Biometrika 27: 260–266.

Baten, W. D. 1936. Influence of position on structure of inflorescences of *Cicuta maculata*. Biometrika 28: 64–83.

Bateson, M. 2002. Recent advances in our understanding of risk-sensitive foraging preferences. Proceedings of the Nutrition Society 61: 509–516.

Bateson, W. 1894. Materials for the study of variation. Macmillan, London.

Bateson, W. 1901. Heredity, differentiation, and other conceptions of biology: a consideration of Professor Karl Pearson's paper "On the principle of homotyposis." Proceedings of the Royal Society of London 69: 193–205.

Bawa, K. S., and C. J. Webb. 1983. Floral variation and sexual differentiation in *Muntingia calabura* (Elaeocarpaceae), a species with hermaphrodite flowers. Evolution 37: 1271–1282.

Bayman, P., L. L. Lebrón, R. L. Tremblay, and D. J. Lodge. 1997. Variation in endophytic fungi from roots and leaves of *Lepanthes* (Orchidaceae). New Phytologist 135: 143–149.

Bechyne, M., and Z. P. Kondra. 1970. Effect of seed pod location on fatty acid composition of seed oil from rapeseed (*Brassica napus* and *B. campestris*). Canadian Journal of Plant Science 50: 151–154.

Beckett, P. H. T., and R. Webster. 1971. Soil variability: a review. Soils Fertility 34: 1–15.

Benedetti-Cecchi, L. 2003. The importance of the variance around the mean effect size of ecological processes. Ecology 84: 2335–2346.

Bennett, J. O., A. H. Krishnan, W. J. Wiebold, and H. B. Krishnan. 2003. Positional effect on protein and oil content and composition of soybeans. Journal of Agricultural and Food Chemistry 51: 6882–6886.

Bentz, J., J. Reeves, P. Barbosa, and B. Francis. 1995. Within-plant variation in nitrogen and sugar content of *Poinsettia* and its effects on the oviposition pattern, survival, and development of *Bemisia argentifolii* (Homoptera: Aleyrodidae). Environmental Entomology 24: 271–277.

Berg, H. 2000. Differential seed dispersal in *Oxalis acetosella*, a cleistogamous perennial herb. Acta Oecologica 21: 109–118.

Berg, H., U. Becker, and D. Matthies. 2005. Phenotypic plasticity in *Carlina vulgaris*: effects of geographical origin, population size, and population isolation. Oecologia 143: 220–231.

Berg, R. L. 1959. A general evolutionary principle underlying the origin of developmental homeostasis. American Naturalist 93: 103–105.

Berg, R. L. 1960. The ecological significance of correlation pleiades. Evolution 14: 171–180.

Berger, C. A., and E. R. Witkus. 1954. The cytology of *Xanthisma texanum* D. C.: 1, Differences in the chromosome number of root and shoot. Bulletin of the Torrey Botanical Club 81: 489–491.

Berná, G., P. Robles, and J. L. Micol. 1999. A mutational analysis of leaf morphogenesis in *Arabidopsis thaliana*. Genetics 152: 729–742.

Bernasconi, G. 2004. Seed paternity in flowering plants: an evolutionary perspective. Perspectives in Plant Ecology, Evolution, and Systematics 6: 149–158.

Bernays, E. A., and W. T. Wcislo. 1994. Sensory capabilities, information processing, and resource specialization. Quarterly Review of Biology 69: 187–204.

Berry, P. E., and R. N. Calvo. 1991. Pollinator limitation and position dependent fruit set in the high Andean orchid *Myrosmodes cochleare* (Orchidaceae). Plant Systematics and Evolution 174: 93–101.

Bertin, N., C. Gary, M. Tchamitchian, and B. E. Vaissiere. 1998. Influence of cultivar, fruit position and seed content on tomato fruit weight during a crop cycle under low and high competition for assimilates. Journal of Horticultural Science and Biotechnology 73: 541–548.

Bertin, R. I. 1982. The ruby-throated hummingbird and its major food plants: ranges, flowering phenology, and migration. Canadian Journal of Zoology 60: 210–219.

Best, L. S., and P. Bierzychudek. 1982. Pollinator foraging on foxglove (*Digitalis purpurea*): a test of a new model. Evolution 36: 70–79.

Biere, A. 1991a. Parental effects in *Lychnis flos-cuculi*: 1, Seed size, germination and seedling performance in a controlled environment. Journal of Evolutionary Biology 3: 447–465.

Biere, A. 1991b. Parental effects in *Lychnis flos-cuculi*: 2, Selection on time of emergence and seedling performance in the field. Journal of Evolutionary Biology 3: 467–486.

Biernaskie, J. M., and R. V. Cartar. 2004. Variation in rate of nectar production depends on floral display size: a pollinator manipulation hypothesis. Functional Ecology 18: 125–129.

Biernaskie, J. M., R. V. Cartar, and T. A. Hurly. 2002. Risk-averse inflorescence departure in hummingbirds and bumble bees: could plants benefit from variable nectar volumes? Oikos 98: 98–104.

Binnie, R. C., and P. E. Clifford. 1999. Sink characteristics of reproductive organs of dwarf bean in relation to likelihood of abscission. Crop Science 39: 1077–1082.

Blake, W. J., M. Kærn, C. R. Cantor, and J. J. Collins. 2003. Noise in eukaryotic gene expression. Nature 422: 633–637.

Blanke, M. M., and F. Lenz. 1989. Fruit photosynthesis. Plant, Cell and Environment 12: 31–46.

Blarer, A., T. Keasar, and A. Shmida. 2002. Possible mechanisms for the formation of flower size preferences by foraging bumblebees. Ethology 108: 341–351.

Blem, C. R., L. B. Blem, J. Felix, and J. van Gelder. 2000. Rufous hummingbird sucrose preference: precision of selection varies with concentration. Condor 102: 235–238.

Bodkin, P. C., D. H. N. Spence, and D. C. Weeks. 1980. Photoreversible control of heterophylly in *Hippuris vulgaris* L. New Phytologist 84: 533–542.

Boerner, R. E. J., and S. D. Koslowsky. 1989. Microsite variations in soil chemistry and nitrogen mineralization in a beech-maple forest. Soil Biology and Biochemistry 21: 795–801.

Bohner, J., and F. Bangerth. 1988. Effects of fruit set sequence and defoliation on cell number, cell size and hormone levels of tomato fruits (*Lycopersicon esculentum* Mill.) within a truss. Plant Growth Regulation 7: 141–155.

Boller, E. F., and R. J. Prokopy. 1976. Bionomics and management of *Rhagoletis*. Annual Review of Entomology 21: 223–246.

Bonaccorso, F. 1975. Foraging and reproductive ecology in a community of bats in Panama. Ph.D. thesis, University of Florida. Original not seen. Quoted by Janzen (1977a).

Bond, T. E. T. 1941. On abnormal flowers of *Primula vulgaris* Huds. grown in Ceylon. New Phytologist 40: 152–156.

Boonman, A., E. Prinsen, F. Gilmer, U. Schurr, A. J. M. Peeters, L. Voesenek, and T. L. Pons. 2007. Cytokinin import rate as a signal for photosynthetic acclimation to canopy light gradients. Plant Physiology 143: 1841–1852.

Boose, D. L. 1997. Sources of variation in floral nectar production rate in *Epilobium canum* (Onagraceae): implications for natural selection. Oecologia 110: 493–500.

Borges, R. M. 2005. Do plants and animals differ in phenotypic plasticity? Journal of Biosciences 30: 41–50.

Borisjuk, L., H. Rolletschek, R. Radchuk, W. Weschke, U. Wobus, and H. Weber. 2004. Seed development and differentiation: a role for metabolic regulation. Plant Biology 6: 375–386.

Bosque, C., and R. Calchi. 2003. Food choice by blue-gray tanagers in relation to protein content. Comparative Biochemistry and Physiology 135A: 321–327.

Bostrack, J. M., and W. F. Millington. 1962. On the determination of leaf form in an aquatic heterophyllous species of *Ranunculus*. Bulletin of the Torrey Botanical Club 89: 1–20.

Bouchard, S., M. J. Vonhof, M. B. Fenton, and G. Monette. 2000. Nutrient preferences of Brazilian hummingbirds. Wilson Bulletin 112: 558–562.

Bowers, M. D., and N. E. Stamp. 1992. Chemical variation within and between individuals of *Plantago lanceolata* (Plantaginaceae). Journal of Chemical Ecology 18: 985–995.

Boyce, C. K., M. A. Zwieniecki, G. D. Cody, C. Jacobsen, S. Wirick, A. H. Knoll, and N. M. Holbrook. 2004. Evolution of xylem lignification and hydrogel transport regulation. Proceedings of the National Academy of Sciences of the USA 101: 17555–17558.

Boyd, R. S., T. Jaffré, and J. W. Odom. 1999. Variation in nickel content in the nickel-hyperaccumulating shrub *Psychotria douarrei* (Rubiaceae) from New Caledonia. Biotropica 31: 403–410.

Bradford, D. F., and C. C. Smith. 1977. Seed predation and seed number in *Scheelea* palm fruits. Ecology 58: 667–673.

Bradshaw, A. D. 1965. Evolutionary significance of phenotypic plasticity in plants. Advances in Genetics 13: 115–155.

Bradshaw, A. D. 1972. Some of the evolutionary consequences of being a plant. Evolutionary Biology 5: 25–48.

Bramble, T., T. J. Herrman, T. Loughin, and F. Dowell. 2002. Single kernel protein variance structure in commercial wheat fields in western Kansas. Crop Science 42: 1488–1492.

Bregitzer, P., S. E. Halbert, and P. G. Lemaux. 1998. Somaclonal variation in the progeny of transgenic barley. Theoretical and Applied Genetics 96: 421–425.

Brennan, E. B., and S. A. Weinbaum. 2001. Performance of adult psyllids in no-choice experiments on juvenile and adult leaves of *Eucalyptus globulus*. Entomologia Experimentalis et Applicata 100: 179–185.

Brennan, E. B., S. A. Weinbaum, J. A. Rosenheim, and R. Karban. 2001. Heteroblasty in *Eucalyptus globulus* (Myricales: Myricaceae) affects ovipositional and settling preferences of *Ctenarytaina eucalypti* and *C. spatulata* (Homoptera: Psyllidae). Environmental Entomology 30: 1144–1149.

Brim, C. A., W. M. Schutz, and F. I. Collins. 1967. Nuclear magnetic resonance analysis for oil in soybeans *Glycine max* (L.) Merrill with implications in selection. Crop Science 7: 220–222.

Brink, D. E. 1982. A bonanza-blank pollinator reward schedule in *Delphinium nelsonii* (Ranunculaceae). Oecologia 52: 292–294.

Brito e Abreu, F., and A. Kacelnik. 1999. Energy budgets and risk-sensitive foraging in starlings. Behavioral Ecology 10: 338–345.

Brocard-Gifford, I. M., T. J. Lynch, and R. R. Finkelstein. 2003. Regulatory networks in seeds integrating developmental, abscisic acid, sugar, and light signaling. Plant Physiology 131: 78–92.

Brody, A. K. 1992a. Oviposition choices by a pre-dispersal seed predator (*Hylemya* sp.): 1, Correspondence with hummingbird pollinators, and the role of plant size, density and floral morphology. Oecologia 91: 56–62.

Brody, A. K. 1992b. Oviposition choices by a pre-dispersal seed predator (*Hyle-*

mya sp.): 2, A positive association between female choice and fruit set. Oecologia 91: 63–67.

Brody, A. K., and N. M. Waser. 1995. Oviposition patterns and larval success of a pre-dispersal seed predator attacking two confamilial host plants. Oikos 74: 447–452.

Brookes, P. C., and D. L. Wigston. 1979. Variation of morphological and chemical characteristics of acorns from populations of *Quercus petraea* (Matt.) Liebl., *Q. robur* L. and their hybrids. Watsonia 12: 315–324.

Brooks, J. R., P. J. Schulte, B. J. Bond, R. Coulombe, J. C. Domec, T. M. Hinckley, N. McDowell, and N. Phillips. 2003. Does foliage on the same branch compete for the same water? Experiments on Douglas-fir trees. Trees 17: 101–108.

Broom, F. D., G. S. Smith, D. B. Miles, and T. G. A. Green 1998. Within and between tree variability in fruit characteristics associated with bitter pit incidence of 'Braeburn' apple. Journal of Horticultural Science and Biotechnology 73: 555–561.

Brouat, C., M. Gibernau, L. Amsellem, and D. McKey. 1998. Corner's rules revisited: ontogenetic and interspecific patterns in leaf-stem allometry. New Phytologist 139: 459–470.

Brown, A. H. D., J. E. Grant, and R. Pullen. 1986. Outcrossing and paternity in *Glycine argyrea* by paired fruit analysis. Biological Journal of the Linnean Society 29: 283–294.

Bruck, D. K., and D. R. Kaplan. 1980. Heterophyllic development in *Muehlenbeckia* (Polygonaceae). American Journal of Botany 67: 337–346.

Bruschi, P., P. Grossoni, and F. Bussotti. 2003. Within- and among-tree variation in leaf morphology of *Quercus petraea* (Matt.) Liebl. natural populations. Trees 17: 164–172.

Bruun, H. H., and P. Poschlod. 2006. Why are small seeds dispersed through animal guts: large numbers or seed size per se? Oikos 113: 402–411.

Buide, M. L. 2004. Intra-inflorescence variation in floral traits and reproductive success of the hermaphrodite *Silene acutifolia*. Annals of Botany 94: 441–448.

Bull, J. J. 1987. Evolution of phenotypic variance. Evolution 41: 303–315.

Bullock, S. H. 1981. Aggregation of *Prunus ilicifolia* (Rosaceae) during dispersal and its effect on survival and growth. Madroño 28: 94–95.

Burns, K. C. 2005. Plastic heteroblasty in beach groundsel (*Senecio lautus*). New Zealand Journal of Botany 43: 665–672.

Burnside, R. R. 1975. Convexity and Jensen's inequality. American Mathematical Monthly 82: 1005.

Byers, R. A., D. L. Gustine, and B. G. Moyer. 1977. Toxicity of β-nitropropionic acid to *Trichoplusia ni*. Environmental Entomology 6: 229–232.

Byrne, M., and S. J. Mazer. 1990. The effect of position on fruit characteristics, and relationships among components of yield in *Phytolacca rivinoides* (Phytolaccaceae). Biotropica 22: 353–365.

Caetano-Anollés, G. 1999. High genome-wide mutation rates in vegetatively propagated bermudagrass. Molecular Ecology 8: 1211–1221.

Calcagno, M. P., J. Coll, J. Lloria, F. Faini, and M. E. Alonso-Amelot. 2002. Evaluation of synergism in the feeding deterrence of some furanocoumarins on *Spodoptera littoralis*. Journal of Chemical Ecology 28: 175–191.

Calderini, D. F., and I. Ortiz-Monasterio. 2003. Grain position affects grain macro-nutrient and micronutrient concentrations in wheat. Crop Science 43: 141–151.

Campbell, C. S., J. A. Quinn, G. P. Cheplick, and T. J. Bell. 1983. Cleistogamy in grasses. Annual Review of Ecology and Systematics 14: 411–441.

Campbell, D. R. 1989. Measurements of selection in a hermaphroditic plant: variation in male and female pollination success. Evolution 43: 318–334.

Campbell, D. R. 1992. Variation in sex allocation and floral morphology in *Ipomopsis aggregata* (Polemoniaceae). American Journal of Botany 79: 516–521.

Campbell, D. R. 1998. Multiple paternity in fruits of *Ipomopsis aggregata* (Polemoniaceae). American Journal of Botany 85: 1022–1027.

Campbell, D. R., N. M. Waser, M. V. Price, E. A. Lynch, and R. J. Mitchell. 1991. Components of phenotypic selection: pollen export and flower corolla width in *Ipomopsis aggregata*. Evolution 45: 1458–1467.

Campbell, J. F. 2002. Influence of seed size on exploitation by the rice weevil, *Sitophilus oryzae*. Journal of Insect Behavior 15: 429–445.

Canto, M. A., R. Pérez, M. Medrano, M. C. Castellanos, and C. M. Herrera. 2007. Intraplant variation in nectar sugar composition in two *Aquilegia* species (Ranunculaceae): contrasting patterns under field and greenhouse conditions. Annals of Botany 99: 653–660.

Caraco, T. 1982. Aspects of risk-aversion in foraging white-crowned sparrows. Animal Behaviour 30: 719–727.

Caraco, T., S. Martindale, and T. S. Whittam. 1980. An empirical demonstration of risk-sensitive foraging preferences. Animal Behaviour 28: 820–830.

Carlquist, S. 1969. Toward acceptable evolutionary interpretations of floral anatomy. Phytomorphology 19: 332–362.

Carroll, D. P., and R. F. Luck. 1984. Within-tree distribution of California red scale, *Aonidiella aurantii* (Maskell) (Homoptera: Diaspididae), and its parasitoid *Comperiella bifasciata* Howard (Hymenoptera: Encyrtidae) on orange trees in the San Joaquin Valley. Environmental Entomology 13: 179–183.

Carroll, S. P., and L. Moore. 1993. Hummingbirds take their vitamins. Animal Behaviour 46: 817–820.

Carromero, W., and J. L. Hamrick. 2005. The mating system of *Verbascum thapsus* (Scrophulariaceae): the effect of plant height. International Journal of Plant Sciences 166: 979–983.

Carsey, K. S., and D. F. Tomback. 1994. Growth form distribution and genetic relationships in tree clusters of *Pinus flexilis*, a bird-dispersed pine. Oecologia 98: 402–411.

Cartar, R. V., and L. M. Dill. 1990. Why are bumble bees risk-sensitive foragers? Behavioral Ecology and Sociobiology 26: 121–127.

Caruso, C. M. 2001. Differential selection on floral traits of *Ipomopsis aggregata* growing in contrasting environments. Oikos 94: 295–302.

Casella, E., and R. Ceulemans. 2002. Spatial distribution of leaf morphological and physiological chracteristics in relation to local radiation regime within the canopies of 3-year-old *Populus* clones in coppiced culture. Tree Physiology 22: 1277–1288.

Castellanos, M. C., M. Medrano, and C. M. Herrera. 2008. Genetic and environmental effects on subindividual variation in seed traits in a European *Aquilegia*. Botany 86: 1125–1132.

Castro, J. 1999. Seed mass versus seedling performance in Scots pine: a maternally dependent trait. New Phytologist 144: 153–161.

Catley, J. L., I. R. Brooking, L. J. Davies, and E. A. Halligan. 2002. Temperature and irradiance effects on *Sandersonia aurantiaca* flower shape and pedicel length. Scientia Horticulturae 93: 157–166.

Causin, H. F., and R. D. Wulff. 2003. Changes in the responses to light quality during ontogeny in *Chenopodium album*. Canadian Journal of Botany 81: 152–163.

Cavers, P. B., and M. G. Steele. 1984. Patterns of change in seed weight over time on individual plants. American Naturalist 124: 324–335.

Chabot, B. F., and D. J. Hicks. 1982. The ecology of leaf life spans. Annual Review of Ecology and Systematics 13: 229–259.

Chabot, B. F., T. W. Jurik, and J. F. Chabot. 1979. Influence of instantaneous and integrated light-flux density on leaf anatomy and photosynthesis. American Journal of Botany 66: 940–945.

Chamberlain, C. J. 1935. Gymnosperms: structure and function. University of Chicago Press, Chicago.

Chang, J. C., and T. S. Lin. 2006. GA$_3$ increases fruit weight in 'Yu Her Pau' litchi. Scientia Horticulturae 108: 442–443.

Chaparro, J. X., D. J. Werner, R. W. Whetten, and D. M. Omalley. 1995. Characterization of an unstable anthocyanin phenotype and estimation of somatic mutation rates in peach. Journal of Heredity 86: 186–193.

Chapman, H. M., D. Parh, and N. Oraguzie. 2000. Genetic structure and colonizing success of a clonal, weedy species, *Pilosella officinarum* (Asteraceae). Heredity 84: 401–409.

Chapman, J. L. 1987. Sporne's advancement index revisited. New Phytologist 106: 319–332.

Charles-Edwards, D. A., H. Stutzel, R. Ferraris, and D. F. Beech. 1987. An analysis of spatial variation in the nitrogen content of leaves from different horizons within a canopy. Annals of Botany 60: 421–426.

Charlesworth, D., and B. Charlesworth. 1987. Inbreeding depression and its evolutionary consequences. Annual Review of Ecology and Systematics 18: 237–268.

Charnov, E. L. 1976. Optimal foraging: the marginal value theorem. Theoretical Population Biology 9: 129–136.

Chelle, M. 2005. Phylloclimate or the climate perceived by individual plant organs: what is it? how to model it? what for? New Phytologist 166: 781–790.

Cheplick, G. P. 1997. Effects of endophytic fungi on the phenotypic plasticity of *Lolium perenne* (Poaceae). American Journal of Botany 84: 34–40.

Cheplick, G. P. 1998. Genotypic variation in the regrowth of *Lolium perenne* following clipping: effects of nutrients and endophytic fungi. Functional Ecology 12: 176–184.

Cheptou, P. O., J. Lepart, and J. Escarré. 2001. Differential outcrossing rates in dispersing and non-dispersing achenes in the heterocarpic plant *Crepis sancta* (Asteraceae). Evolutionary Ecology 15: 1–13.

Chittka, L., and J. D. Thomson, eds. 2001. Cognitive ecology of pollination: animal behavior and floral evolution. Cambridge University Press, Cambridge.

Chittka, L., J. D. Thomson, and N. M. Waser. 1999. Flower constancy, insect psychology, and plant evolution. Naturwissenschaften 86: 361–377.

Christensen, K. M., T. G. Whitham, and R. P. Balda. 1991. Discrimination among pinyon pine trees by Clark's nutcrackers: effects of cone crop size and cone characters. Oecologia 86: 402–407.

Chung, J. C., and D. M. Waller. 1986. Patterns of insect predation on seeds of smooth sumac (*Rhus glabra* L.). American Midland Naturalist 116: 315–322.

Cipollini, M. L., and E. W. Stiles. 1991. Seed predation by the bean weevil *Acanthoscelides obtectus* on *Phaseolus* species: consequences for seed size, early growth and reproduction. Oikos 60: 205–214.

Clancy, K. M. 1992. Response of western spruce budworm (Lepidoptera: Tortricidae) to increased nitrogen in artificial diets. Environmental Entomology 21: 331–344.

Clark, C. J., J. R. Poulsen, B. M. Bolker, E. F. Connor, and V. T. Parker. 2005. Comparative seed shadows of bird-, monkey-, and wind-dispersed trees. Ecology 86: 2684–2694.

Clark, L. 1991. Odor detection thresholds in tree swallows and cedar waxwings. Auk 108: 177–180.

Clark, R. W. 1968. JBS: The life and work of J. B. S. Haldane. Coward-McCann, New York.

Clarke, G. M. 1993. The genetic basis of developmental stability: 1, Relationships between stability, heterozygosity and genomic coadaptation. Genetica 89: 15–23.

Clausen, J., D. D. Keck, and W. M. Hiesey. 1941. Regional differentiation in plant species. American Naturalist 75: 231–250.

Clausen, J. J., and T. T. Kozlowski. 1965. Heterophyllous shoots in *Betula papyrifera*. Nature 205: 1030–1031.

Clay, K. 1983a. The differential establishment of seedlings from chasmogamous and cleistogamous flowers in natural populations of the grass *Danthonia spicata* (L.) Beauv. Oecologia 57: 183–188.

Clay, K. 1983b. Variation in the degree of cleistogamy within and among species of the grass *Danthonia*. American Journal of Botany 70: 835–843.

Clay, K. 1990. Fungal endophytes of grasses. Annual Review of Ecology and Systematics 21: 275–297.

Clay K., and C. Schardl. 2002. Evolutionary origins and ecological consequences of endophyte symbiosis with grasses. American Naturalist 160, suppl.: S99–S127.

Clearwater, M. J., and K. S. Gould. 1993. Comparative leaf development of juvenile and adult *Pseudopanax crassifolius*. Canadian Journal of Botany 72: 658–670.

Clements, F. E., and F. L. Long. 1923. Experimental pollination: an outline of the ecology of flowers and insects. Carnegie Institution of Washington, Washington, DC.

Cnaani, J., J. D. Thomson, and D. R. Papaj. 2006. Flower choice and learning in foraging bumblebees: effects of variation in nectar volume and concentration. Ethology 112: 278–285.

Cohen, D. 1966. Optimizing reproduction in a randomly varying environment. Journal of Theoretical Biology 12: 119–129.

Collins, F. I., and J. L. Carter. 1956. Variability in chemical composition of seed from different portions of the soybean plant. Agronomy Journal 48: 216–219.

Condrashoff, S. F. 1964. Bionomics of the aspen leaf miner, *Phyllocnistis populiella* Cham. (Lepidoptera: Gracillariidae). Canadian Entomologist 96: 857–874.

Conner, J. K., and S. Rush. 1996. Effects of flower size and number on pollinator visitation to wild radish, *Raphanus raphanistrum*. Oecologia 105: 509–516.

Conner, J. K., and A. Sterling. 1995. Testing hypotheses of functional relationships: a comparative survey of correlation patterns among floral traits in five insect-pollinated plants. American Journal of Botany 82: 1399–1406.

Conner, J. K., and S. Via. 1993. Patterns of phenotypic and genetic correlations among morphological and life-history traits in wild radish, *Raphanus raphanistrum*. Evolution 47: 704–711.

Connor, D. J., V. O. Sadras, and A. J. Hall. 1995. Canopy nitrogen distribution and the photosynthetic performance of sunflower crops during grain filling: a quantitative analysis. Oecologia 101: 274–281.

Cook, A. D., P. R. Atsatt, and C. A. Simon. 1971. Doves and dove weed: multiple defenses against avian predation. BioScience 21: 277–281.

Cook, S. A., and M. P. Johnson. 1968. Adaptation to heterogeneous environments: I, Variation in heterophylly in *Ranunculus flammula* L. Evolution 22: 496–516.

Cope, J. M., and C. W. Fox. 2003. Oviposition decisions in the seed beetle, *Callosobruchus maculatus* (Coleoptera: Bruchidae): effects of seed size on superparasitism. Journal of Stored Products Research 39: 355–365.

Corbet, S. A. 1978. Bee visits and the nectar of *Echium vulgare* L. and *Sinapis alba* L. Ecological Entomology 3: 25–37.

Corbet, S. A., I. Cuthill, M. Fallows, T. Harrison, and G. Hartley. 1981. Why do nectar-foraging bees and wasps work upwards on inflorescences? Oecologia 51: 79–83.

Corradini, P., C. Edelin, A. Bruneau, and A. Bouchard. 2002. Architectural and genotypic variation in the clonal shrub *Taxus canadensis* as determined from random amplified polymorphic DNA and amplified fragment length polymorphism. Canadian Journal of Botany 80: 205–219.

Cottrell, J. E., and J. E. Dale. 1984. Variation in size and development of spikelets within the ear of barley. New Phytologist 97: 565–573.

Courtney, S. P., and M. I. Manzur. 1985. Fruiting and fitness in *Crataegus monogyna*: the effects of frugivores and seed predators. Oikos 44: 398–406.

Cowart, N. M., and J. H. Graham. 1999. Within- and among-individual variation in fluctuating asymmetry of leaves in the fig (*Ficus carica* L.). International Journal of Plant Sciences 160: 116–121.

Crane, J. C. 1964. Growth substances in fruit setting and development. Annual Review of Plant Physiology 15: 303–326.

Credland, P. F., K. M. Dick, and A. W. Wright. 1986. Relationships between larval density, adult size and egg production in the cowpea seed beetle, *Callosobruchus maculatus*. Ecological Entomology 11: 41–50.

Crespan, M. 2004. Evidence on the evolution of polymorphism of microsatellite markers in varieties of *Vitis vinifera* L. Theoretical and Applied Genetics 108: 231–237.

Cresswell, J. E. 1998. Stabilizing selection and the structural variability of flowers within species. Annals of Botany 81: 463–473.

Cresswell, J. E. 1999. The influence of nectar and pollen availability on pollen transfer by individual flowers of oil-seed rape (*Brassica napus*) when pollinated by bumblebees (*Bombus lapidarius*). Journal of Ecology 87: 670–677.

Cresswell, J. E. 2000. Manipulation of female architecture in flowers reveals a narrow optimum for pollen deposition. Ecology 81: 3244–3249.

Critchfield, W. B. 1960. Leaf dimorphism in *Populus trichocarpa*. American Journal of Botany 47: 699–711.

Critchfield, W. B. 1970. Shoot growth and leaf dimorphism in Boston ivy (*Parthenocissus tricuspidata*). American Journal of Botany 57: 535–542.

Crome, F. H. J. 1975. The ecology of fruit pigeons in tropical Northern Queensland. Australian Wildlife Research 2: 155–185.

Cronquist, A. 1981. An integrated system of classification of flowering plants. Columbia University Press, New York.

Crowley, G. M., and S. T. Garnett. 2001. Food value and tree selection by glossy black-cockatoos *Calyptorhynchus lathami*. Austral Ecology 26: 116–126.

Cruz, A., B. Pérez, A. Velasco, and J. M. Moreno. 2003. Variability in seed germination at the interpopulation, intrapopulation and intraindividual levels of the shrub *Erica australis* in response to fire-related cues. Plant Ecology 169: 93–103.

Cui, K. H., S. B. Peng, Y. Z. Xing, S. B. Yu, C. G. Xu, and Q. Zhang. 2003. Molecular dissection of the genetic relationships of source, sink and transport tissue with yield traits in rice. Theoretical and Applied Genetics 106: 649–658.

Cullen Speirs, D., T. N. Sherratt, and S. F. Hubbard. 1991. Parasitoid diets: does superparasitism pay? Trends in Ecology and Evolution 6: 22–25.

Culley, T. M., and M. R. Klooster. 2007. The cleistogamous breeding system: a review of its frequency, evolution, and ecology in angiosperms. Botanical Review 73: 1–30.

Curtis, J. D., and N. R. Lersten. 1978. Heterophylly in *Populus grandidentata* (Salicaceae) with emphasis on resin glands and extrafloral nectaries. American Journal of Botany 65: 1003–1010.

Curtis, W. M. 1931. Variation in the flowers of *Lychnis dioica*, Linn. New Phytologist 30: 69–72.

D'Amato, F. 1997. Role of somatic mutations in the evolution of higher plants. Caryologia 50: 1–15.

Damuth, J., and I. L. Heisler. 1988. Alternative formulations of multilevel selection. Biology and Philosophy 3: 407–430.

Danell, K., L. Edenius, and P. Lundberg. 1991. Herbivory and tree stand composition: moose patch use in winter. Ecology 72: 1350–1357.

Darwin, C. 1859. On the origin of species. Murray, London.

Darwin, C. 1883. The variation of animals and plants under domestication. 2nd edition, revised. Appleton, New York.

Davis, J. M., M. P. Gordon, and B. A. Smit. 1991. Assimilate movement dictates remote sites of wound-induced gene expression in poplar leaves. Proceedings of the National Academy of Sciences of the USA 88: 2393–2396.

Davis, T. A., and C. Ramanujacharyulu. 1971. Statistical analysis of bilateral symmetry in plant organs. Shankhya, series B, 33: 159–290.

de Boer, N. J. 1999. Pyrrolizidine alkaloid distribution in *Senecio jacobaea* rosettes minimises losses to generalist feeding. Entomologia Experimentalis et Applicata 91: 169–173.

de Candolle, A. P., and K. Sprengel. 1821. Elements of the philosophy of plants. Blackwood, Edinburgh.

Deckert, R. J., and R. L. Peterson. 2000. Distribution of foliar fungal endophytes of *Pinus strobus* between and within host trees. Canadian Journal of Forest Research 30: 1436–1442.

de Jong, G. 1990. Quantitative genetics of reaction norms. Journal of Evolutionary Biology 3: 447–468.

de Jong, T. J., N. M. Waser, and P. G. L. Klinkhamer. 1993. Geitonogamy: the neglected side of selfing. Trends in Ecology and Evolution 8: 321–325.

de Kogel, W. J., M. van der Hoek, and C. Mollema. 1997. Oviposition preference of western flower thrips for cucumber leaves from different positions along the plant stem. Entomologia Experimentalis et Applicata 82: 283–288.

de la Bandera, M. C., and A. Traveset. 2006. Reproductive ecology of *Thymelaea velutina* (Thymelaeaceae): factors contributing to the maintenance of heterocarpy. Plant Systematics and Evolution 256: 97–112.

Delesalle, V. A., and S. J. Mazer. 1996. Nutrient levels and salinity affect gender and floral traits in the autogamous *Spergularia marina*. International Journal of Plant Sciences 157: 621–631.

Dengler, N. G. 1999. Anisophylly and dorsiventral shoot symmetry. International Journal of Plant Sciences 160, suppl.: S67–S80.

Dengler, N. G., and A. A. Sánchez-Burgos. 1988. Effect of light level on the expression of anisophylly in *Paradrymonia ciliosa* (Gesneriaceae). Botanical Gazette 149: 158–165.

Denslow, J. S. 1987. Fruit removal rates from aggregated and isolated bushes of the red elderberry, *Sambucus pubens*. Canadian Journal of Botany 65: 1229–1235.

Dermen, H., and R. N. Stewart. 1973. Ontogenetic study of floral organs of peach (*Prunus persica*) utilizing cytochimeral plants. American Journal of Botany 60: 283–291.

Deschamp, P. A., and T. J. Cooke. 1984. Causal mechanisms of leaf dimorphism in the aquatic angiosperm *Callitriche heterophylla*. American Journal of Botany 71: 319–329.

Deschamp, P. A., and T. J. Cooke. 1985. Leaf dimorphism in the aquatic angiosperm *Callitriche heterophylla*. American Journal of Botany 72: 1377–1387.

de Schepper, S., L. Leus, T. Eeckhaut, E. Van Bockstaele, P. Debergh, and M. De Loose. 2004. Somatic polyploid petals: regeneration offers new roads for breeding Belgian pot azaleas. Plant Cell Tissue and Organ Culture 76: 183–188.

de Silva, H. N., and R. D. Ball. 1997. Mixed model analysis of within-plant variation of fruit weight with implications for sampling of kiwifruit vines. Annals of Botany 79: 411–418.

de Silva, H. N., A. J. Hall, W. M. Cashmore, and D. S. Tustin. 2000. Variation of fruit size and growth within an apple tree and its influence on sampling methods for estimating the parameters of mid-season size distributions. Annals of Botany 86: 493–501.

de Soyza, A. G., and D. T. Kincaid. 1991. Patterns in leaf morphology and photosynthesis in shoots of Sassafras albidum (Lauraceae). American Journal of Botany 78: 89–98.

de Soyza, A. G., D. T. Kincaid, and C. R. Ramirez. 1990. Variability of leaf chlorophyll content in a population of sassafras. Bulletin of the Torrey Botanical Club 117: 167–172.

De Steven, D. 1983. Reproductive consequences of insect seed predation in Hamamelis virginiana. Ecology 64: 89–98.

de Verno, L. L., Y. S. Park, J. M. Bonga, and J. D. Barrett. 1999. Somaclonal variation in cryopreserved embryogenic clones of white spruce (Picea glauca [Moench] Voss.). Plant Cell Reports 18: 948–953.

Devlin, B., and A. G. Stephenson. 1985. Sex differential flower longevity, nectar secretion, and pollinator foraging in a protandrous species. American Journal of Botany 72: 303–310.

DeWitt, T. J., and S. M. Scheiner, eds. 2004a. Phenotypic plasticity: functional and conceptual approaches. Oxford University Press, Oxford.

DeWitt, T. J., and S. M. Scheiner. 2004b. Phenotypic variation from single genotypes: a primer. In T. J. DeWitt and S. M. Scheiner, eds., Phenotypic plasticity: functional and conceptual approaches, pp. 1–9. Oxford University Press, Oxford.

Díaz, M., A. P. Møller, and F. J. Pulido. 2003. Fruit abortion, developmental selection and developmental stability in Quercus ilex. Oecologia 135: 378–385.

Dickason, E. A. 1960. Mortality factors for the vetch bruchid, Bruchus brachialis. Journal of Economic Entomology 53: 555–558.

Diggle, P. K. 1991. Labile sex expression in andromonoecious Solanum hirtum: pattern of variation in floral structure. Canadian Journal of Botany 69: 2033–2043.

Diggle, P. K. 1994. The expression of andromonoecy in Solanum hirtum (Solanaceae): phenotypic plasticity and ontogenic contingency. American Journal of Botany 81: 1354–1365.

Diggle, P. K. 1995. Architectural effects and the interpretation of patterns of fruit and seed development. Annual Review of Ecology and Systematics 26: 531–552.

Diggle, P. K. 1997. Ontogenetic contingency and floral morphology: the effects of architecture and resource limitation. International Journal of Plant Sciences 158, suppl.: S99–S107.

Diggle, P. K. 2003. Architectural effects on floral form and function: a review. In T. F. Stuessy, E. Hörandl, and V. Mayer, eds., Deep morphology: toward a renaissance of morphology in plant systematics, pp. 63–80. Koeltz, Königstein, Germany.

Dixon, M. D., W. C. Johnson, and C. S. Adkisson. 1997. Effects of weevil larvae on acorn use by blue jays. Oecologia 111: 201–208.

Doehlert, D. C., M. S. McMullen, J. L. Jannink, S. Panigrahi, H. Z. Gu, and

N. R. Riveland. 2004. Evaluation of oat kernel size uniformity. Crop Science 44: 1178–1186.

Dogterom, M. H., M. L. Winston, and A. Mukai. 2000. Effect of pollen load size and source (self, outcross) on seed and fruit production in highbush blueberry cv. 'Bluecrop' (*Vaccinium corymbosum*; Ericaceae). American Journal of Botany 87: 1584–1591.

Dohm, M. R. 2002. Repeatability estimates do not always set an upper limit to heritability. Functional Ecology 16: 273–280.

Domínguez, C. A., L. E. Eguiarte, J. Núñez-Farfán, and R. Dirzo. 1998. Flower morphometry of *Rhizophora mangle* (Rhizophoraceae): geographical variation in Mexican populations. American Journal of Botany 85: 637–643.

Donohue, K., E. H. Pyle, D. Messiqua, M. S. Heschel, and J. Schmitt. 2001. Adaptive divergence in plasticity in natural populations of *Impatiens capensis* and its consequences for performance in novel habitats. Evolution 55: 692–702.

Donovan, L. A., S. A. Dudley, D. M. Rosenthal, and F. Ludwig. 2007. Phenotypic selection on leaf water use efficiency and related ecophysiological traits for natural populations of desert sunflowers. Oecologia 152: 13–25.

Doust, A. N. 2001. The developmental basis of floral variation in *Drimys winteri* (Winteraceae). International Journal of Plant Sciences 162: 697–717.

Drake, J. M. 2005. Population effects of increased climate variation. Proceedings of the Royal Society London B 272: 1823–1827.

Dreisig, H. 1985. Movement patterns of a clear-wing hawkmoth, *Hemaris fuciformis*, foraging at red catchfly, *Viscaria vulgaris*. Oecologia 67: 360–366.

Dronamraju, K. R. 1961. Non-genetic polymorphism in *Bauhinia acuminata* L. Journal of Genetics 57: 299–311.

Dronamraju, K. R. 1987. On some aspects of the life and work of John Burdon Sanderson Haldane, F. R. S., in India. Notes and Records of the Royal Society of London 41: 211–237.

Duda, J. J., D. C. Freeman, M. L. Brown, J. H. Graham, A. J. Krzysik, J. M. Emlen, J. C. Zak, and D. A. Kovacic. 2004. Estimating disturbance effects from military training using developmental instability and physiological measures of plant stress. Ecological Indicators 3: 251–262.

Dudash, M. R. 1991. Plant size effects on female and male function in hermaphroditic *Sabatia angularis* (Gentianaceae). Ecology 72: 1004–1012.

Duffield, G. E., R. C. Gibson, P. M. Gilhooly, A. J. Hesse, C. R. Inkley, F. S. Gilbert, and C. J. Barnard. 1993. Choice of flowers by foraging honey bees (*Apis mellifera*): possible morphological cues. Ecological Entomology 18: 191–197.

Dukas, R. 1998. Constraints on information processing and their effects on behavior. In R. Dukas, ed., Cognitive ecology, pp. 89–127. University of Chicago Press, Chicago.

Dukas, R., and L. A. Real. 1993. Learning constraints and floral choice behavior in bumble bees. Animal Behaviour 46: 637–644.

Dungan, J. L., J. N. Perry, M. R. T. Dale, P. Legendre, S. Citron-Pousty, M. J. Fortin, A. Jakomulska, M. Miriti, and M. S. Rosenberg. 2002. A balanced view of scale in spatial statistical analysis. Ecography 25: 626–640.

Dyson, W. G., and G. A. Herbin. 1970. Variation in leaf wax alkanes in cypress trees grown in Kenya. Phytochemistry 9: 585–589.

Dziminski, M. A., and R. A. Alford. 2005. Patterns and fitness consequences of intraclutch variation in egg provisioning in tropical Australian frogs. Oecologia 146: 98–109.

Dziminski, M. A., and J. D. Roberts. 2006. Fitness consequences of variable maternal provisioning in quacking frogs (Crinia georgiana). Journal of Evolutionary Biology 19: 144–155.

East, E. M. 1913. Inheritance of flower size in crosses between species of Nicotiana. Botanical Gazette 55: 177–188.

Edmunds, G. F., and D. N. Alstad. 1978. Coevolution in insect herbivores and conifers. Science 199: 941–945.

Edwards, P. B., W. J. Wanjura, W. V. Brown, and J. M. Dearn. 1990. Mosaic resistance in plants. Nature 347: 434.

Ehlers, B. K., J. M. Olesen, and J. Ågren. 2002. Floral morphology and reproductive success in the orchid Epipactis helleborine: regional and local across-habitat variation. Plant Systematics and Evolution 236: 19–32.

Einum, S., and I. A. Fleming. 2004. Environmental unpredictability and offspring size: conservative versus diversified bet-hedging. Evolutionary Ecology Research 6: 443–455.

Ellegren, H. 2000. Microsatellite mutations in the germline: implications for evolutionary inference. Trends in Genetics 16: 551–558.

Ellmore, G. S., A. E. Zanne, and C. M. Orians. 2006. Comparative sectoriality in temperate hardwoods: hydraulics and xylem anatomy. Botanical Journal of the Linnean Society 150: 61–71.

Ellstrand, N. C. 1983. Floral formula inconstancy within and among plants and populations of Ipomopsis aggregata (Polemoniaceae). Botanical Gazette 144: 119–123.

Ellstrand, N. C. 1984. Multiple paternity within the fruits of the wild radish, Raphanus sativus. American Naturalist 123: 819–828.

Ellstrand, N. C., E. M. Lord, and K. J. Eckard. 1984. The inflorescence as a metapopulation of flowers: position dependent differences in function and form in the cleistogamous species Collomia grandiflora Dougl. ex Lindl. (Polemoniaceae). Botanical Gazette 145: 329–333.

Ellstrand, N. C., and R. J. Mitchell. 1988. Spatial and temporal patterns of floral inconstancy in plants and populations of Ipomopsis aggregata (Polemoniaceae). Botanical Gazette 149: 209–212.

Ellstrand, N. C., and M. L. Roose. 1987. Patterns of genotypic diversity in clonal plant species. American Journal of Botany 74: 123–131.

Ellsworth, D. S., and P. B. Reich. 1993. Canopy structure and vertical patterns of photosynthesis and related leaf traits in a deciduous forest. Oecologia 96: 169–178.

Elowitz, M. B., A. J. Levine, E. D. Siggia, and P. S. Swain. 2002. Stochastic gene expression in a single cell. Science 297: 1183–1186.

Endler, J. A. 1986. Natural selection in the wild. Princeton University Press, Princeton, NJ.

Englund, R. 1993. Fruit removal in *Viburnum opulus*: copious seed predation and sporadic massive seed dispersal in a temperate shrub. Oikos 67: 503–510.

Erhardt, A. 1991. Nectar sugar and amino acid preferences of *Battus philenor* (Lepidoptera, Papilionidae). Ecological Entomology 16: 425–434.

Esau, K. 1977. Anatomy of seed plants. 2nd edition. Wiley, New York.

Eskildsen, L. I., J. M. Olesen, and C. G. Jones. 2004. Feeding response of the Aldabra giant tortoise (*Geochelone gigantea*) to island plants showing heterophylly. Journal of Biogeography 31: 1785–1790.

Espinosa-Garcia, F. J., and J. H. Langenheim. 1990. The endophytic fungal community in leaves of a coastal redwood population: diversity and spatial patterns. New Phytologist 116: 89–97.

Esselman, E. J., L. Jianqiang, D. J. Crawford, J. L. Windus, and A. D. Wolfe. 1999. Clonal diversity in the rare *Calamagrostis porteri* ssp. *insperata* (Poaceae): comparative results for allozymes and random amplified polymorphic DNA (RAPD) and intersimple sequence repeat (ISSR) markers. Molecular Ecology 8: 443–451.

Evans, A. S., and M. Marshall. 1996. Developmental instability in *Brassica campestris* (Cruciferae): fluctuating asymmetry of foliar and floral traits. Journal of Evolutionary Biology 9: 717–736.

Fægri, K., and L. van der Pijl. 1966. The principles of pollination ecology. Pergamon Press, Oxford.

Faeth, S. H. 1985. Host leaf selection by leaf miners: interactions among three trophic levels. Ecology 66: 870–875.

Faeth, S. H. 1990. Aggregation of a leafminer, *Cameraria* sp. nov. (Davis): consequences and causes. Journal of Animal Ecology 59: 569–586.

Faeth, S. H. 1991. Effect of oak leaf size on abundance, dispersion, and survival of the leafminer *Cameraria* sp. (Lepidoptera, Gracillariidae). Environmental Entomology 20: 196–204.

Falconer, D. S., and T. F. C. MacKay. 1996. Introduction to quantitative genetics. 4th edition. Addison Wesley Longman, Harlow, Essex, UK.

Fay, P. A., and T. G. Whitham. 1990. Within-plant distribution of a galling adelgid (Homoptera: Adelgidae): the consequences of conflicting survivorship, growth, and reproduction. Ecological Entomology 15: 245–254.

Fedriani, J. M. 2005. Do frugivorous mice choose where or what to feed on? Journal of Mammalogy 86: 576–586.

Feeny, P. 1970. Seasonal changes in oak leaf tannins and nutrients as a cause of spring feeding by winter moth caterpillars. Ecology 51: 565–581.

Feinsinger, P. 1978. Ecological interactions between plants and hummingbirds in a successional tropical community. Ecological Monographs 48: 269–287.

Feinsinger, P. 1983. Variable nectar secretion in a *Heliconia* species pollinated by hermit hummingbirds. Biotropica 15: 48–52.

Feldman, R., D. F. Tomback, and J. Koehler. 1999. Cost of mutualism: competition, tree morphology, and pollen production in limber pine clusters. Ecology 80: 324–329.

Fenster, C. B. 1991. Selection on floral morphology by hummingbirds. Biotropica 23: 98–101.

Fergason, V. L., J. L. Helm, and M. S. Zuber. 1966. Effect of kernel position on amylose starch content: distribution of amylose within corn endosperm (*Zea mays* L.). Crop Science 6: 273–275.

Fernando, D. D., and D. D. Cass. 1996. Genotypic differentiation in *Butomus umbellatus* (Butomaceae) using isozymes and random amplified polymorphic DNAs. Canadian Journal of Botany 74: 647–652.

Fick, G. N., and D. C. Zimmerman. 1973. Variability in oil content among heads and seeds within heads of sunflowers (*Helianthus annuus* L.). Journal of the American Oil Chemists' Society 50: 529–531.

Field, C. 1983. Allocating leaf nitrogen for the maximization of carbon gain: leaf age as a control on the allocation program. Oecologia 56: 341–347.

Fisher, R. A. 1918. The correlation between relatives on the supposition of Mendelian inheritance. Transactions of the Royal Society of Edinburgh 52: 399–433.

Flatt, T. 2005. The evolutionary genetics of canalization. Quarterly Review of Biology 80: 287–316.

Fleck, S., Ü. Niinemets, A. Cescatti, and J. D. Tenhunen. 2003. Three-dimensional lamina architecture alters light-harvesting efficiency in *Fagus*: a leaf-scale analysis. Tree Physiology 23: 577–589.

Fleming, T. H. 1992. How do fruit- and nectar-feeding birds and mammals track their food resources? In M. D. Hunter, T. Ohgushi, and P. W. Price, eds., Effects of resource distribution on animal-plant interactions, pp. 355–391. Academic Press, New York.

Fletcher, B. S. 1987. The biology of dacine fruit flies. Annual Review of Entomology 32: 115–144.

Font Quer, P. 1979. Diccionario de botánica. Editorial Labor, Barcelona.

Ford, E. D., and P. J. Newbould. 1971. The leaf canopy of a coppiced deciduous woodland: 1, Development and structure. Journal of Ecology 59: 843–862.

Forsberg, J. 1987. Size discrimination among conspecific host plants in two pierid butterflies, *Pieris napi* L. and *Pontia daplidice* L. Oecologia 72: 52–57.

Fortin, M., and Y. Mauffette. 2002. The suitability of leaves from different canopy layers for a generalist herbivore (Lepidoptera: Lasiocampidae) foraging on sugar maple. Canadian Journal of Forest Research 32: 379–389.

Foster, M. S. 1990. Factors influencing bird foraging preferences among conspecific fruit trees. Condor 92: 844–854.

Fourré, J. L., P. Berger, L. Niquet, and P. André. 1997. Somatic embryogenesis and somaclonal variation in Norway spruce: morphogenetic, cytogenetic and molecular approaches. Theoretical and Applied Genetics 94: 159–169.

Fox, C. W., J. D. Martin, M. S. Thakar, and T. A. Mousseau. 1996. Clutch size manipulations in two seed beetles: consequences for progeny fitness. Oecologia 108: 88–94.

Fox, C. W., and T. A. Mousseau. 1995. Determinants of clutch size and seed preference in a seed beetle, *Stator beali* (Coleoptera: Bruchidae). Environmental Entomology 24: 1557–1561.

Franke, E., S. Jackson, and S. Nicolson. 1998. Nectar sugar preferences and absorption in a generalist African frugivore, the Cape white-eye *Zosterops pallidus*. Ibis 140: 501–506.

Frankie, G. W., and R. Coville. 1979. An experimental study of the foraging behavior of selected solitary bee species in the Costa Rican dry forest (Hymenoptera: Apoidea). Journal of the Kansas Entomological Society 52: 591–602.

Franks, T., R. Botta, and M. R. Thomas. 2002. Chimerism in grapevines: implications for cultivar identity, ancestry, and genetic improvement. Theoretical and Applied Genetics 104: 192–199.

Frazee, J. E., and R. J. Marquis. 1994. Environmental contribution to floral trait variation in *Chamaecrista fasciculata* (Fabaceae: Caesalpinoideae). American Journal of Botany 81: 206–215.

Freeman, C. E., and K. C. Head. 1990. Temperature and sucrose composition of floral nectars in *Ipomopsis longiflora* under field conditions. Southwestern Naturalist 35: 423–426.

Freeman, C. E., and D. H. Wilken. 1987. Variation in nectar sugar composition at the intraplant level in *Ipomopsis longiflora* (Polemoniaceae). American Journal of Botany 74: 1681–1689.

Freeman, D. C., M. L. Brown, J. J. Duda, J. H. Graham, J. M. Emlen, A. J. Krzysik, H. Balbach, D. A. Kovacic, and J. C. Zak. 2005. Leaf fluctuating asymmetry, soil disturbance and plant stress: a multiple year comparison using two herbs, *Ipomoea pandurata* and *Cnidoscolus stimulosus*. Ecological Indicators 5: 85–95.

Freeman, D. C., J. M. Emlen, J. H. Graham, R. L. Mara, M. Tracy, and C. L. Alados. 1996. Developmental instability as a bioindicator of ecosystem health. In J. R. Barrow, E. D. McArthur, R. E. Sosebee, and R. J. Tausch, eds., Proceedings: shrubland ecosystem dynamics in a changing environment, pp. 170–177. General Technical Report INT-GTR-338. USDA Forest Service Intermountain Research Station, Ogden, UT.

Freeman, D. C., J. H. Graham, and J. M. Emlen. 1993. Developmental stability in plants: symmetries, stress and epigenesis. Genetica 89: 97–119.

Freeman, D. C., J. H. Graham, M. Tracy, J. M. Emlen, and C. L. Alados. 1999. Developmental instability as a means of assessing stress in plants: a case study using electromagnetic fields and soybeans. International Journal of Plant Sciences 160, suppl.: S157–S166.

Friedmann, F., and T. Cadet. 1976. Observations sur l'hétérophyllie dans les Îles Mascareignes. Adansonia, series 2, 15: 423–440.

Fry, A. 1902. Note on variation in leaves of mulberry trees. Biometrika 1: 258–259.

Fryer, H. C., L. Ascham, A. B. Cardwell, J. C. Frazier, and W. W. Willis. 1954. Effect of fruit cluster position on the ascorbic acid content of tomatoes. Proceedings of the American Society for Horticultural Science 64: 360–364.

Fuentes, M. 1995. The effect of unripe fruits on ripe fruit removal by birds in *Pistacia terebinthus*: flag or handicap? Oecologia 101: 55–58.

Fuentes, M., and E. W. Schupp. 1998. Empty seeds reduce seed predation by birds in *Juniperus osteosperma*. Evolutionary Ecology 12: 823–827.

Fukuta, N., K. Fukuzono, H. Kawaide, H. Abe, and M. Nakayama. 2006. Physical restriction of pods causes seed size reduction of a brassinosteroid-deficient faba bean (*Vicia faba*). Annals of Botany 97: 65–69.

Furnier, G. R., P. Knowles, M. A. Clyde, and B. P. Dancik. 1987. Effects of avian seed dispersal on the genetic structure of whitebark pine populations. Evolution 41: 607–612.

Gain, E. 1904. Étude biométrique sur les variations de la fleur et sur l'hétérostylie de *Pulmonaria officinalis* L. Biometrika 3: 398–458.

Galetto, L., G. Bernardello, and C. A. Sosa. 1998. The relationship between floral nectar composition and visitors in *Lycium* (Solanaceae) from Argentina and Chile: what does it reflect? Flora 193: 303–314.

García, D., J. M. Gómez, R. Zamora, and J. A. Hódar. 2000. Do empty *Juniperus communis* seeds defend filled seeds against predation by *Apodemus sylvaticus*? Écoscience 7: 214–221.

García, D., R. Zamora, J. M. Gómez, and J. A. Hódar. 1999. Bird rejection of unhealthy fruit reinforces the mutualism between juniper and its avian dispersers. Oikos 85: 536–544.

García, M. B. 2003. Sex allocation in a long-lived monocarpic plant. Plant Biology 5: 203–209.

Gardener, C. J., J. G. McIvor, and A. Jansen. 1993. Passage of legume and grass seeds through the digestive tract of cattle and their survival in faeces. Journal of Applied Ecology 30: 63–74.

Gargiullo, M. B., and E. W. Stiles. 1991. Chemical and nutritional differences between two bird-dispersed fruits: *Ilex opaca* and *Ilex verticillata*. Journal of Chemical Ecology 17: 1091–1106.

Garrido, J. L. 2003. Semillas y plántulas de *Helleborus foetidus* L. (Ranunculaceae): variación geográfica, ecología y evolución. Ph.D. thesis, Facultad de Ciencias Experimentales, Universidad de Jaén.

Garrido, J. L., P. J. Rey, X. Cerdá, and C. M. Herrera. 2002. Geographical variation in diaspore traits of an ant-dispersed plant (*Helleborus foetidus*): are ant community composition and diaspore traits correlated? Journal of Ecology 90: 446–455.

Garriz, P. I., H. L. Alvarez, and A. J. Alvarez. 1994. Influence of altered irradiance on fruits and leaves of mature pear trees. Biologia Plantarum 39: 229–234.

Garrod, J. F., and G. P. Harris. 1974. Studies on the glasshouse carnation: effects of temperature and growth substances on petal number. Annals of Botany 38: 1025–1031.

Gaudeul, M., and I. Till-Bottraud. 2004. Reproductive ecology of the endangered alpine species *Eryngium alpinum* L. (Apiaceae): phenology, gene dispersal and reproductive success. Annals of Botany 93: 711–721.

Gavrilets, S., and A. Hastings. 1994. A quantitative-genetic model for selection on developmental noise. Evolution 48: 1478–1486.

Gavrilets, S., and S. M. Scheiner. 1993a. The genetics of phenotypic plasticity: 5, Evolution of reaction norm shape. Journal of Evolutionary Biology 6: 31–48.

Gavrilets, S., and S. M. Scheiner. 1993b. The genetics of phenotypic plasticity: 6, Theoretical predictions for directional selection. Journal of Evolutionary Biology 6: 49–68.

Gawleta, N., Y. Zimmermann, and T. Eltz. 2005. Repellent foraging scent recognition across bee families. Apidologie 36: 325–330.

Gegear, R. J., and T. M. Laverty. 2001. The effect of variation among floral traits on the flower constancy of pollinators. In L. Chittka and J. D. Thomson, eds., Cognitive ecology of pollination, pp. 1–20. Cambridge University Press, Cambridge.

Génard, M. 1992. Influence du nombre de feuilles et de la répartition des fruits sur la production et la qualité des pêches. Canadian Journal of Plant Science 72: 517–525.

Génard, M., and F. Baret. 1994. Spatial and temporal variation of light inside peach-trees. Journal of the American Society for Horticultural Science 119: 669–677.

Génard, M., and C. Bruchou. 1992. Multivariate analysis of within-tree factors accounting for the variation of peach fruit quality. Scientia Horticulturae 52: 37–51.

Ghent, A. W. 1973. Gravity and the distribution of leaf shape in the trees of *Sassafras albidum*. New Phytologist 72: 1141–1158.

Gianoli, E. 1999. Within-plant distribution of *Rhopalosiphum padi* on wheat seedlings is affected by induced responses. Entomologia Experimentalis et Applicata 93: 227–230.

Gibson, S. I. 2005. Control of plant development and gene expression by sugar signaling. Current Opinion in Plant Biology 8: 93–102.

Giles, S., and A. Lill. 1999. The effect of fruit abundance, conspicuousness and sugar concentration on fruit colour choice by captive silvereyes. Ethology, Ecology, and Evolution 11: 229–242.

Gill, D. E. 1986. Individual plants as genetic mosaics: ecological organisms versus evolutionary individuals. In M. Crawley, ed., Plant ecology, pp. 321–343. Blackwell, Oxford.

Gill, D. E., L. Chao, S. L. Perkins, and J. B. Wolf. 1995. Genetic mosaicism in plants and clonal animals. Annual Review of Ecology and Systematics 26: 423–444.

Gill, D. E., and T. G. Halverson. 1984. Fitness variation among branches within trees. In B. Shorrocks, ed., Evolutionary ecology, pp. 105–116. 23rd Symposium of the British Ecological Society. Blackwell, Oxford.

Gillan, T. L., and D. H. S. Richardson. 1997. The chalcid seed wasp, *Megastigmus nigrovariegatus* Hymenoptera: Torymidae, on *Rosa rugosa* Thunb. in Nova Scotia. Canadian Entomologist 129: 809–814.

Gillespie, J. H. 1977. Natural selection for variances in offspring numbers: a new evolutionary principle. American Naturalist 111: 1010–1014.

Giurfa, M., and J. A. Núñez. 1992. Honeybees mark with scent and reject recently visited flowers. Oecologia 89: 113–117.

Givnish, T. J. 1988. Adaptation to sun and shade: a whole-plant perspective. Australian Journal of Plant Physiology 15: 63–92.

Givnish, T. J., K. J. Sytsma, J. F. Smith, and W. J. Hahn. 1994. Thorn-like prickles and heterophylly in *Cyanea*: adaptations to extinct avian browsers on Hawaii? Proceedings of the National Academy of Sciences of the USA 91: 2810–2814.

Glaettli, M., L. Pescatore, and J. Goudet. 2006. Proximity-dependent pollen performance in *Silene vulgaris*. Annals of Botany 98: 431–437.

Glendinning, D. R. 1963. The inheritance of bean size, pod size and number of beans per pod in cocoa (*Theobroma cacao* L.), with a note on bean shape. Euphytica 12: 311–322.

Godoy, J. A., and P. Jordano. 2001. Seed dispersal by animals: exact identification of source trees with endocarp DNA microsatellites. Molecular Ecology 10: 2275–2283.

Goethe, J. W. von. 1817. My discovery of a worthy forerunner. In Goethe's botanical writings, edited and translated by Bertha Mueller, pp. 176–181. University of Hawaii Press, Honolulu, HI, 1952.

Gómez, J. M., F. Perfectti, and J. P. M. Camacho. 2006. Natural selection on *Erysimum mediohispanicum* flower shape: insights into the evolution of zygomorphy. American Naturalist 168: 531–545.

Goodspeed, T. H., and R. E. Clausen. 1915. Factors influencing flower size in *Nicotiana* with special reference to questions of inheritance. American Journal of Botany 2: 332–374.

Gorchov, D. L. 1985. Fruit ripening asynchrony is related to variable seed number in *Amelanchier* and *Vaccinium*. American Journal of Botany 72: 1939–1943.

Gottsberger, G., T. Arnold, and H. F. Linskens. 1990. Variation in floral nectar amino acids with aging of flowers, pollen contamination, and flower damage. Israel Journal of Botany 39: 167–176.

Gould, S. J., and R. F. Johnston. 1972. Geographic variation. Annual Review of Ecology and Systematics 3: 457–498.

Goulson, D. 1999. Foraging strategies of insects for gathering nectar and pollen, and implications for plant ecology and evolution. Perspectives in Plant Ecology, Evolution, and Systematics 2: 185–209.

Goulson, D. 2003. Bumblebees: their behaviour and ecology. Oxford University Press, Oxford.

Goulson, D., J. W. Chapman, and W. O. H. Hughes. 2001. Discrimination of unrewarding flowers by bees: direct detection of rewards and use of repellent scent marks. Journal of Insect Behavior 14: 669–678.

Goulson, D., S. A. Hawson, and J. C. Stout. 1998. Foraging bumblebees avoid flowers already visited by conspecifics or by other bumblebee species. Animal Behaviour 55: 199–206.

Grafius, J. E. 1978. Multiple characters and correlated response. Crop Science 18: 931–934.

Graham, B. F., and F. H. Bormann. 1966. Natural root grafts. Botanical Review 32: 255–292.

Graham, L. E., M. E. Cook, and J. S. Busse. 2000. The origin of plants: body plan changes contributing to a major evolutionary radiation. Proceedings of the National Academy of Sciences of the USA 97: 4535–4540.

Green, S., T. L. Green, and Y. Heslop-Harrison. 1979. Seasonal heterophylly and leaf gland features in *Triphyophyllum* (Dioncophyllaceae), a new carnivorous plant genus. Botanical Journal of the Linnean Society 78: 99–116.

Green, T. R., and C. A. Ryan. 1972. Wound-induced proteinase inhibitor in plant leaves: a possible defense mechanism against insects. Science 175: 776–777.

Greene, D. F., and E. A. Johnson. 1992. Can the variation in samara mass and terminal velocity on an individual plant affect the distribution of dispersal distances? American Naturalist 139: 825–838.

Gregoire, T. G., H. T. Valentine, and G. M. Furnival. 1995. Sampling methods to estimate foliage and other characteristics of individual trees. Ecology 76: 1181–1194.

Greig-Smith, P. W., and D. R. Crocker. 1986. Mechanisms of food size selection by bullfinches (*Pyrrhula pyrrhula* L.) feeding on sunflower seeds. Animal Behaviour 34: 843–859.

Greig-Smith, P. W., and M. F. Wilson. 1985. Influences of seed size, nutrient composition and phenolic content on the preferences of bullfinches feeding in ash trees. Oikos 44: 47–54.

Grigal, D. F., R. E. McRoberts, and L. F. Ohmann. 1991. Spatial variation in chemical properties of forest floor and surface mineral soil in the north central United States. Soil Science 151: 282–290.

Grindeland, J. M., N. Sletvold, and R. A. Ims. 2005. Effects of floral display size and plant density on pollinator visitation rate in a natural population of *Digitalis purpurea*. Functional Ecology 19: 383–390.

Gripenberg, S., and T. Roslin. 2005. Host plants as islands: resource quality and spatial setting as determinants of insect distribution. Annales Zoologici Fennici 42: 335–345.

Gronquist, M., A. Bezzerides, A. Attygalle, J. Meinwald, M. Eisner, and T. Eisner. 2001. Attractive and defensive functions of the ultraviolet pigments of a flower (*Hypericum calycinum*). Proceedings of the National Academy of Sciences of the USA 98: 13745–13750.

Grossmueller, D. W., and R. C. Lederhouse. 1985. Oviposition site selection: an aid to rapid growth and development in the tiger swallowtail butterfly, *Papilio glaucus*. Oecologia 66: 68–73.

Grubb, P. J., and D. F. R. P. Burslem. 1998. Mineral nutrient concentrations as a function of seed size within seed crops: implications for competition among seedlings and defence against herbivory. Journal of Tropical Ecology 14: 177–185.

Guitián, J., M. Medrano, and J. E. Oti. 2004. Variation in floral sex allocation in *Polygonatum odoratum* (Liliaceae). Annals of Botany 94: 433–440.

Guitián, J., I. Munilla, and P. Guitián. 1994. Influencia de los depredadores de aves en el consumo de frutos de *Crataegus monogyna* por zorzales y mirlos. Ardeola 41: 45–54.

Gurrieri, F., J. M. Audergon, G. Albagnac, and M. Reich. 2001. Soluble sugars and carboxylic acids in ripe apricot fruit as parameters for distinguishing different cultivars. Euphytica 117: 183–189.

Gutiérrez, D., R. Menéndez, and J. R. Obeso. 1996. Effect of ovule position on seed maturation and seed weight in *Ulex europaeus* and *Ulex gallii* (Fabaceae). Canadian Journal of Botany 74: 848–853.

Gutschick, V. P. 1999. Biotic and abiotic consequences of differences in leaf structure. New Phytologist 143: 3–18.

Habegger, R., and W. H. Schnitzler. 2000. Aroma compounds in the essential oil of carrots (*Daucus carota* L. ssp *sativus*): 2, Intra-leaf distribution. Journal of Applied Botany 74: 229–232.

Habib, R., D. Tisneagostini, M. P. Vanniere, and P. Monestiez. 1991. Geostatistical method for independent sampling in kiwifruit vine to estimate yield components. New Zealand Journal of Crop and Horticultural Science 19: 329–335.

Hainsworth, F. R., and L. L. Wolf. 1976. Nectar characteristics and food selection by hummingbirds. Oecologia 25: 101–113.

Haldane, J. B. S. 1932. The causes of evolution. Longmans Green, London. Reprint. Princeton University Press, Princeton, NJ, 1990.

Haldane, J. B. S. 1955. The measurement of variation. Evolution 9: 484.

Haldane, J. B. S. 1957. Karl Pearson, 1857–1957. Biometrika 44: 303–313.

Haldane, J. B. S. 1959. The theory of natural selection to-day. Nature 183: 710–713.

Halket, A. C. 1932. A note on the occurrence of abnormal flowers of *Nasturtium officinale* R. Br. New Phytologist 31: 284–286.

Hall, G. D., and J. H. Langenheim. 1986. Within-tree spatial variation in the leaf monoterpenes of *Sequoia sempervirens*. Biochemical Systematics and Evolution 14: 625–632.

Hallé, F., R. A. A. Oldeman, and P. B. Tomlinson. 1978. Tropical trees and forests: an architectural analysis. Springer-Verlag, Berlin.

Halpern, S. L. 2005. Sources and consequences of seed size variation in *Lupinus perennis* (Fabaceae): adaptive and non-adaptive hypotheses. American Journal of Botany 92: 205–213.

Harder, L. D. 1986. Effects of nectar concentration and flower depth on flower handling efficiency of bumble bees. Oecologia 69: 309–315.

Harder, L. D. 1988. Choice of individual flowers by bumble bees: interaction of morphology, time and energy. Behaviour 104: 60–77.

Harder, L. D., and S. C. H. Barrett. 1996. Pollen dispersal and mating patterns in animal-pollinated plants. In D. G. Lloyd and S. C. H. Barrett, eds., Floral biology, pp. 140–190. Chapman and Hall, New York.

Harder, L. D., C. Y. Jordan, W. E. Gross, and M. B. Routley. 2004. Beyond floricentrism: the pollination function of inflorescences. Plant Species Biology 19: 137–148.

Harder, L. D., and L. A. Real. 1987. Why are bumble bees risk averse? Ecology 68: 1104–1108.

Harding, P. L. 1936. Distribution of total soluble solids and catalase in different parts of Jonathan apples. Journal of Agricultural Research 53: 43–48.

Hare, J. D. 1992. Effects of plant variation on herbivore–natural enemy interactions. In R. S. Fritz and E. L. Simms, eds., Plant resistance to herbivores and pathogens, pp. 278–298. University of Chicago Press, Chicago.

Harlow, B. A., R. A. Duursma, and J. D. Marshall. 2005. Leaf longevity of western red cedar (*Thuja plicata*) increases with depth in the canopy. Tree Physiology 25: 557–562.

Harper, J. L. 1977. Population biology of plants. Academic Press, London.

Harper, J. L., P. H. Lovell, and K. G. Moore. 1970. The shapes and sizes of seeds. Annual Review of Ecology and Systematics 1: 327–356.

Harris, J. A. 1909a. The leaves of *Podophyllum*. Botanical Gazette 47: 438–444.

Harris, J. A. 1909b. Note on variation in *Adoxa*. Biometrika 7: 218–222.

Harris, J. A. 1910. A quantitative study of the morphology of the fruit of the blood-root, *Sanguinaria canadensis*. Biometrika 7: 305–351.

Harris, J. A. 1911. On the correlation between somatic characters and fertility: illustrations from the involucral whorl of *Hibiscus*. Biometrika 8: 52–65.

Harris, J. A. 1916. A contribution to the problem of homotyposis: data from the legume *Cercis canadensis*. Biometrika 11: 201–214.

Hata, K., R. Atari, and K. Sone. 2002. Isolation of endophytic fungi from leaves of *Pasania edulis* and their within-leaf distributions. Mycoscience 43: 369–373.

Hay, M. J. M., and N. R. Sackville Hamilton. 1996. Influence of xylem vascular architecture on the translocation of phosphorus from nodal roots in a genotype of *Trifolium repens* during undisturbed growth. New Phytologist 132: 575–582.

Heard, S. B., M. A. Campbell, M. L. Bonine, and S. D. Hendrix. 1999. Developmental instability in fragmented populations of prairie phlox: a cautionary tale. Conservation Biology 13: 274–281.

Heard, T. 1995. Oviposition preferences and larval performance of a flower-feeding weevil, *Coelocephalapion aculeatum*, in relation to host development. Entomologia Experimentalis et Applicata 76: 195–201.

Hegde, S. G., K. N. Ganeshaiah, and R. Uma Shaanker. 1991. Fruit preference criteria by avian frugivores: their implications for the evolution of clutch size in *Solanum pubescens*. Oikos 60: 20–26.

Heinrich, B. 1975. Energetics of pollination. Annual Review of Ecology and Systematics 6: 139–170.

Heinrich, B. 1979. Resource heterogeneity and patterns of movement in foraging bumblebees. Oecologia 40: 235–245.

Heinrich, B., C. C. Joerg, S. S. Madden, and E. W. Sanders. 1997. Black-capped chickadees and red-breasted nuthatches "weigh" sunflower seeds. Auk 114: 298–299.

Heisler, I. L., and J. Damuth. 1987. A method for analyzing selection in hierarchically structured populations. American Naturalist 130: 582–602.

Helenurm, K., and B. A. Schaal. 1996. Genetic and maternal effects on offspring fitness in *Lupinus texensis* (Fabaceae). American Journal of Botany 83: 1596–1608.

Hendrix, S. D. 1984. Variation in seed weight and its effects on germination in *Pastinaca sativa* L. (Umbelliferae). American Journal of Botany 71: 795–802.

Hendrix, S. D., and I. F. Sun. 1989. Inter- and intraspecific variation in seed mass in seven species of umbellifer. New Phytologist 112: 445–451.

Herrera, C. M. 1981a. Are tropical fruits more rewarding to dispersers than temperate ones? American Naturalist 118: 896–907.

Herrera, C. M. 1981b. Fruit variation and competition for dispersers in natural populations of *Smilax aspera*. Oikos 36: 51–58.

Herrera, C. M. 1984a. The annual cycle of *Osyris quadripartita*, a hemiparasitic dioecious shrub of Mediterranean scrublands. Journal of Ecology 72: 1065–1078.

Herrera, C. M. 1984b Selective pressures on fruit seediness: differential predation of fly larvae on the fruits of *Berberis hispanica*. Oikos 42: 166–170.

Herrera, C. M. 1984c. A study of avian frugivores, bird-dispersed plants, and their interaction in Mediterranean scrublands. Ecological Monographs 54: 1–23.

Herrera, C. M. 1985. Predispersal reproductive biology of female *Osyris quadripartita* (Santalaceae), a hemiparasitic dioecious shrub of Mediterranean scrublands. Botanical Journal of the Linnean Society 90: 113–127.

Herrera, C. M. 1987a. Components of pollinator "quality": comparative analysis of a diverse insect assemblage. Oikos 50: 79–90.

Herrera, C. M. 1987b. Vertebrate-dispersed plants of the Iberian Peninsula: a study of fruit characteristics. Ecological Monographs 57: 305–331.

Herrera, C. M. 1988. The fruiting ecology of *Osyris quadripartita*: individual variation and evolutionary potential. Ecology 69: 233–249.

Herrera, C. M. 1989a. Frugivory and seed dispersal by carnivorous mammals, and associated fruit characteristics, in undisturbed Mediterranean habitats. Oikos 55: 250–262.

Herrera, C. M. 1989b. Pollinator abundance, morphology, and flower visitation rate: analysis of the "quantity" component in a plant-pollinator system. Oecologia 80: 241–248.

Herrera, C. M. 1989c. Vertebrate frugivores and their interaction with invertebrate fruit predators: supporting evidence from a Costa Rican dry forest. Oikos 54: 185–188.

Herrera, C. M. 1993. Selection on floral morphology and environmental determinants of fecundity in a hawk moth–pollinated violet. Ecological Monographs 63: 251–275.

Herrera, C. M. 1995a. Microclimate and individual variation in pollinators: flowering plants are more than their flowers. Ecology 76: 1516–1524.

Herrera, C. M. 1995b. Plant-vertebrate seed dispersal systems in the Mediterranean: ecological, evolutionary and historical determinants. Annual Review of Ecology and Systematics 26: 705–727.

Herrera, C. M. 1996. Floral traits and plant adaptation to insect pollinators: a devil's advocate approach. In D. G. Lloyd and S. C. H. Barrett, eds., Floral biology, pp. 65–87. Chapman and Hall, New York.

Herrera, C. M. 2000. Individual differences in progeny viability in *Lavandula latifolia*: a long-term field study. Ecology 81: 3036–3047.

Herrera, C. M. 2002a. Censusing natural microgametophyte populations: variable spatial mosaics and extreme fine-graininess in winter-flowering *Helleborus foetidus* (Ranunculaceae). American Journal of Botany 89: 1570–1578.

Herrera, C. M. 2002b. Seed dispersal by vertebrates. In C. M. Herrera and O. Pellmyr, eds., Plant-animal interactions: an evolutionary approach, pp. 185–208. Blackwell Science, Oxford.

Herrera, C. M. 2004. Distribution ecology of pollen tubes: fine-grained, labile spatial mosaics in southern Spanish Lamiaceae. New Phytologist 161: 473–484.

Herrera, C. M. 2005. Plant generalization on pollinators: species property or local phenomenon? American Journal of Botany 92: 13–20.

Herrera, C. M., M. C. Castellanos, and M. Medrano. 2006. Geographical context of floral evolution: towards an improved research program in floral diversification. In L. D. Harder and S. C. H. Barrett, eds., The ecology and evolution of flowers, pp. 278–294. Oxford University Press, Oxford.

Herrera, C. M., X. Cerdá, M. B. García, J. Guitián, M. Medrano, P. Rey, and A. M. Sánchez-Lafuente. 2002. Floral integration, phenotypic covariance structure and pollinator variation in bumblebee-pollinated *Helleborus foetidus*. Journal of Evolutionary Biology 15: 108–121.

Herrera, C. M., and P. Jordano. 1981. *Prunus mahaleb* and birds: the high-efficiency seed dispersal system of a temperate fruiting tree. Ecological Monographs 51: 203–218.

Herrera, C. M., P. Jordano, J. Guitián, and A. Traveset. 1998. Annual variability in seed production by woody plants and the masting concept: reassessment of principles and relationship to pollination and seed dispersal. American Naturalist 152: 576–594.

Herrera, C. M., P. Jordano, L. López-Soria, and J. A. Amat. 1994. Recruitment of a mast-fruiting, bird-dispersed tree: bridging frugivore activity and seedling establishment. Ecological Monographs 64: 315–344.

Herrera, C. M., and O. Pellmyr, eds. 2002. Plant-animal interactions: an evolutionary approach. Blackwell, Oxford.

Herrera, C. M., R. Pérez, and C. Alonso. 2006. Extreme intraplant variation in nectar sugar composition in an insect-pollinated perennial herb. American Journal of Botany 93: 575–581.

Herrera, C. M., and R. C. Soriguer. 1983. Intra- and inter-floral heterogeneity of nectar production in *Helleborus foetidus* L. (Ranunculaceae). Botanical Journal of the Linnean Society 86: 253–260.

Herrera, L. G., D. Leblanc, and J. Nassar. 2000. Sugar discrimination and gustatory thresholds in captive-born frugivorous Old World bats. Mammalia 64: 135–143.

Heschel, M. S., and C. Riginos. 2005. Mechanisms of selection for drought stress tolerance and avoidance in *Impatiens capensis* (Balsaminaceae). American Journal of Botany 92: 37–44.

Heslop-Harrison, J. 1959. Variability and environment. Evolution 13: 145–147.

Hespenheide, H. A. 1966. The selection of seed size by finches. Wilson Bulletin 78: 191–197.

Hessing, M. B. 1988. Geitonogamous pollination and its consequences in *Geranium caespitosum*. American Journal of Botany 75: 1324–1333.

Hiesey, W. M., J. Clausen, and D. D. Keck. 1942. Relations between climate and intraspecific variation in plants. American Naturalist 76: 5–22.

Hills, M. J. 2004. Control of storage-product synthesis in seeds. Current Opinion in Plant Biology 7: 302–308.

Hirabayashi, Y., H. S. Ishii, and G. Kudo. 2006. Significance of nectar distribution for bumblebee behaviour within inflorescences, with reference to inflorescence architecture and display size. Écoscience 13: 351–359.

Hirano, H. Y., and Y. Sano. 1998. Enhancement of Wx gene expression and the accumulation of amylose in response to cool temperatures during seed development in rice. Plant and Cell Physiology 39: 807–812.

Hirano, H. Y., and Y. Sano. 2000. Comparison of Waxy gene regulation in the endosperm and pollen in *Oryza sativa* L. Genes and Genetic Systems 75: 245–249.

Hirose, T., and M. J. A. Werger. 1987. Maximizing daily canopy photosynthesis with respect to the leaf nitrogen allocation pattern in the canopy. Oecologia 72: 520–526.

Hochwender, C. G., V. L. Sork, and R. J. Marquis. 2003. Fitness consequences of herbivory on *Quercus alba*. American Midland Naturalist 150: 246–253.

Hocquigny, S., F. Pelsy, V. Dumas, S. Kindt, M. C. Heloir, and D. Merdinoglu. 2004. Diversification within grapevine cultivars goes through chimeric states. Genome 47: 579–589.

Hódar, J. A., and R. T. Palo. 1997. Feeding by vertebrate herbivores in a chemically heterogeneous environment. Écoscience 4: 304–310.

Hodge, A. 2004. The plastic plant: root responses to heterogeneous supplies of nutrients. New Phytologist 162: 9–24.

Hodges, C. M., and L. L. Wolf. 1981. Optimal foraging in bumblebees: why is nectar left behind in flowers? Behavioural Ecology and Sociobiology 9: 41–44.

Hodges, S. A. 1995. The influence of nectar production on hawkmoth behavior, self-pollination, and seed production in *Mirabilis multiflora* (Nyctaginaceae). American Journal of Botany 82: 197–204.

Holbrook, K. M., and T. B. Smith. 2000. Seed dispersal and movement patterns in two species of *Ceratogymna* hornbills in a West African tropical lowland forest. Oecologia 125: 249–257.

Hollinger, D. Y. 1989. Canopy organization and foliage photosynthetic capacity in a broad-leaved evergreen montane forest. Functional Ecology 3: 53–62.

Hollinger, D. Y. 1996. Optimality and nitrogen allocation in a tree canopy. Tree Physiology 16: 627–634.

Honěk, A., and Z. Martinková. 2002. Factors of between- and within-plant distribution of *Metopolophium dirhodum* (Hom., Aphididae) on small grain cereals. Journal of Applied Entomology 126: 378–383.

Hopkirk, G., D. J. Beever, and C. M. Triggs. 1986. Variation in soluble solids concentration in kiwifruit at harvest. New Zealand Journal of Agricultural Research 29: 475–484.

Horn, C. N. 1988. Developmental heterophylly in the genus *Heteranthera* (Pontederiaceae). Aquatic Botany 31: 197–209.

Horovitz, A., L. Meiri, and A. Beiles. 1976. Effects of ovule positions in fabaceous flowers on seed set and outcrossing rates. Botanical Gazette 137: 250–254.

Hossaert, M., and M. Valéro. 1988. Effect of ovule position in the pod on patterns of seed formation in two species of *Lathyrus* (Leguminosae: Papilionoideae). American Journal of Botany 75: 1714–1731.

Housley, T. L., and D. M. Peterson. 1982. Oat stem vascular size in relation to kernel number and weight: 1, Controlled environment. Crop Science 22: 259–263.

Howe, H. F. 1979. Fear and frugivory. American Naturalist 114: 925–931.

Howe, H. F. 1980. Monkey dispersal and waste of a Neotropical fruit. Ecology 61: 944–959.

Howe, H. F., and W. M. Richter. 1982. Effects of seed size on seedling size in *Virola surinamensis*: a within and between tree analysis. Oecologia 53: 347–351.

Howe, H. F., and G. A. Vande Kerckhove. 1980. Nutmeg dispersal by tropical birds. Science 210: 925–927.

Hoy, C. W., G. P. Head, and F. R. Hall. 1998. Spatial heterogeneity and insect adaptation to toxins. Annual Review of Entomology 43: 571–594.

Hrncir, M., S. Jarau, R. Zucchi, and F. G. Barth. 2004. On the origin and properties of scent marks deposited at the food source by a stingless bee, *Melipona seminigra*. Apidologie 35: 3–13.

Huang, S. Q., L. L. Tang, J. F. Sun, and Y. Lu. 2006. Pollinator response to female and male floral display in a monoecious species and its implications for the evolution of floral dimorphism. New Phytologist 171: 417–424.

Huether, C. A. 1968. Exposure of natural genetic variability underlying the pentamerous corolla constancy in *Linanthus androsaceus* ssp. *androsaceus*. Genetics 60: 123–146.

Huether, C. A. 1969. Constancy of the pentamerous corolla phenotype in natural populations of *Linanthus*. Evolution 23: 572–588.

Hunter, M. D., T. Ohgushi, and P. W. Price, eds. 1992. Effects of resource distribution on animal-plant interactions. Academic Press, New York.

Hurly, T. A., and M. D. Oseen. 1999. Context-dependent, risk-sensitive foraging preferences in wild rufous hummingbirds. Animal Behaviour 58: 59–66.

Hutchinson, J. M. C. 2005. Is more choice always desirable? Evidence and arguments from leks, food selection, and environmental enrichment. Biological Reviews 80: 73–92.

Hutchison, B. A., D. R. Matt, R. T. McMillen, L. J. Gross, S. J. Tajchman, and J. M. Norman. 1986. The architecture of a deciduous forest canopy in eastern Tennessee, U.S.A. Journal of Ecology 74: 635–646.

Ide, J. Y. 2004. Selection of age classes of *Sasa* leaves by caterpillars of the skipper butterfly *Thoressa varia* using albo-margination of overwintered leaves. Journal of Ethology 22: 99–103.

Ide, J. Y. 2006. Inter- and intra-shoot distributions of the ramie moth caterpillar, *Arcte coerulea* (Lepidoptera: Noctuidae), in ramie shrubs. Applied Entomology and Zoology 41: 49–55.

Ikonen, A. 2002. Preferences of six leaf beetle species among qualitatively different leaf age classes of three salicaceous host species. Chemoecology 12: 23–28.

Imbert, E. 2002. Ecological consequences and ontogeny of seed heteromorphism. Perspectives in Plant Ecology, Evolution and Sytematics 5: 13–36.

Imbert, E., J. Escarré, and J. Lepart. 1996. Achene dimorphism and among-population variation in *Crepis sancta* (Asteraceae). International Journal of Plant Sciences 157: 309–315.

Infante, J. M., A. Mauchamp, R. Fernández-Alés, R. Joffre, and S. Rambal. 2001. Within-tree variation in transpiration in isolated evergreen oak trees: evidence in support of the pipe model theory. Tree Physiology 21: 409–414.

Inouye, B. D. 2005. The importance of the variance around the mean effect size of ecological processes: comment. Ecology 86: 262–265.

Inouye, D. W., and G. D. Waller. 1984. Responses of honey bees (*Apis mellifera*) to amino acid solutions mimicking floral nectars. Ecology 65: 618–625.

Irwin, R. E. 2000. Morphological variation and female reproductive success in two sympatric *Trillium* species: evidence for phenotypic selection in *Trillium erectum* and *Trillium grandiflorum*. American Journal of Botany 87: 205–214.

Ishihara, M., and M. Shimada. 1993. Female-biased sex ratio in a wild bruchid seed predator, *Kytorhinus sharpianus*: 1, Larval competition and other factors. Ecological Entomology 18: 54–60.

Ishii, H. S., and S. Sakai. 2002. Temporal variation in floral display size and individual floral sex allocation in racemes of *Narthecium asiaticum* (Liliaceae). American Journal of Botany 89: 441–446.

Iwasa, Y., T. Kubo, N. van Dam, and T. J. de Jong. 1996. Optimal level of chemical defense decreasing with leaf age. Theoretical Population Biology 50: 124–148.

Iwasa, Y., Y. Suzuki, and H. Matsuda. 1984. Theory of oviposition strategy of parasitoids: 1, Effect of mortality and limited egg number. Theoretical Population Biology 26: 205–227.

Izhaki, I., and U. N. Safriel. 1990. The effect of some Mediterranean scrubland frugivores upon germination patterns. Journal of Ecology 78: 56–65.

Izhaki, I., P. B. Walton, and U. N. Safriel. 1991. Seed shadows generated by frugivorous birds in an eastern Mediterranean scrub. Journal of Ecology 79: 575–590.

Jackson, J. E., R. O. Sharples, and J. W. Palmer. 1971. The influence of shade and within-tree position on apple fruit size, colour and storage quality. Journal of Horticultural Science 46: 277–287.

Jackson, R. B., J. H. Manwaring, and M. M. Caldwell. 1990. Rapid physiological adjustment of roots to localized soil enrichment. Nature 344: 58–60.

Jackson, S., S. W. Nicolson, and C. N. Lotz. 1998. Sugar preferences and "side bias" in Cape sugarbirds and lesser double-collared sunbirds. Auk 115: 156–165.

Jacquemyn, H., R. Brys, and M. Hermy. 2001. Within and between plant variation in seed number, seed mass and germinability of *Primula elatior*: effect of population size. Plant Biology 3: 561–568.

Jakobsen, H. B., and K. Kristjánsson. 1994. Influence of temperature and floret age on nectar secretion in *Trifolium repens* L. Annals of Botany 74: 327–334.

Jansen, S., P. Baas, P. Gasson, F. Lens, and E. Smets. 2004. Variation in xylem structure from tropics to tundra: evidence from vestured pits. Proceedings of the National Academy of Sciences of the USA 101: 8833–8837.

Jansen, S., P. Baas, P. Gasson, and E. Smets. 2003. Vestured pits: do they promote safer water transport? International Journal of Plant Sciences 164: 405–413.

Jansen, S., P. Baas, and E. Smets. 2001. Vestured pits: their occurrence and systematic importance in eudicots. Taxon 50: 135–167.

Janzen, D. H. 1969. Seed-eaters versus seed size, number, toxicity and dispersal. Evolution 23: 1–27.

Janzen, D. H. 1976. Effect of defoliation on fruit-bearing branches of the Kentucky cofee tree, *Gymnocladus dioicus* (Leguminosae). American Midland Naturalist 95: 474–478.

Janzen, D. H. 1977a. Promising directions of study in tropical animal-plant interactions. Annals of the Missouri Botanical Garden 64: 706–736.

Janzen, D. H. 1977b. Variation in seed size within a crop of a Costa Rican *Mucuna andreana* (Leguminosae). American Journal of Botany 64: 347–349.

Janzen, D. H. 1977c. Variation in seed weight in Costa Rican *Cassia grandis* (Leguminosae). Tropical Ecology 18: 177–186.

Janzen, D. H. 1978. Inter- and intra-crop variation in seed weight of Costa Rican *Ateleia herbert-smithii* Pitt. (Leguminosae). Brenesia 14/15: 311–323.

Janzen, D. H. 1982a. Differential seed survival and passage rates in cows and horses, surrogate Pleistocene dispersal agents. Oikos 38: 150–156.

Janzen, D. H. 1982b. Variation in average seed size and fruit seediness in a fruit crop of a guanacaste tree (Leguminosae: *Enterolobium cyclocarpum*). American Journal of Botany 69: 1169–1178.

Janzen, D. H. 1982c. Weight of seeds in 1–3 seeded fruits of *Lonchocarpus costaricensis* (Leguminosae), a Costa Rican wind-dispersed tree. Brenesia 20: 363–368.

Janzen, D. H. 1987. How moths pass the dry season in a Costa Rican dry forest. Insect Science and Its Application 8: 489–500.

Jarlan, A., D. de Oliveira, and J. Gingras. 1997a. Effects of *Eristalis tenax* (Diptera: Syrphidae) pollination on characteristics of greenhouse sweet pepper fruits. Journal of Economic Entomology 90: 1650–1654.

Jarlan, A., D. de Oliveira, and J. Gingras. 1997b. Pollination by *Eristalis tenax* (Diptera: Syrphidae) and seed set of greenhouse sweet pepper. Journal of Economic Entomology 90: 1646–1649.

Jellum, M. D. 1967. Fatty acid composition of corn (*Zea mays* L.) oil as influenced by kernel position on ear. Crop Science 7: 593–595.

Jeng, T. L., T. H. Tseng, C. S. Wang, C. L. Chen, and J. M. Sung. 2003. Starch biosynthesizing enzymes in developing grains of rice cultivar Tainung 67 and its sodium azide–induced rice mutant. Field Crops Research 84: 261–269.

Jeng, T. L., C. S. Wang, C. L. Chen, and J. M. Sung. 2003. Effects of grain position on the panicle on starch biosynthetic enzyme activity in developing grains of rice cultivar Tainung 67 and its NaN_3-induced mutant. Journal of Agricultural Science 141: 303–311.

Jenik, P. D., and Irish V. F. 2000. Regulation of cell proliferation patterns by homeotic genes during *Arabidopsis* floral development. Development 127: 1267–1276.

Jennions, M. D. 1996. The allometry of fluctuating asymmetry in southern African plants: flowers and leaves. Biological Journal of the Linnean Society 59: 127–142.

Jerling, L. 1985. Are plants and animals alike? A note on evolutionary plant population ecology. Oikos 45: 150–153.

Johnson, E. D. 1926. A comparison of the juvenile and adult leaves of *Eucalyptus globulus*. New Phytologist 25: 202–212.

Johnson, L. S., J. M. Marzluff, and R. P. Balda. 1987. Handling of pinyon pine seed by the Clark's nutcracker. Condor 89: 117–125.

Johnson, S. N., D. A. Elston, and S. E. Hartley. 2003. Influence of host plant heterogeneity on the distribution of a birch aphid. Ecological Entomology 28: 533–541.

378   LITERATURE CITED

Jones, C. G., R. F. Hopper, J. S. Coleman, and V. A. Krischik. 1993. Control of systemically induced herbivore resistance by vascular plant architecture. Oecologia 93: 452–456.

Jones, C. S. 1993. Heterochrony and heteroblastic leaf development in two subspecies of Cucurbita argyrosperma (Cucurbitaceae). American Journal of Botany 80: 778–795.

Jones, K. N., J. S. Reithel, and R. E. Irwin. 1998. A trade-off between the frequency and duration of bumblebee visits to flowers. Oecologia 117: 161–168.

Jordano, P. 1984. Seed weight variation and differential avian dispersal in blackberries Rubus ulmifolius. Oikos 43: 149–153.

Jordano, P. 1987. Avian fruit removal: effects of fruit variation, crop size, and insect damage. Ecology 68: 1711–1723.

Jordano, P. 1988. Polinización y variabilidad de la producción de semillas en Pistacia lentiscus L. (Anacardiaceae). Anales del Jardín Botánico de Madrid 45: 213–231.

Jordano, P. 1989. Pre-dispersal biology of Pistacia lentiscus (Anacardiaceae): cumulative effects on seed removal by birds. Oikos 55: 375–386.

Jordano, P. 1990. Utilización de los frutos de Pistacia lentiscus (Anacardiaceae) por el Verderón Común (Carduelis chloris). In L. Arias, P. Recuerda, and T. Redondo, eds., Principios en etología, pp. 145–153. Publicaciones del Monte de Piedad, Córdoba, Spain.

Jordano, P. 1991. Gender variation and expression of monoecy in Juniperus phoenicea (L.) (Cupressaceae). Botanical Gazette 152: 476–485.

Jordano, P. 1995a. Angiosperm fleshy fruits and seed dispersers: a comparative analysis of adaptation and constraints in plant-animal interactions. American Naturalist 145: 163–191.

Jordano, P. 1995b. Frugivore-mediated selection on fruit and seed size: birds and St. Lucie's cherry, Prunus mahaleb. Ecology 76: 2627–2639.

Josens, R. B., and W. M. Farina. 2001. Nectar feeding by the hovering hawk moth Macroglossum stellatarum: intake rate as a function of viscosity and concentration of sucrose solutions. Journal of Comparative Physiology 187A: 661–665.

Josens, R. B., W. M. Farina, and F. Roces. 1998. Nectar feeding by the ant Camponotus mus: intake rate and crop filling as a function of sucrose concentration. Journal of Insect Physiology 44: 579–585.

Juenger, T., M. Purugganan, and T. F. C. Mackay. 2000. Quantitative trait loci for floral morphology in Arabidopsis thaliana. Genetics 156: 1379–1392.

Kacelnik, A., and M. Bateson. 1996. Risky theories: the effects of variance on foraging decisions. American Zoologist 36: 402–434.

Kacelnik, A., and M. Bateson. 1997. Risk-sensitivity: crossroads for theories of decision-making. Trends in Cognitive Sciences 1: 304–309.

Kadmon, R. 1992. Dynamics of forager arrivals and nectar renewal in flowers of Anchusa strigosa. Oecologia 92: 552–555.

Kadmon, R., and A. Shmida. 1992. Departure rules used by bees foraging for nectar: a field test. Evolutionary Ecology 6: 142–151.

Kadmon, R., A. Shmida, and R. Selten. 1991. Within-plant foraging behavior of bees and its relationship to nectar distribution in *Anchusa strigosa*. Israel Journal of Botany 40: 283–294.

Kærn, M., T. C. Elston, W. J. Blake, and J. J. Collins. 2005. Stochasticity in gene expression: from theories to phenotypes. Nature Reviews Genetics 6: 451–464.

Kainulainen, P., J. Tarhanen, K. Tiilikkala, and J. K. Holopainen. 1998. Foliar and emission composition of essential oil in two carrot varieties. Journal of Agricultural and Food Chemistry 46: 3780–3784.

Kalisz, S. 1986. Variable selection on the timing of germination in *Collinsia verna* (Scrophulariaceae). Evolution 40: 479–491.

Kane, M. E., and L. S. Albert. 1982. Environmental and growth regulator effects on heterophylly and growth of *Proserpinaca intermedia* (Haloragaceae). Aquatic Botany 13: 73–85.

Kang, H., and R. B. Primack. 1991. Temporal variation of flower and fruit size in relation to seed yield in celandine poppy (*Chelidonium majus*: Papaveraceae). American Journal of Botany 78: 711–722.

Kappel, F., and R. MacDonald. 2007. Early gibberellic acid sprays increase firmness and fruit size of 'Sweetheart' sweet cherry. Journal of the American Pomological Society 61: 38–43.

Kappel, F., and G. H. Neilsen. 1994. Relationship between light microclimate, fruit growth, fruit quality, specific leaf weight and N and P content of spur leaves of 'Bartlett' and 'Anjou' pear. Scientia Horticulturae 59: 187–196.

Karban, R. 1992. Plant variation: its effects on populations of herbivorous insects. In R. S. Fritz and E. L. Simms, eds., Plant resistance to herbivores and pathogens, pp. 195–215. University of Chicago Press, Chicago.

Karban, R., A. A. Agrawal, and M. Mangel. 1997. The benefits of induced defenses against herbivores. Ecology 78: 1351–1355.

Karban, R., and I. T. Baldwin. 1997. Induced responses to herbivory. University of Chicago Press, Chicago.

Karban, R., and S. Courtney. 1987. Intraspecific host plant choice: lack of consequences for *Streptanthus tortuosus* (Cruciferae) and *Euchloe hyantis* (Lepidoptera: Pieridae). Oikos 48: 243–248.

Karlsson, M. G., R. D. Heins, J. E. Erwin, R. D. Berghage, W. H. Carlson, and J. A. Biernbaum. 1989. Irradiance and temperature effects on time of development and flower size in chrysanthemum. Scientia Horticulturae 39: 257–267.

Karowe, D. N., and M. M. Martin. 1989. The effects of quantity and quality of diet nitrogen on the growth, efficiency of food utilization, nitrogen budget, and metabolic rate of fifth-instar *Spodoptera eridania* larvae (Lepidoptera: Noctuidae). Journal of Insect Physiology 35: 699–708.

Karrenberg, S., and M. Suter. 2003. Phenotypic trade-offs in the sexual reproduction of Salicaceae from flood plains. American Journal of Botany 90: 749–754.

Karron, J. D., R. J. Mitchell, and J. M. Bell. 2006. Multiple pollinator visits to *Mimulus ringens* (Phrymaceae) flowers increase mate number and seed set within fruits. American Journal of Botany 93: 1306–1312.

Karron, J. D., R. J. Mitchell, K. G. Holmquist, J. M. Bell, and B. Funk. 2004. The influence of floral display size on selfing rates in *Mimulus ringens*. Heredity 92: 242–248.

Kato, M. 1988. Bumblebee visits to *Impatiens* spp.: pattern and efficiency. Oecologia 76: 364–370.

Katsoyannos, B. I., and I. S. Pittara. 1983. Effect of size of artificial oviposition substrates and presence of natural host fruits on the selection of oviposition site by *Dacus oleae*. Entomologia Experimentalis et Applicata 34: 326–332.

Kausik, S. B. 1938. Morphology of abnormal flowers in some angiosperms. New Phytologist 37: 396–408.

Kay, Q. O. N. 1976. Preferential pollination of yellow-flowered morphs of *Raphanus raphanistrum* by *Pieris* and *Eristalis* spp. Nature 261: 230–232.

Kay, Q. O. N. 1982. Intraspecific discrimination by pollinators and its role in evolution. In J. A. Armstrong, J. M. Powell, and A. J. Richards, eds., Pollination and evolution, pp. 9–28. Royal Botanic Gardens, Sydney.

Kearns, C. A., and D. W. Inouye. 1993. Techniques for pollination biologists. University Press of Colorado, Niwot, CO.

Kearsley, M. J. C., and T. G. Whitham. 1989. Developmental changes in resistance to herbivory: implications for individuals and populations. Ecology 70: 422–434.

Kearsley, M. J. C., and T. G. Whitham. 1997. The developmental stream of cottonwoods affects ramet growth and resistance to galling aphids. Ecology 79: 178–191.

Kelber, A. 2003. Sugar preferences and feeding strategies in the hawkmoth *Macroglossum stellatarum*. Journal of Comparative Physiology 189A: 661–666.

Kelly, D., and V. L. Sork. 2002. Mast seeding in perennial plants: why, how, where? Annual Review of Ecology and Systematics 33: 427–447.

Kenrick, P., and P. R. Crane. 1997. The origin and early evolution of plants on land. Nature 389: 33–39.

Kenta, T., N. Inari, T. Nagamitsu, K. Goka, and T. Hiura. 2007. Commercialized European bumblebee can cause pollination disturbance: an experiment on seven native plant species in Japan. Biological Conservation 134: 298–309.

Kester, K. M., S. C. Peterson, F. Hanson, D. M. Jackson, and R. F. Severson. 2002. The roles of nicotine and natural enemies in determining larval feeding site distributions of *Manduca sexta* L. and *Manduca quinquemaculata* (Haworth) on tobacco. Chemoecology 12: 1–10.

Khudamrongsawat, J., R. Tayyar, and J. S. Holt. 2004. Genetic diversity of giant reed (*Arundo donax*) in the Santa Ana River, California. Weed Science 52: 395–405.

Kigel, J. 1992. Diaspore heteromorphism and germination in populations of the ephemeral *Hedypnois rhagadioloides* (L.) F. W. Schmidt (Asteraceae) inhabiting a geographic range of increasing aridity. Acta Oecologica 13: 45–53.

Kincaid, D. T., P. J. Anderson, and S. A. Mori. 1998. Leaf variation in a tree of *Pourouma tomentosa* (Cecropiaceae) in French Guiana. Brittonia 50: 324–338.

King, L. M., and B. A. Schaal. 1990. Genotypic variation within asexual lineages of *Taraxacum officinale*. Proceedings of the National Academy of Sciences of the USA 87: 998–1002.

Kingsolver, J. G., H. E. Hoekstra, J. M. Hoekstra, D. Berrigan, S. N. Vignieri, C. E.

Hill, A. Hoang, P. Gibert, and P. Beerli. 2001. The strength of phenotypic selection in natural populations. American Naturalist 157: 245–261.

Kingsolver, J. G., and D. W. Pfennig. 2007. Patterns and power of phenotypic selection in nature. BioScience 57: 561–572.

Kinnaird, M. F., T. G. O'Brien, and S. Suryadi. 1996. Population fluctuation in Sulawesi red-knobbed hornbills: tracking figs in space and time. Auk 113: 431–440.

Kitajima, K., S. S. Mulkey, and S. J. Wright. 1997. Seasonal leaf phenotypes in the canopy of a tropical dry forest: photosynthetic characteristics and associated traits. Oecologia 109: 490–498.

Kitamura, S., T. Yumoto, P. Poonswad, P. Chuailua, K. Plongmai, T. Maruhashi, and N. Noma. 2002. Interactions between fleshy fruits and frugivores in a tropical seasonal forest in Thailand. Oecologia 133: 559–572.

Klein, I., T. M. DeJong, S. A. Weinbaum, and T. T. Muraoka 1991. Specific leaf weight and nitrogen allocation responses to light exposure within walnut trees. Hortscience 26: 183–185.

Klekowski, E. J. 1988a. Mutation, developmental selection, and plant evolution. Columbia University Press, New York.

Klekowski, E. J. 1988b. Progressive cross- and self-sterility associated with aging in fern clones and perhaps other plants. Heredity 61: 247–253.

Klekowski, E. J., and P. J. Godfrey. 1989. Ageing and mutation in plants. Nature 340: 389–391.

Klekowski, E. J., N. Kazarinova-Fukshansky, and H. Mohr. 1985. Shoot apical meristems and mutation: stratified meristems and angiosperm evolution. American Journal of Botany 72: 1788–1800.

Kliber, A., and C. G. Eckert. 2004. Sequential decline in allocation among flowers within inflorescences: proximate mechanisms and adaptive significance. Ecology 85: 1675–1687.

Klinkhamer, P. G. L., and C. A. M. Van der Veen–Van Wijk. 1999. Genetic variation in floral traits of Echium vulgare. Oikos 85: 515–522.

Kluge, A. G., and W. C. Kerfoot. 1973. The predictability and regularity of character divergence. American Naturalist 107: 426–442.

Kondra, Z. P., and R. K. Downey. 1970. Glucosinolate content of rapeseed (Brassica napus L. and B. campestris L.) meal as influenced by pod position on plant. Crop Science 10: 54–56.

Koo, J., H. Roh, and J. Choe. 2003. Oviposition preference and offspring performance in Mechoris ursulus Roelofs (Coleoptera: Attelabidae). Journal of Ethology 21: 37–43.

Koops, M. A., J. A. Hutchings, and B. K. Adams. 2003. Environmental predictability and the cost of imperfect information: influences on offspring size variability. Evolutionary Ecology Research 5: 29–42.

Korine, C., and E. K. V. Kalko. 2005. Fruit detection and discrimination by small fruit-eating bats (Phyllostomidae): echolocation call design and olfaction. Behavioral Ecology and Sociobiology 59: 12–23.

Korn, R. W. 2001. Analysis of shoot apical organization in six species of the Cupressaceae based on chimeric behavior. American Journal of Botany 88: 1945–1952.

Kouki, J. 1993. Herbivory modifies the production of different leaf types in the yellow water-lily, *Nuphar lutea* (Nymphaeaceae). Functional Ecology 7: 21–26.

Kovalchuk, I., O. Kovalchuk, and B. Hohn. 2000. Genome-wide variation of the somatic mutation frequency in transgenic plants. EMBO Journal 19: 4431–4438.

Kozlov, M. V., and P. Niemelä. 1999. Difference in needle length: a new and objective indicator of pollution impact on Scots pine (*Pinus sylvestris*). Water, Air, and Soil Pollution 116: 365–370.

Kozlowski, T. T., and J. J. Clausen. 1966. Shoot growth characteristics of heterophyllous woody plants. Canadian Journal of Botany 44: 827–843.

Krannitz, P. G. 1997a. Seed weight variability of antelope bitterbrush (*Purshia tridentata*: Rosaceae). American Midland Naturalist 138: 306–321.

Krannitz, P. G. 1997b. Variation in magnesium and nitrogen content in seeds of antelope bitterbrush (*Purshia tridentata*, Rosaceae). American Journal of Botany 84: 1738–1742.

Kreher, S. A., S. A. Foré, and B. S. Collins. 2000. Genetic variation within and among patches of the clonal species, *Vaccinium stamineum* L. Molecular Ecology 9: 1247–1252.

Krischik, V., E. S. McCloud, and J. A. Davidson. 1989. Selective avoidance by vertebrate frugivores of green holly berries infested with a cecidomyiid fly (Diptera: Cecidomyiidae). American Midland Naturalist 121: 350–354.

Kudo, G. 2003. Anther arrangement influences pollen deposition and removal in hermaphrodite flowers. Functional Ecology 17: 349–355.

Kudo, G., T. Maeda, and K. Narita. 2001. Variation in floral sex allocation and reproductive success within inflorescences of *Corydalis ambigua* (Fumariaceae): pollination efficiency or resource limitation? Journal of Ecology 89: 48–56.

Kuijt, J. 1980. A note on heterophylly and branching patterns in the *Amyema* complex (Loranthaceae). Blumea 26: 403–410.

Kull, O. 2002. Acclimation of photosynthesis in canopies: models and limitations. Oecologia 133: 267–279.

Kull, O., and Ü. Niinemets. 1993. Variations in leaf morphometry and nitrogen concentration in *Betula pendula* Roth., *Corylus avellana* L. and *Lonicera xylosteum* L. Tree Physiology 12: 311–318.

Kuwabara, A., K. Ikegami, T. Koshiba, and T. Nagata. 2003. Effects of ethylene and abscisic acid upon heterophylly in *Ludwigia arcuata* (Onagraceae). Planta 217: 880–887.

Kuwabara, A., H. Tsukaya, and T. Nagata. 2001. Identification of factors that cause heterophylly in *Ludwigia arcuata* Walt. (Onagraceae). Plant Biology 3: 98–105.

Labra, M., C. Savini, M. Bracale, N. Pelucchi, L. Colombo, M. Bardini, and F. Sala. 2001. Genomic changes in transgenic rice (*Oryza sativa* L.) plants produced by infecting calli with *Agrobacterium tumefaciens*. Plant Cell Reports 20: 325–330.

Lacey, E. P., and C. Marshall. 1992. Carbon integration in two *Plantago* species. American Journal of Botany 79: 1108–1112.

Lacey, E. P., L. Real, J. Antonovics, and D. G. Heckel. 1983. Variance models in the study of life histories. American Naturalist 122: 114–131.

Lacointe, A., E. Deleens, T. Ameglio, B. Saint-Joanis, C. Lelarge, M. Vandame, G. C. Song, and F. A. Daudet. 2004. Testing the branch autonomy theory: a

$^{13}C/^{14}C$ double-labelling experiment on differentially shaded branches. Plant Cell and Environment 27: 1159–1168.

Laird, R. A., and L. W. Aarssen. 2005. Size inequality and the tragedy of the commons phenomenon in plant competition. Plant Ecology 179: 127–131.

Lake, J. A., W. P. Quick, D. J. Beerling, and F. I. Woodward. 2001. Signals from mature to new leaves. Nature 411: 154.

Lake, J. A., F. I. Woodward, and W. P. Quick. 2002. Long-distance CO_2 signalling in plants. Journal of Experimental Botany 53: 183–193.

Lambert, F. 1989. Fig-eating by birds in a Malaysian lowland rain forest. Journal of Tropical Ecology 5: 401–412.

Lande, R. 1977. On comparing coefficients of variation. Systematic Zoology 26: 214–217.

Lande, R., and S. J. Arnold. 1983. The measurement of selection on correlated characters. Evolution 37: 1210–1226.

Lanza, J., G. C. Smith, S. Sack, and A. Cash. 1995. Variation in nectar volume and composition of *Impatiens capensis* at the individual, plant, and population levels. Oecologia 102: 113–119.

Larkin, P. J., and W. R. Scowcroft. 1981. Somaclonal variation: a novel source of variability from cell cultures for plant improvement. Theoretical and Applied Genetics 60: 197–214.

Laska, M., E. Carrera Sánchez, J. A. Rodríguez Rivera, and E. Rodríguez Luna. 1996. Gustatory thresholds for food-associated sugars in the spider monkey (*Ateles geoffroyi*). American Journal of Primatology 39: 189–193.

Laska, M. S., and E. W. Stiles. 1994. Effects of fruit crop size on intensity of fruit removal in *Viburnum prunifolium* (Caprifoliaceae). Oikos 69: 199–202.

Laverty, T. M. 1994. Costs to foraging bumble bees of switching plant species. Canadian Journal of Zoology 72: 43–47.

Lawton, J. H. 1983. Plant architecture and the diversity of phytophagous insects. Annual Review of Entomology 28: 23–39.

Leal, D. B., and S. C. Thomas. 2003. Vertical gradients and tree-to-tree variation in shoot morphology and foliar nitrogen in an old-growth *Pinus strobus* stand. Canadian Journal of Forest Research 33: 1304–1314.

Leary, R. F., and F. W. Allendorf. 1989. Fluctuating asymmetry as an indicator of stress: implications for conservation biology. Trends in Ecology and Evolution 4: 214–217.

Lee, T. D. 1988. Patterns of fruit and seed production. In J. Lovett-Doust and L. Lovett-Doust, eds., Plant reproductive ecology, pp. 179–202. Oxford University Press, Oxford.

Lehmann, N. L., and R. Sattler. 1994. Floral development and homeosis in *Actaea rubra* (Ranunculaceae). International Journal of Plant Sciences 155: 658–671.

Lehrer, M., G. A. Horridge, S. W. Zhang, and R. Gadagkar. 1995. Shape vision in bees: innate preference for flower-like patterns. Philosophical Transactions of the Royal Society of London B 347: 123–137.

Leishman, M. R., I. J. Wright, A. T. Moles, and M. Westoby. 2000. The evolutionary ecology of seed size. In M. Fenner, ed., Seeds: the ecology of regeneration in plant communities, 2nd edition, pp. 31–57. CABI Publishing, Wallingford, UK.

Lemaire, G., B. Onillon, G. Gosse, M. Chartier, and J. M. Allirand. 1991. Nitrogen distribution within a lucerne canopy during regrowth: relation with light distribution. Annals of Botany 68: 483–488.

León, J., E. Rojo, and J. J. Sánchez-Serrano. 2001. Wound signalling in plants. Journal of Experimental Botany 52: 1–9.

Lepczyk, C. A., K. G. Murray, K. Winnett-Murray, P. Bartell, E. Geyer, and T. Work. 2000. Seasonal fruit preferences for lipids and sugars by American robins. Auk 117: 709–717.

Lerner, I. M. 1954. Genetic homeostasis. Oliver and Boyd, Edinburgh.

Le Roux, X., H. Sinoquet, and M. Vandame. 1999. Spatial distribution of leaf dry weight per area and leaf nitrogen concentration in relation to local radiation regime within an isolated tree crown. Tree Physiology 19: 181–188.

Les, D. H., and D. J. Sheridan. 1990. Biochemical heterophylly and flavonoid evolution in North American *Potamogeton* (Potamogetonaceae). American Journal of Botany 77: 453–465.

Lessells, C. M., and P. T. Boag. 1987. Unrepeatable repeatabilities: a common mistake. Auk 104: 116–121.

Leuning, R., R. N. Cromer, and S. Rance. 1991. Spatial distributions of foliar nitrogen and phosphorus in crowns of *Eucalyptus grandis*. Oecologia 88: 504–510.

Levey, D. J. 1986. Methods of seed processing by birds and seed deposition patterns. In A. Estrada and T. H. Fleming, eds., Frugivores and seed dispersal, pp. 147–158. Junk, Dordrecht, Netherlands.

Levey, D. J. 1987. Sugar-tasting ability and fruit selection in tropical fruit-eating birds. Auk 104: 173–179.

Levey, D. J., and W. H. Karasov. 1992. Digestive modulation in a seasonal frugivore, the American robin (*Turdus migratorius*). American Journal of Physiology 262: G711-G718.

Levey, D. J., T. C. Moermond, and J. S. Denslow. 1984. Fruit choice in Neotropical birds: the effect of distance between fruits on preference patterns. Ecology 65: 844–850.

Levin, D. A. 1978. Some genetic consequences of being a plant. In P. F. Brussard, ed., Ecological genetics: the interface, pp. 189–212. Springer-Verlag, New York.

Levin, D. A., and H. W. Kerster. 1973. Assortative pollination for stature in *Lythrum salicaria*. Evolution 27: 144–152.

Levy, F. 1988. Effects of pollen source and time of pollination on seed production and seed weight in *Phacelia dubia* and *P. maculata* (Hydrophyllaceae). American Midland Naturalist 119: 193–198.

Lewis, W. H., R. L. Oliver, and T. J. Luikart. 1971. Multiple genotypes in individuals of *Claytonia virginica*. Science 172: 564–565.

Lewontin, R. C. 1966. On the measurement of relative variability. Systematic Zoology 15: 141–142.

Li, N., and R. C. Jackson. 1961. Cytology of supernumerary chromosomes in *Haplopappus spinulosus* ssp. *cotula*. American Journal of Botany 48: 419–426.

Lightfoot, D. C., and W. G. Whitford. 1989. Interplant variation in creosotebush foliage characteristics and canopy arthropods. Oecologia 81: 166–175.

Ligon, J. D., and D. J. Martin. 1974. Piñon seed assessment by the piñon jay, *Gymnorhinus cyanocephalus*. Animal Behaviour 22: 421–429.

Lin, S. K., M. C. Chang, Y. G. Tsai, and H. S. Lur. 2005. Proteomic analysis of the expression of proteins related to rice quality during caryopsis development and the effect of high temperature on expression. Proteomics 5: 2140–2156.

Linhart, Y. B., and D. F. Tomback. 1985. Seed dispersal by nutcrackers causes multi-trunk growth form in pines. Oecologia 67: 107–110.

Lipow, S. R., and R. Wyatt. 1999. Diallel crosses reveal patterns of variation in fruit-set, seed mass, and seed number in *Asclepias incarnata*. Heredity 83: 310–318.

Littell, R. C., G. A. Milliken, W. W. Stroup, and R. D. Wolfinger. 1996. SAS System for mixed models. SAS Institute, Cary, NC.

Liu, Z. H., F. M. Cheng, W. D. Cheng, and G. P. Zhang. 2005. Positional variations in phytic acid and protein content within a panicle of japonica rice. Journal of Cereal Science 41: 297–303.

Llorens, L., J. Peñuelas, and B. Emmett. 2002. Developmental instability and gas exchange responses of a heathland shrub to experimental drought and warming. International Journal of Plant Sciences 163: 959–967.

Lloyd, D. G. 1984. Variation strategies of plants in heterogeneous environments. Biological Journal of the Linnean Society 21: 357–385.

Lockhart, C. S. 1996. Aquatic heterophylly as a survival strategy in *Melaleuca quinquenervia* (Myrtaceae). Canadian Journal of Botany 74: 243–246.

López-Almansa, J. C., and L. Gil. 2003. Empty samara and parthenocarpy in *Ulmus minor* s.l. in Spain. Silvae Genetica 52: 241–243.

López-Calleja, M. V., F. Bozinovic, and C. Martínez del Río. 1997. Effects of sugar concentration on hummingbird feeding and energy use. Comparative Biochemistry and Physiology 118A: 1291–1299.

Lord, E. M. 1980. An anatomical basis for the divergent floral forms in the cleistogamous species *Lamium amplexicaule* L. (Labiatae). American Journal of Botany 67: 1430–1441.

Lord, E. M. 1981. Cleistogamy: a tool for the study of floral morphogenesis, function and evolution. Botanical Review 47: 421–449.

Lötscher, M., and M. J. M. Hay. 1996. Distribution of mineral nutrition from nodal roots of *Trifolium repens*: genotypic variation in intra-plant allocation of ^{32}P and ^{45}Ca. Physiologia Plantarum 97: 269–276.

Lötscher, M., and M. J. M. Hay. 1997. Genotypic differences in physiological integration, morphological plasticity and utilization of phosphorus induced by variation in phosphate supply in *Trifolium repens*. Journal of Ecology 85: 341–350.

Lowndes, A. G. 1931. Note on individual variation in *Paris quadrifolia* L. New Phytologist 30: 298–299.

Lucas, F. C. 1898. Variation in the number of ray-flowers in the white daisy. American Naturalist 32: 509–511.

Luyssaert, S., M. Van Meirvenne, and N. Lust. 2001. Cadmium variability in leaves of a *Salix fragilis*: simulation and implications for leaf sampling. Canadian Journal of Forest Research 31: 313–321.

Lynch, J., and A. González. 1993. Canopy nutrient allocation in relation to incident light in the tropical fruit tree *Borojoa patinoi* (Cuatr). Journal of the American Society for Horticultural Science 118: 777–785.

Lynch, M., and B. Walsh. 1998. Genetics and analysis of quantitative traits. Sinauer, Sunderland, MA.

Lyndon, R. F. 1979. Aberrations in flower development in *Silene*. Canadian Journal of Botany 57: 233–235.

Maad, J. 2000. Phenotypic selection in hawkmoth-pollinated *Platanthera bifolia*: targets and fitness surfaces. Evolution 54: 112–123.

MacArthur, R. H., and E. R. Pianka. 1966. On optimal use of a patchy environment. American Naturalist 100: 603–609.

Macedo, C. A., and J. H. Langenheim. 1989. Intraplant and interplant leaf sesquiterpene variability in *Copaifera langsdorfii*: relation to microlepidopteran herbivory. Biochemical Systematics and Ecology 17: 551–557.

MacGregor, S. D., and T. G. O'Connor. 2004. Response of *Acacia tortilis* to utilization by elephants in a semi-arid African savanna. South African Journal of Wildlife Research 34: 55–66.

Macnair, M. R., and Q. J. Cumbes. 1990. The pattern of sexual resource allocation in the yellow monkey flower, *Mimulus guttatus*. Proceedings of the Royal Society London B 242: 101–107.

Maffei, M., F. Chialva, and T. Sacco. 1989. Glandular trichomes and essential oils in developing peppermint leaves. New Phytologist 111: 707–716.

Mal, T. K., and J. Lovett-Doust. 2005. Phenotypic plasticity in vegetative and reproductive traits in an invasive weed, *Lythrum salicaria* (Lythraceae) in response to soil moisture. American Journal of Botany 92: 819–825.

Manasse, R. S., and M. L. Stanton. 1991. The influence of the mating system on seed size variation in *Crinum erubescens* (Amaryllidaceae). Evolution 45: 883–890.

Mandák, B. 1997. Seed heteromorphism and plant life cycle: a review of literature. Preslia 69: 129–159.

Manzur, M. I., and S. P. Courtney. 1984. Influence of insect damage in fruits of hawthorn on bird foraging and seed dispersal. Oikos 43: 265–270.

Marañón, T. 1987. Ecología del polimorfismo somático de semillas y la sinaptospermia en *Aegilops neglecta* Req. ex Bertol. Anales del Jardín Botánico de Madrid 44: 97–107.

Marañón, T. 1989. Variations in seed size and germination in three *Aegilops* species. Seed Science and Technology 17: 583–588.

Marchetti, S., A. Giordano, and C. Chiabà. 1995. Within-plot and within-plant variation for seed content of soya bean protease inhibitors. Journal of the Science of Food and Agriculture 68: 465–469.

Marcotrigiano, M. 1986. Experimentally synthesized plant chimeras: 3, Qualitative and quantitative characteristics of the flowers of interspecific *Nicotiana* chimeras. Annals of Botany 57: 435–442.

Marcotrigiano, M. 1997. Chimeras and variegation: patterns of deceit. Hortscience 32: 773–784.

Marcotrigiano, M. 2000. Herbivory could unlock mutations sequestered in stratified shoot apices of genetic mosaics. American Journal of Botany 87: 355–361.

Marcotrigiano, M. 2001. Genetic mosaics and the analysis of leaf development. International Journal of Plant Sciences 162: 513–525.

Marden, J. H. 1984a. Intrapopulation variation in nectar secretion in *Impatiens capensis*. Oecologia 63: 418–422.

Marden, J. H. 1984b. Remote perception of floral nectar by bumblebees. Oecologia 64: 232–240.

Markham, J. H. 2002. A hierarchical analysis of seed production by *Alnus rubra*. American Midland Naturalist 148: 246–252.

Markow, T. A. 1995. Evolutionary ecology and developmental instability. Annual Review of Entomology 40: 105–120.

Marquis, R. J. 1988. Intra-crown variation in leaf herbivory and seed production in striped maple, *Acer pensylvanicum* L. (Aceraceae). Oecologia 77: 51–55.

Marquis, R. J. 1996. Plant architecture, sectoriality and plant tolerance to herbivores. Vegetatio 127: 85–97

Marshall, C. 1996. Sectoriality and physiological organisation in herbaceous plants: an overview. Vegetatio 127: 9–16.

Marshall, D. L. 1991. Nonrandom mating in wild radish: variation in pollen donor success and effects of multiple paternity among one- to six-donor pollinations. American Journal of Botany 78: 1404–1418.

Marshall, D. L., D. A. Levin, and N. L. Fowler. 1985. Plasticity in yield components in response to fruit predation and date of fruit initiation in three species of *Sesbania* (Leguminosae). Journal of Ecology 73: 71–81.

Martín, C., E. Uberhuaga, and C. Pérez. 2002. Application of RAPD markers in the characterisation of *Chrysanthemum* varieties and the assessment of somaclonal variation. Euphytica 127: 247–253.

Martin, N. H. 2004. Flower size preferences of the honeybee (*Apis mellifera*) foraging on *Mimulus guttatus* (Scrophulariaceae). Evolutionary Ecology Research 6: 777–782.

Martinez Arias, A., and P. Hayward. 2006. Filtering transcriptional noise during development: concepts and mechanisms. Nature Reviews Genetics 7: 34–44.

Martínez del Río, C., W. H. Karasov, and D. J. Levey. 1989. Physiological basis and ecological consequences of sugar preferences in cedar waxwings. Auk 106: 64–71.

Martínez-Gómez, P., and T. M. Gradziel. 2003. Sexual polyembryony in almond. Sexual Plant Reproduction 16: 135–139.

Masters, M. T. 1869. Vegetable teratology. Ray Society, London.

Masuda, M., T. Yahara, and M. Maki. 2001. An ESS model for the mixed production of cleistogamous and chasmogamous flowers in a facultative cleistogamous plant. Evolutionary Ecology Research 3: 429–439.

Mata, A., and C. Bosque. 2004. Sugar preferences, absorption efficiency and water influx in a Neotropical nectarivorous passerine, the bananaquit (*Coereba flaveola*). Comparative Biochemistry and Physiology 139A: 395–404.

Mather, K. 1953. Genetical control of stability in development. Heredity 7: 297–336.

Matilla, A., M. Gallardo, and M. I. Puga-Hermida. 2005. Structural, physiological and molecular aspects of heterogeneity in seeds: a review. Seed Science Research 15: 63–76.

Maxwell, C. D., A. Zobel, and D. Woodfine. 1994. Somatic polymorphism in the achenes of *Tragopogon dubius*. Canadian Journal of Botany 72: 1282–1288.

May, P. G. 1985. Nectar uptake rates and optimal nectar concentrations of two butterfly species. Oecologia 66: 381–386.

Mazer, S. J., and K. A. Dawson. 2001. Size-dependent sex allocation within flowers of the annual herb *Clarkia unguiculata* (Onagraceae): ontogenetic and among-plant variation. American Journal of Botany 88: 819–831.

Mazer, S. J., and V. A. Delesalle. 1996. Floral trait variation in *Spergularia marina* (Caryophyllaceae): ontogenetic, maternal family, and population effects. Heredity 77: 269–281.

Mazer, S. J., and D. E. Lowry. 2003. Environmental, genetic, and seed mass effects on winged seed production in the heteromorphic *Spergularia marina* (Caryophyllaceae). Functional Ecology 17: 637–650.

Mazer, S. J., A. A. Snow, and M. L. Stanton. 1986. Fertilization dynamics and parental effects upon fruit development in *Raphanus raphanistrum*: consequences for seed size variation. American Journal of Botany 73: 500–511.

Mazer, S. J., and N. T. Wheelwright. 1993. Fruit size and shape: allometry at different taxonomic levels in bird-dispersed plants. Evolutionary Ecology 7: 556–575.

McAdams, H. H., and A. Arkin. 1999. It's a noisy business! Genetic regulation at the nanomolar scale. Trends in Genetics 15: 65–69.

McArdle, B. H., and K. J. Gaston. 1992. Comparing population variabilities. Oikos 64: 610–612.

McArdle, B. H., and K. J. Gaston. 1995. The temporal variability of densities: back to basics. Oikos 74: 165–171.

McClintock, B. 1984. The significance of responses of the genome to challenge. Science 226: 792–801.

McClure, M., D. T. Quiring, and J. J. Turgeon. 1998. Proximate and ultimate factors influencing oviposition site selection by endoparasites on conifer seed cones: two sympatric dipteran species on larch. Entomologia Experimentalis et Applicata 87: 1–13.

McCrea, R. H. 1924. Abnormal flower of the honeysuckle (*Lonicera periclymenum* L.). New Phytologist 23: 159–160.

McGinley, M. A. 1989. Within and among plant variation in seed mass and pappus size in *Tragopogon dubius*. Canadian Journal of Botany 67: 1298–1304.

McGinley, M. A., C. C. Smith, P. F. Elliott, and J. J. Higgins. 1990. Morphological constraints on seed mass in lodgepole pine. Functional Ecology 4: 183–192.

McKone, M. J., R. Ostertag, J. T. Rauscher, D. A. Heiser, and F. L. Russe. 1995. An exception to Darwin's syndrome: floral position, protogyny, and insect visitation in *Besseya bullii* (Scrophulariaceae). Oecologia 101: 68–74.

McLellan, T. 1993. The roles of heterochrony and heteroblasty in the diversification of leaf shapes in *Begonia dregei* (Begoniaceae). American Journal of Botany 80: 796–804.

McPheeters, K., and R. M. Skirvin. 1983. Histogenic layer manipulation in chimeral 'Thornless Evergreen' trailing blackberry. Euphytica 32: 351–360.

McPherson, J. M. 1988. Preferences of cedar waxwings in the laboratory for fruit species, colour and size: a comparison with field observations. Animal Behaviour 36: 961–969.

Medrano, M., P. Guitián, and J. Guitián. 2000. Patterns of fruit and seed set within inflorescences of *Pancratium maritimum* (Amaryllidaceae): nonuniform pollination, resource limitation, or architectural effects? American Journal of Botany 87: 493–501.

Medrano, M., C. M. Herrera, and S. C. H. Barrett. 2005. Herkogamy and mating patterns in the self-compatible daffodil *Narcissus longispathus*. Annals of Botany 95: 1105–1111.

Mehlman, D. W. 1993. Seed size and seed packaging variation in *Baptisia lanceolata* (Fabaceae). American Journal of Botany 80: 735–742.

Meir, P., B. Kruijt, M. Broadmeadow, E. Barbosa, O. Kull, F. Carswell, A. Nobre, and P. G. Jarvis. 2002. Acclimation of photosynthetic capacity to irradiance in tree canopies in relation to leaf nitrogen concentration and leaf mass per unit area. Plant, Cell, and Environment 25: 343–357.

Mena-Alí, J. I., and O. J. Rocha. 2005. Effect of ovule position within the pod on the probability of seed production in *Bauhinia ungulata* (Fabaceae). Annals of Botany 95: 449–455.

Menadue, Y., and R. K. Crowden. 1990. Leaf polymorphism in *Ranunculus nanus* Hook. (Ranunculaceae). New Phytologist 114: 265–274.

Méndez, M. 1997. Sources of variation in seed mass in *Arum italicum*. International Journal of Plant Sciences 158: 298–305.

Merilä, J., and M. Björklund. 1995. Fluctuating asymmetry and measurement error. Systematic Biology 44: 97–101.

Merritt, S. Z. 1996. Within-plant variation in concentrations of amino acids, sugar, and sinigrin in phloem sap of black mustard, *Brassica nigra* (L.) Koch (Cruciferae). Journal of Chemical Ecology 22: 1133–1145.

Meyer, G. A. 1998. Pattern of defoliation and its effect on photosynthesis and growth of goldenrod. Functional Ecology 12: 270–279.

Meyer, S. T., F. Roces, and R. Wirth. 2006. Selecting the drought stressed: effects of plant stress on intraspecific and within-plant herbivory patterns of the leaf-cutting ant *Atta colombica*. Functional Ecology 20: 973–981.

Meyerowitz, E. M. 2002. Plants compared to animals: the broadest comparative study of development. Science 295: 1482–1485.

Michaels, H. J., B. Benner, A. P. Hartgerink, T. D. Lee, S. Rice, M. F. Willson, and R. I. Bertin. 1988. Seed size variation: magnitude, distribution, and ecological correlates. Evolutionary Ecology 2: 157–166.

Midgley, J. J., R. M. Cowling, and B. B. Lamont. 1991. Relationship of follicle size and seed size in *Hakea* (Proteaceae): isometry, allometry and adaptation. South African Journal of Botany 57: 107–110.

Miles, D. B., G. S. Smith, and S. A. Miller. 1996. Within plant sampling procedures: fruit variation in kiwifruit vines. Annals of Botany 78: 289–294.

Millener, L. H. 1961. Day length as related to vegetative development in *Ulex europaeus*: 1, The experimental approach. New Phytologist 60: 339–354.

Millet, E. 1986. Relationships between grain weight and the size of floret cavity in the wheat spike. Annals of Botany 58: 417–423.

Miner, B. G., and J. R. Vonesh. 2004. Effects of fine grain environmental variability on morphological plasticity. Ecology Letters 7: 794–801.

Minorsky, P. V. 2003. Heterophylly in aquatic plants. Plant Physiology 133: 1671–1672.

Mitchell, M. G. E., J. A. Antos, and G. A. Allen. 2004. Modules of reproduction in females of the dioecious shrub *Oemleria cerasiformis*. Canadian Journal of Botany 82: 393–400.

Mitchell, R. 1975. The evolution of oviposition tactics in the bean weevil, *Calloso-bruchus maculatus* (F.). Ecology 56: 696–702.

Mitchell, R. 1976. Bruchid beetles and seed packaging by palo verde. Ecology 58: 644–651.

Mitchell, R. J. 1993. Adaptive significance of *Ipomopsis aggregata* nectar production: observation and experiment in the field. Evolution 47: 25–35.

Mitchell, R. J. 1994. Effects of floral traits, pollinator visitation, and plant size on *Ipomopsis aggregata* fruit production. American Naturalist 143: 870–889.

Mitchell, R. J., J. D. Karron, K. G. Holmquist, and J. M. Bell. 2004. The influence of *Mimulus ringens* floral display size on pollinator visitation patterns. Functional Ecology 18: 116–124.

Mitchell, R. J., and D. C. Paton. 1990. Effects of nectar volume and concentration on sugar intake rates of Australian honeyeaters (Meliphagidae). Oecologia 83: 238–246.

Mitchell, R. J., and N. M. Waser. 1992. Adaptive significance of *Ipomopsis aggregata* nectar production: pollination success of single flowers. Ecology 73: 633–638.

Mitchell, R. S. 1971. Comparative leaf structure of aquatic *Polygonum* species. American Journal of Botany 58: 342–360.

Mitchell-Olds, T., and R. G. Shaw. 1987. Regression analysis of natural selection: statistical inference and biological interpretation. Evolution 41: 1149–1161.

Mitton, J. B., and M. C. Grant. 1984. Associations among protein heterozygosity, growth rate, and developmental homeostasis. Annual Review of Ecology and Systematics 15: 479–499.

Miyaji, K. I., W. S. da Silva, and P. D. T. Alvim. 1997. Longevity of leaves of a tropical tree, *Theobroma cacao*, grown under shading, in relation to position within the canopy and time of emergence. New Phytologist 135: 445–454.

Mizukami, Y. 2001. A matter of size: developmental control of organ size in plants. Current Opinion in Plant Biology 4: 533–539.

Mo, J., M. T. Tanton, and F. L. Bygrave. 1997. Within-tree distribution of attack by *Hypsipyla robusta* Moore (Lepidoptera: Pyralidae) in Australian red cedar (*Toona australis* [F. Muell.] Harmes). Forest Ecology and Management 96: 147–154.

Moegenburg, S. M. 1996. *Sabal palmetto* seed size: causes of variation, choices of predators, and consequences for seedlings. Oecologia 106: 539–543.

Moermond, T. C., and J. S. Denslow. 1983. Fruit choice in Neotropical birds: effects of fruit type and accessibility on selectivity. Journal of Animal Ecology 52: 407–420.

Mohana, G. S., R. Uma Shaanker, K. N. Ganeshaiah, and S. Dayanandan. 2001. Genetic relatedness among developing seeds and intra fruit seed abortion in *Dalbergia sissoo* (Fabaceae). American Journal of Botany 88: 1181–1188.

Møller, A. P. 1995. Bumblebee preference for symmetrical flowers. Proceedings of the National Academy of Sciences of the USA 92: 2288–2292.

Møller, A. P. 1997. Developmental stability and fitness: a review. American Naturalist 149: 916–932.

Møller, A. P., and M. Eriksson. 1994. Patterns of fluctuating asymmetry in flowers: implications for sexual selection in plants. Journal of Evolutionary Biology 7: 97–113.

Møller, A. P., and M. Eriksson. 1995. Pollinator preference for symmetrical flowers and sexual selection in plants. Oikos 73: 15–22.

Møller, A. P., and J. A. Shykoff. 1999. Morphological developmental stability in plants: patterns and causes. International Journal of Plant Sciences 160 (suppl. 6): S135–S146.

Møller, A. P., and G. Sorci. 1998. Insect preference for symmetrical artificial flowers. Oecologia 114: 37–42.

Monestiez, P., R. Habib, and J. M. Audergon. 1990. Geostatistics, spatial dependencies in a tree: a new approach in fruit tree studies. Acta Horticulturae 276: 257–263.

Montalvo, A. M., and R. G. Shaw. 1994. Quantitative genetics of sequential life-history and juvenile traits in the partially selfing perennial, *Aquilegia caerulea*. Evolution 48: 828–841.

Montgomerie, R. D. 1984. Nectar extraction by hummingbirds: response to different floral characters. Oecologia 63: 229–236.

Moore, B. D., and W. J. Foley. 2005. Tree use by koalas in a chemically complex landscape. Nature 435: 488–490.

Moore, L. V., J. H. Myers, and R. Eng. 1988. Western tent caterpillars prefer the sunny side of the tree, but why? Oikos 51: 321–326.

Morandin, L. A., T. M. Laverty, and P. G. Kevan. 2001. Effect of bumble bee (Hymenoptera: Apidae) pollination intensity on the quality of greenhouse tomatoes. Journal of Economic Entomology 94: 172–179.

Morse, D. H. 1978. Size-related foraging differences of bumble bee workers. Ecological Entomology 3: 189–192.

Mosjidis, J. A., and D. M. Yermanos. 1985. Plant position effect on seed weight, oil content, and oil composition in sesame. Euphytica 34: 193–199.

Mothershead, K., and R. J. Marquis. 2000. Fitness impacts of herbivory through indirect effects on plant-pollinator interactions in *Oenothera macrocarpa*. Ecology 81: 30–40.

Mouissie, A. M., C. E. J. Van der Veen, G. F. Veen, and R. Van Diggelen. 2005. Ecological correlates of seed survival after ingestion by fallow deer. Functional Ecology 19: 284–290.

Mulkey, S. S., A. P. Smith, S. J. Wright, J. L. Machado, and R. Dudley. 1992. Contrasting leaf phenotypes control seasonal variation in water loss in a tropical forest shrub. Proceedings of the National Academy of Sciences of the USA 89: 9084–9088.

Müller, H. 1883. The fertilisation of flowers. Translated by D'Arcy W. Thompson. Macmillan, London.

Muñiz, M., G. Nombela, and L. Barrios. 2002. Within-plant distribution and infestation pattern of the B- and Q-biotypes of the whitefly, *Bemisia tabaci*, on tomato and pepper. Entomologia Experimentalis et Applicata 104: 369–373.

Munshi, S. K., B. Kaushal, and R. K. Bajaj. 2003. Compositional changes in seeds influenced by their positions in different whorls of mature sunflower head. Journal of the Science of Food and Agriculture 83: 1622–1626.

Munshi, S. K., and A. Kumari. 1994. Physical characteristics of siliqua and lipid composition of seeds located at different positions in mature mustard inflorescence. Journal of the Science of Food and Agriculture 64: 289–293.

Murcia, C. 1990. Effect of floral morphology and temperature on pollen receipt and removal in *Ipomoea trichocarpa*. Ecology 71: 1098–1109.

Murdoch, A. J., and R. H. Ellis. 2000. Dormancy, viability and longevity. In M. Fenner, ed., Seeds: the ecology of regeneration in plant communities, 2nd edition, pp. 183–214. CABI Publishing, Wallingford, UK.

Murphy, H. T., and J. Lovett-Doust. 2004. Landscape-level effects on developmental instability: fluctuating asymmetry across the range of honey locust, *Gleditsia triacanthos* (Fabaceae). International Journal of Plant Sciences 165: 795–803.

Murray, J. R., and W. P. Hackett. 1991. Dihydroflavonol reductase activity in relation to differential anthocyanin accumulation in juvenile and mature phase *Hedera helix* L. Plant Physiology 97: 343–351.

Murray, K. G., K. Winnett-Murray, E. A. Cromie, M. Minor, and E. Meyers. 1993. The influence of seed packaging and fruit color on feeding preferences of American robins. Vegetatio 107/108: 217–226.

Murray, X. J., D. M. Holcroft, N. C. Cook, and S. J. E. Wand. 2005. Postharvest quality of 'Laetitia' and 'Songold' (*Prunus salicina* Lindell) plums as affected by preharvest shading treatments. Postharvest Biology and Technology 37: 81–92.

Nahrung, H. F., and G. R. Allen. 2003. Intra-plant host selection, oviposition preference and larval survival of *Chrysophtharta agricola* (Chapuis) (Coleoptera: Chrysomelidae: Paropsini) between foliage types of a heterophyllous host. Agricultural and Forest Entomology 5: 155–162.

Nakamura, R. R. 1988. Seed abortion and seed size variation within fruits of *Phaseolus vulgaris*: pollen donors and resource limitation effects. American Journal of Botany 75: 1003–1010.

Nalepa, C. A., and E. E. Grissell. 1993. Host seed size and adult size, emergence, and morphology of *Megastigmus aculeatus nigroflavus* (Hymenoptera: Torymidae). Environmental Entomology 22: 1313–1317.

Naranjo, S. E., and H. M. Flint. 1995. Spatial distribution of adult *Bemisia tabaci* (Homoptera: Aleyrodidae) in cotton and development and validation of fixed-precision sampling plans for estimating population. Environmental Entomology 24: 261–270.

Nátrová, Z., and L. Nátr. 1993. Limitation of kernel yield by the size of conducting tissue in winter wheat varieties. Field Crops Research 31: 121–130.

Navarro, L. 1996. Biología reproductiva y conservación de dos endemismos del noroccidente Ibérico: *Petrocoptis grandiflora* Rothm. y *Petrocoptis viscosa* Rothm. (Caryophyllaceae). Ph.D. thesis, University of Santiago de Compostela.

Nelson, C. J., J. N. Seiber, and L. P. Brower. 1981. Seasonal and intraplant variation of cardenolide content in the California milkweed, *Asclepias eriocarpa*, and implications for plant defense. Journal of Chemical Ecology 7: 981–1010.

Nelson, S. L., D. V. Masters, S. R. Humphrey, and T. H. Kunz. 2005. Fruit choice and calcium block use by Tongan fruit bats in American Samoa. Journal of Mammalogy 86: 1205–1209.

Niemelä, P., J. Tuomi, and S. Sirén. 1984. Selective herbivory on mosaic leaves of variegated *Acer pseudoplatanus*. Experientia 40: 1433–1434.

Niesenbaum, R. A. 1994. Spatial and temporal variation in pollen tube numbers in *Lindera benzoin* (spicebush). Canadian Journal of Botany 72: 268–271.

Niinemets, Ü., and O. Kull. 1998. Stoichiometry of foliar carbon constituents varies along light gradients in temperate woody canopies: implications for foliage morphological plasticity. Tree Physiology 18: 467–479.

Niinemets, Ü., O. Kull, and J. D. Tenhunen. 1998. An analysis of light effects on foliar morphology, physiology, and light interception in temperate deciduous woody species of contrasting shade tolerance. Tree Physiology 18: 681–696.

Niinemets, Ü., and A. Lukjanova. 2003. Total foliar area and average leaf age may be more strongly associated with branching frequency than with leaf longevity in temperate conifers. New Phytologist 158: 75–89.

Niinemets, Ü., A. Portsmuth, and L. Truus. 2002. Leaf structural and photosynthetic characteristics, and biomass allocation to foliage in relation to foliar nitrogen content and tree size in three *Betula* species. Annals of Botany 89: 191–204.

Niinemets, Ü., F. Valladares, and R. Ceulemans. 2003. Leaf-level phenotypic variability and plasticity of invasive *Rhododendron ponticum* and non-invasive *Ilex aquifolium* co-occurring at two contrasting European sites. Plant Cell and Environment 26: 941–956.

Niklas, K. J. 1994. Plant allometry. University of Chicago Press, Chicago.

Niu, G. H., R. D. Heins, A. Cameron, and W. Carlson. 2001. Temperature and daily light integral influence plant quality and flower development of *Campanula carpatica* 'Blue Clips' and 'Deep Blue Clips', and *Campanula* 'Birch Hybrid'. Hortscience 36: 664–668.

Njoku, E. 1956. Studies in the morphogenesis of leaves: 11, The effect of light intensity on leaf shape in *Ipomoea caerulea*. New Phytologist 55: 91–110.

Norton, B. J. 1975. Biology and philosophy: the methodological foundations of biometry. Journal of the History of Biology 8: 85–93.

Nothnagl, M., A. Kosiba, and R. U. Larsen. 2004. Predicting the effect of irradiance and temperature on the flower diameter of greenhouse grown *Chrysanthemum*. Scientia Horticulturae 99: 319–329.

Obeso, J. R. 1986. Ecología de la comunidad de aves en un medio transicional de la Sierra de Cazorla (Jaén). Ph.D. thesis, University of Oviedo.

Obeso, J. R. 1993. Seed mass variation in the perennial herb *Asphodelus albus*: sources of variation and position effect. Oecologia 93: 571–575.

Obeso, J. R. 1996. Producción de frutos y semillas en *Ilex aquifolium* L. (Aquifoliaceae). Anales del Jardín Botánico de Madrid 54: 533–539.

Obeso, J. R. 1998a. Effects of defoliation and girdling on fruit production in *Ilex aquifolium*. Functional Ecology 12: 486–491.

Obeso, J. R. 1998b. Patterns of variation in *Ilex aquifolium* fruit traits related to fruit consumption by birds and seed predation by rodents. Écoscience 5: 463–469.

Obeso, J. R. 2002. The costs of reproduction in plants. New Phytologist 155: 321–348.

Obeso, J. R. 2004a. A hierarchical perspective in allocation to reproduction from whole plant to fruit and seed level. Perspectives in Plant Ecology, Evolution, and Systematics 6: 217–225.

Obeso, J. R. 2004b. Seed provisioning within holly fruits: test of the hierarchical model. Evolutionary Ecology 18: 133–144.

Obeso, J. R., and C. M. Herrera. 1994. Inter- and intraspecific variation in fruit traits in co-occurring vertebrate-dispersed plants. International Journal of Plant Sciences 155: 382–387.

O'Connell, L. M., and K. Ritland. 2004. Somatic mutations at microsatellite loci in western redcedar (*Thuja plicata*: Cupressaceae). Journal of Heredity 95: 172–176.

O'Connell, L. M., J. Russell, and K. Ritland. 2004. Fine-scale estimation of outcrossing in western redcedar with microsatellite assay of bulked DNA. Heredity 93: 443–449.

Okane, I., A. Nakagiri, and T. Ito. 1998. Endophytic fungi in leaves of ericaceous plants. Canadian Journal of Botany 76: 657–663.

Oke, T. R. 1987. Boundary layer climates. 2nd edition. Methuen, London.

Orians, C. M. 2005. Herbivores, vascular pathways, and systemic induction: facts and artifacts. Journal of Chemical Ecology 31: 2231–2242.

Orians, C. M., M. Ardón, and B. A. Mohammad. 2002. Vascular architecture and patchy nutrient availability generate within-plant heterogeneity in plant traits important to herbivores. American Journal of Botany 89: 270–278.

Orians, C. M., and C. G. Jones. 2001. Plants as resource mosaics: a functional model for predicting patterns of within-plant resource heterogeneity to consumers based on vascular architecture and local environmental variability. Oikos 94: 493–504.

Orians, C. M., J. Pomerleau, and R. Ricco. 2000. Vascular architecture generates fine scale variation in systemic induction of proteinase inhibitors in tomato. Journal of Chemical Ecology 26: 471–485.

Orians, C. M., S. D. P. Smith, and L. Sack. 2005. How are leaves plumbed inside a branch? Differences in leaf-to-leaf hydraulic sectoriality among six temperate tree species. Journal of Experimental Botany 56: 2267–2273.

Orians, C. M., M. M. I. van Vuuren, N. L. Harris, B. A. Babst, and G. S. Ellmore. 2004. Differential sectoriality in long-distance transport in temperate tree species: evidence from dye flow, ^{15}N transport, and vessel element pitting. Trees 18: 501–509.

Ott, J. R., L. Real, and E. Silverfine. 1985. The effect of nectar variance on bumblebee patterns of movement and potential gene dispersal. Oikos 45: 333–340.

Otto, S. P., and I. M. Hastings. 1998. Mutation and selection within the individual. Genetica 102: 507–524.

Otto, S. P., and M. E. Orive. 1995. Evolutionary consequences of mutation and selection within an individual. Genetics 141: 1173–1187.

Ozga, J. A., and D. M. Reinecke. 2003. Hormonal interactions in fruit development. Journal of Plant Growth Regulation 22: 73–81.

Palá-Paúl, J., J. J. Brophy, R. J. Goldsack, L. M. Copeland, M. J. Pérez-Alonso, and A. Velasco-Negueruela. 2003. Essential oil composition of the seasonal heterophyllous leaves of *Eryngium vesiculosum*. Australian Journal of Botany 51: 497–501.

Palmer, A. R. 1996. Waltzing with asymmetry. BioScience 46: 518–532.

Palmer, A. R., and C. Strobeck. 1986. Fluctuating asymmetry: measurement, analysis, patterns. Annual Review of Ecology and Systematics 17: 391–421.

Palombi, M. A., and C. Damiano. 2002. Comparison between RAPD and SSR molecular markers in detecting genetic variation in kiwifruit (*Actinidia deliciosa* A. Chev.). Plant Cell Reports 20: 1061–1066.

Pappers, S. M., T. J. de Jong, P. G. L. Klinkhamer, and E. Meelis. 1999. Effects of nectar content on the number of bumblebee approaches and the length of visitation sequences in *Echium vulgare* (Boraginaceae). Oikos 87: 580–586.

Parciak, W. 2002. Environmental variation in seed number, size, and dispersal of a fleshy-fruited plant. Ecology 83: 780–793.

Parr, J. C., and R. Thurston. 1972. Toxicity of nicotine in synthetic diets to larvae of the tobacco hornworm. Annals of the Entomological Society of America 65: 1185–1188.

Parra-Tabla, V., and S. H. Bullock. 2000. Phenotypic natural selection on flower biomass allocation in the tropical tree *Ipomoea wolcottiana* Rose (Convolvulaceae). Plant Systematics and Evolution 221: 167–177.

Parra-Tabla, V., and S. H. Bullock. 2005. Ecological and selective effects of stigma-anther separation in the self-incompatible tropical tree *Ipomoea wolcottiana* (Convolvulaceae). Plant Systematics and Evolution 252: 85–95.

Parsons, P. A. 1990. Fluctuating asymmetry: an epigenetic measure of stress. Biological Reviews 65: 131–145.

Parsons, P. A. 1992. Fluctuating asymmetry: a biological monitor of environmental and genomic stress. Heredity 68: 361–364.

Patzak, J. 2003. Assessment of somaclonal variability in hop (*Humulus lupulus* L.) in vitro meristem cultures and clones by molecular methods. Euphytica 131: 343–350.

Paul, J., and F. Roces. 2003. Fluid intake rates in ants correlate with their feeding habits. Journal of Insect Physiology 49: 347–357.

Paxman, G. J. 1956. Differentiation and stability in the development of *Nicotiana rustica*. Annals of Botany 20: 331–347.

Pearl, R. 1936. Karl Pearson, 1857–1936. Journal of the American Statistical Association 31: 653–664.

Pearsall, W. H., and A. M. Hanby. 1925. The variation of leaf form in *Potamogeton perfoliatus*. New Phytologist 24: 112–120.

Pearson, K. 1901. Mathematical contributions to the theory of evolution: 9, On the principle of homotyposis and its relation to heredity, to the variability of the individual, and to that of the race; part 1, homotyposis in the vegetable kingdom. Philosophical Transactions of the Royal Society of London A 197: 285–379.

Pearson, K. 1903. Mathematical contributions to the theory of evolution: 11, On the influence of natural selection on the variability and correlation of organs. Philosophical Transactions of the Royal Society of London A 200: 1–66.

Pearson, S., A. Parker, S. R. Adams, P. Hadley, and D. R. May. 1995. Effects of temperature on the flower size of pansy (*Viola* × *wittrockinna* Gams). Journal of Horticultural Science 70: 183–190.

Peat, J., J. Tucker, and D. Goulson. 2005. Does intraspecific size variation in bumblebees allow colonies to efficiently exploit different flowers? Ecological Entomology 30: 176–181.

Pelletier, B., J. W. Fyles, and P. Dutilleul. 1999. Tree species control and spatial structure of forest floor properties in a mixed-species stand. Écoscience 6: 79–91.

Pellis, A., I. Laureysens, and R. Ceulemans. 2004. Growth and production of a short rotation coppice culture of poplar: 1, Clonal differences in leaf characteristics in relation to biomass production. Biomass and Bioenergy 27: 9–19.

Pellmyr, O. 1987. Temporal patterns of ovule allocation, fruit set, and seed predation in *Anemonopsis macrophylla* (Ranunculaceae). Botanical Magazine (Tokyo) 100: 175–183.

Pérez, F. J., and M. Gómez. 2000. Possible role of soluble invertase in the gibberellic acid berry-sizing effect in Sultana grape. Plant Growth Regulation 30: 111–116.

Perez, S. M., and K. D. Waddington. 1996. Carpenter bee (*Xylocopa micans*) risk indifference and a review of nectarivore risk-sensitivity studies. American Zoologist 36: 435–446.

Pérez-Pérez, J. M., J. Serrano-Cartagena, and J. L. Micol. 2002. Genetic analysis of natural variations in the architecture of *Arabidopsis thaliana* vegetative leaves. Genetics 162: 893–915.

Perfectti, F., and J. P. M. Camacho. 1999. Analysis of genotypic differences in developmental stability in *Annona cherimola*. Evolution 53: 1396–1405.

Perica, S. 2001. Seasonal fluctuation and intracanopy variation in leaf nitrogen level in olive. Journal of Plant Nutrition 24: 779–787.

Pertoldi, C., T. N. Kristensen, D. H. Andersen, and V. Loeschcke. 2006. Developmental instability as an estimator of genetic stress. Heredity 96: 122–127.

Philippi, T. 1993. Bet-hedging germination of desert annuals: beyond the first year. American Naturalist 142: 474–487.

Philippi, T., and J. Seger. 1989. Hedging one's evolutionary bets, revisited. Trends in Ecology and Evolution 4: 41–44.

Phillips, R. L., S. M. Keppler, and P. Olhoft. 1994. Genetic variability of plant tissue cultures: breakdown of normal controls. Proceedings of the National Academy of Science of the USA 91: 5222–5226.

Pierce, B. J., S. R. McWilliams, A. R. Place, and M. A. Huguenin. 2004. Diet preferences for specific fatty acids and their effect on composition of fat reserves in migratory red-eyed vireos (*Vireo olivaceus*). Comparative Biochemistry and Physiology 138A: 503–514.

Pigliucci, M. 1996. How organisms respond to environmental changes: from phenotypes to molecules (and vice versa). Trends in Ecology and Evolution 11: 168–173.

Pigliucci, M. 1998. Developmental phenotypic plasticity: where internal programming meets the external environment. Current Opinion in Plant Biology 1: 87–91.

Pigliucci, M. 2001. Phenotypic plasticity: beyond nature and nurture. Johns Hopkins University Press, London.

Pigliucci, M. 2005. Evolution of phenotypic plasticity: where are we going now? Trends in Ecology and Evolution 20: 481–486.

Pigliucci, M., J. Whitton, and C. D. Schlichting. 1995. Reaction norms of *Arabidop-*

sis: 1, Plasticity of characters and correlations across water, nutrient and light gradients. Journal of Evolutionary Biology 8: 421–438.

Pineda-Krch, M., and A. G. B. Poore. 2004. Spatial interactions within modular organisms: genetic heterogeneity and organism fitness. Theoretical Population Biology 66: 25–36.

Pivnick, K. A., and J. N. McNeil. 1985. Effects of nectar concentration on butterfly feeding: measured feeding rates for *Thymelicus lineola* (Lepidoptera: Hesperiidae) and a general feeding model for adult Lepidoptera. Oecologia 66: 226–237.

Platenkamp, G. A. J., and R. G. Shaw. 1993. Environmental and genetic maternal effects on seed characters in *Nemophila menziesii*. Evolution 47: 540–555.

Pleasants, J. M. 1981. Bumblebee response to variation in nectar availability. Ecology 62: 1648–1661.

Pleasants, J. M. 1983. Nectar production patterns in *Ipomopsis aggregata* (Polemoniaceae). American Journal of Botany 70: 1468–1475.

Pleasants, J. M. 1989. Optimal foraging by nectarivores: a test of the marginal-value theorem. American Naturalist 134: 51–71.

Pleasants, J. M., and M. Zimmerman. 1979. Patchiness in the dispersion of nectar resources: evidence for hot and cold spots. Oecologia 41: 283–288.

Plowright, C. M. S., and R. C. Plowright. 1997. The advantage of short tongues in bumble bees (*Bombus*): analyses of species distributions according to flower corolla depth, and of working speeds on white clover. Canadian Entomologist 129: 51–59.

Poethig, R. S. 1989. Genetic mosaics and cell lineage analysis in plants. Trends in Genetics 5: 273–277.

Poethig, R. S. 1997. Leaf morphogenesis in flowering plants. Plant Cell 9: 1077–1087.

Pons, T. L., and N. P. R. Anten. 2004. Is plasticity in partitioning of photosynthetic resources between and within leaves important for whole-plant carbon gain in canopies? Functional Ecology 18: 802–811.

Pons, T. L., and M. Bergkotte. 1996. Nitrogen allocation in response to partial shading of a plant: possible mechanisms. Physiologia Plantarum 98: 571–577.

Porté, A., and D. Loustau. 1998. Variability of the photosynthetic characteristics of mature needles within the crown of a 25-year-old *Pinus pinaster*. Tree Physiology 18: 223–232.

Poulton, J., and A. A. Winn. 2002. Costs of canalization and plasticity in response to neighbors in *Brassica rapa*. Plant Species Biology 17: 109–118.

Powell, G. R., K. J. Tosh, and J. E. MacDonald. 1982. Indeterminate shoot extension and heterophylly in *Acer saccharum*. Canadian Journal of Forest Research 12: 166–170.

Powell, J. S., and K. F. Raffa. 1999. Sources of variation in concentration and composition of foliar monoterpenes in tamarack (*Larix laricina*) seedlings: roles of nutrient availability, time of season, and plant architecture. Journal of Chemical Ecology 25: 1771–1797.

Pratt, T. K., and E. W. Stiles. 1985. The influence of fruit size and structure on composition of frugivore assemblages in New Guinea. Biotropica 17: 314–321.

Preston, K. A. 1998. The effects of developmental stage and source leaf position on integration and sectorial patterns of carbohydrate movement in an annual plant, *Perilla frutescens* (Lamiaceae). American Journal of Botany 85: 1695–1703.

Price, E. A. C., M. J. Hutchings, and C. Marshall. 1996. Causes and consequences of sectoriality in the clonal herb *Glechoma hederacea*. Vegetatio 127: 41–54.

Primack, R. B. 1987. Relationships among flowers, fruits, and seeds. Annual Review of Ecology and Systematics 18: 409–430.

Prokopy, R. J. 1969. Visual responses of European cherry fruit flies *Rhagoletis cerasi* L. (Diptera, Trypetidae). Polskie Pismo Entomologiczne 39: 539–566.

Prokopy, R. J. 1977. Attraction of *Rhagoletis* flies (Diptera: Tephritidae) to red spheres of different sizes. Canadian Entomologist 109: 593–596.

Prokopy, R. J., and G. L. Bush. 1973. Ovipositional responses to different sizes of artificial fruit by flies of *Rhagoletis pomonella* species group. Annals of the Entomological Society of America 66: 927–929.

Prokopy, R. J., and E. D. Owens. 1983. Visual detection of plants by herbivorous insects. Annual Review of Entomology 28: 337–364.

Pyke, G. H. 1978a. Optimal foraging in bumblebees and coevolution with their plants. Oecologia 36: 281–293.

Pyke, G. H. 1978b. Optimal foraging: movement patterns of bumblebees between inflorescences. Theoretical Population Biology 13: 72–98.

Pyke, G. H. 1981. Optimal nectar production in a hummingbird-pollinated plant. Theoretical Population Biology 20: 326–343.

Pyke, G. H. 1984. Optimal foraging theory: a critical review. Annual Review of Ecology and Systematics 15: 523–575.

Queitsch, C., T. A. Sangster, and S. Lindquist. 2002. Hsp90 as a capacitor of phenotypic variation. Nature 417: 618–624.

Rabasa, S. G., D. Gutiérrez, and A. Escudero. 2005. Egg laying by a butterfly on a fragmented host plant: a multi-level approach. Ecography 28: 629–639.

Raghu, S., R. A. I. Drew, and A. R. Clarke. 2004. Influence of host plant structure and microclimate on the abundance and behavior of a tephritid fly. Journal of Insect Behavior 17: 179–190.

Raguso, R. A. 2001. Floral scent, olfaction, and scent-driven foraging behavior. In L. Chittka and J. D. Thomson, eds., Cognitive ecology of pollination, pp. 83–105. Cambridge University Press, Cambridge.

Rahman, M. H., and O. P. Rajora. 2001. Microsatellite DNA somaclonal variation in micropropagated trembling aspen (*Populus tremuloides*). Plant Cell Reports 20:531–536.

Rajala, A., and P. Peltonen-Sainio. 2004. Intra-plant variation for progress of cell division in developing oat grains: a preliminary study. Agricultural and Food Science 13: 163–169.

Ram, H. Y. M., and S. Rao. 1982. In vitro induction of aerial leaves and of precocious flowering in submerged shoots of *Limnophila indica*. Planta 155: 521–523.

Raser, J. M., and E. K. O'Shea. 2004. Control of stochasticity in eukaryotic gene expression. Science 304: 1811–1814.

Raser, J. M., and E. K. O'Shea. 2005. Noise in gene expression: origins, consequences, and control. Science 309: 2010–2013.

Raske, A. G., and D. G. Bryant. 1977. Distribution, survival, and intra-tree movement of late-instar birch casebearer larvae on white birch (Lepidoptera: Coleophoridae). Canadian Entomologist 109: 1297–1306.

Raspé, O., C. Findlay, and A. L. Jacquemart. 2000. Biological flora of the British Isles: 214, *Sorbus aucuparia* L. Journal of Ecology 88: 910–930.

Raupp, M. J., and R. F. Denno. 1983. Leaf age as a predictor of herbivore distribution and abundance. In R. F. Denno and M. S. McClure, eds., Variable plants and herbivores in natural and managed systems, pp. 91–124. Academic Press, New York.

Ravindra, N. S., R. N. Kulkarni, M. C. Gayathri, and S. Ramesh. 2004. Somaclonal variation for some morphological traits, herb yield, essential oil content and essential oil composition in an Indian cultivar of rose-scented geranium. Plant Breeding 123: 84–86.

Ravishankar, K. V., R. Uma Shaanker, and K. N. Ganeshaiah. 1995. War of hormones over resource allocation to seeds: strategies and counterstrategies of offspring and maternal parent. Journal of Biosciences 20: 89–103.

Ray, T. S. 1987. Cyclic heterophylly in *Syngonium* (Araceae). American Journal of Botany 74: 16–26.

Ray, T. S. 1990. Metamorphosis in the Araceae. American Journal of Botany 77: 1599–1609.

Reader, T., I. MacLeod, P. T. Elliott, O. J. Robinson, and A. Manica. 2005. Interorder interactions between flower-visiting insects: foraging bees avoid flowers previously visited by hoverflies. Journal of Insect Behavior 18: 51–57.

Real, L. A. 1980. On uncertainty and the law of diminishing returns in evolution and behavior. In J. E. R. Staddon, ed., Limits to action: the allocation of individual behavior, pp. 37–64. Academic Press, New York.

Real, L. A. 1981. Uncertainty and pollinator-plant interactions: the foraging behavior of bees and wasps on artificial flowers. Ecology 62: 20–26.

Real, L. A., and T. Caraco. 1986. Risk and foraging in stochastic environments. Annual Review of Ecology and Systematics 17: 371–390.

Real, L. A., and B. J. Rathcke. 1988. Patterns of individual variability in floral resources. Ecology 69: 728–735.

Redmon, S. G., T. G. Forrest, and G. P. Markin. 2000. Biology of *Bruchidius villosus* (Coleoptera: Bruchidae) on scotch broom in North Carolina. Florida Entomologist 83: 242–253.

Rehill, B. J., and J. C. Schultz. 2001. *Hormaphis hamamelidis* and gall size: a test of the plant vigor hypothesis. Oikos 95: 94–104.

Reich, P. B., C. Uhl, M. B. Walters, L. Prugh, and D. S. Ellsworth. 2004. Leaf demography and phenology in Amazonian rain forest: a census of 40,000 leaves of 23 tree species. Ecological Monographs 74: 3–23.

Reid, N. 1989. Dispersal of mistletoes by honeyeaters and flowerpeckers: components of seed dispersal quality. Ecology 70: 137–145.

Rey, P. J. 1995. Spatio-temporal variation in fruit and frugivorous bird abundance in olive orchards. Ecology 76: 1625–1635.

Rey, P. J., and J. E. Gutiérrez. 1997. Elección de fruto y conducta de alimentación de aves frugívoras en olivares y acebuchares: una estrategia óptima basada en la razón beneficio/tiempo de manipulación. Ardeola 44: 27–39.

Rey, P. J., C. M. Herrera, J. Guitián, X. Cerdá, A. M. Sánchez-Lafuente, M. Medrano, and J. L. Garrido. 2006. The geographic mosaic in pre-dispersal interactions and selection on *Helleborus foetidus* (Ranunculaceae). Journal of Evolutionary Biology 19: 21–34.

Rhodes, J. D., J. F. Thain, and D. C. Wildon. 1999. Evidence for physically distinct systemic signalling pathways in the wounded tomato plant. Annals of Botany 84: 109–116.

Riaz, S., K. E. Garrison, G. S. Dangl, J. M. Boursiquot, and C. P. Meredith. 2002. Genetic divergence and chimerism within ancient asexually propagated wine-grape cultivars. Journal of the American Society for Horticultural Science 127: 508–514.

Richards, J. H., and D. W. Lee. 1986. Light effects on leaf morphology in water hyacinth (*Eichhornia crassipes*). American Journal of Botany 73: 1741–1747.

Rinderer, T. E., B. D. Marx, M. Gries, and S. Tingek. 1996. A scientific note on stratified foraging by Sabahan bees on the yellow flame tree (*Peltophorum pterocarpum*). Apidologie 27: 423–425.

Rinne, R. W., and R. G. Langston. 1960. Studies on lateral movement of phosphorus 32 in peppermint. Plant Physiology 35: 216–219.

Roach, D. A. 1987. Variation in seed and seedling size in *Anthoxanthum odoratum*. American Midland Naturalist 117: 258–264.

Robbins, W. J. 1960. Further observations on juvenile and adult *Hedera*. American Journal of Botany 47: 485–491.

Robbins, W. W. 1908. Variation in flower-heads of *Gaillardia aristata*. Biometrika 6: 106–108.

Robe, W. E., and H. Griffiths. 1998. Adaptations for an amphibious life: changes in leaf morphology, growth rate, carbon and nitrogen investment, and reproduction during adjustment to emersion by the freshwater macrophyte *Littorella uniflora*. New Phytologist 140: 9–23.

Robertson, G. P., M. A. Huston, F. C. Evans, and J. M. Tiedje. 1988. Spatial variability in a successional plant-community: patterns of nitrogen availability. Ecology 69: 1517–1524.

Robertson, I. C., and A. C. Ulappa. 2004. Distance between pollen donor and recipient influences fruiting success in slickspot peppergrass, *Lepidium papilliferum*. Canadian Journal of Botany 82: 1705–1710.

Robinson, D. 1994. The responses of plants to non-uniform supply of nutrients. New Phytologist 127: 635–674.

Rocha, O. J., and A. G. Stephenson. 1990. Effect of ovule position on seed production, seed weight, and progeny performance in *Phaseolus coccineus* L. (Leguminosae). American Journal of Botany 77: 1320–1329.

Rocha, O. J., and A. G. Stephenson. 1991. Order of fertilization within the ovary in *Phaseolus coccineus* L. (Leguminosae). Sexual Plant Reproduction 4: 126–131.

Rodríguez, I., A. Gumbert, N. H. de Ibarra, J. Kunze, and M. Giurfa. 2004. Symmetry is in the eye of the "beeholder": innate preference for bilateral symmetry in flower-naïve bumblebees. Naturwissenschaften 91: 374–377.

Rodríguez-Burruezo, A., J. Prohens, and F. Nuez. 2002. Genetic analysis of quantitative traits in pepino (*Solanum muricatum*) in two growing seasons. Journal of the American Society for Horticultural Science 127: 271–278.

Rodríguez-Burruezo, A., J. Prohens, and E. Nuez. 2003. Wild relatives can contribute to the improvement of fruit quality in pepino (*Solanum muricatum*). Euphytica 129: 311–318.

Rodríguez Larrinaga, A. 2004. Factores que afectan a la selección de fruto por parte de aves frugívoras del género *Turdus*. Ph. D. thesis, Universidad de Santiago de Compostela.

Roff, D., and D. Réale. 2004. The quantitative genetics of fluctuating asymmetry: a comparison of two models. Evolution 58: 47–58.

Rogers, R. W., and H. T. Clifford. 1993. The taxonomic and evolutionary significance of leaf longevity. New Phytologist 123: 811–821.

Rohlf, F. J., A. J. Gilmartin, and G. Hart. 1983. The Kluge-Kerfoot phenomenon—a statistical artifact. Evolution 37: 180–202.

Rohloff, J. 1999. Monoterpene composition of essential oil from peppermint (*Mentha × piperita* L.) with regard to leaf position using solid-phase microextraction and gas chromatography/mass spectrometry analysis. Journal of Agricultural and Food Chemistry 47: 3782–3786.

Roiloa, S. R., P. Alpert, N. Tharayil, G. Hancock, and P. C. Bhowmik. 2007. Greater capacity for division of labour in clones of *Fragaria chiloensis* from patchier habitats. Journal of Ecology 95: 397–405.

Roldán Serrano, A., and J. M. Guerra-Sanz. 2006. Quality fruit improvement in sweet pepper culture by bumblebee pollination. Scientia Horticulturae 110: 160–166.

Rolland, F., E. Baena-Gonzalez, and J. Sheen. 2006. Sugar sensing and signaling in plants: conserved and novel mechanisms. Annual Review of Plant Biology 57: 675–709.

Ros, M., D. Sorensen, R. Waagepetersen, M. Dupont-Nivet, M. SanCristobal, J. C. Bonnet, and J. Mallard. 2004. Evidence for genetic control of adult weight plasticity in the snail *Helix aspersa*. Genetics 168: 2089–2097.

Roslin, T., S. Gripenberg, J. P. Salminen, M. Karonen, R. B. O'Hara, K. Pihlaja, and P. Pulkkinen. 2006. Seeing the trees for the leaves: oaks as mosaics for a host-specific moth. Oikos 113: 106–120.

Rossi, R. E., D. J. Mulla, A. G. Journel, and E. H. Franz. 1992. Geostatistical tools for modeling and interpreting ecological spatial dependence. Ecological Monographs 62: 277–314.

Rottenberg, A., E. Nevo, and D. Zohary. 2000. Genetic variability in sexually dimorphic and monomorphic populations of *Populus euphratica* (Salicaceae). Canadian Journal of Forest Research 30: 482–486.

Roubik, D. W., J. D. Ackerman, C. Copenhaver, and B. H. Smith. 1982. Stratum, tree, and flower selection by tropical bees: implications for the reproductive biology of outcrossing *Cochlospermum vitifolium* in Panama. Ecology 63: 712–720.

Rowe, E. C., and G. Cadisch. 2002. Implications of heterogeneity on procedures for estimating plant ^{15}N recovery in hedgerow intercrop systems. Agroforestry Systems 54: 61–70.

Rowe, W. J., and D. A. Potter. 1996. Vertical stratification of feeding by Japanese beetles within linden tree canopies: selective foraging or height per se? Oecologia 108: 459–466.

Roy, S. K. 1959. Regulation of morphogenesis in an oleaceous tree, *Nyctanthes arbor-tristis*. Nature 183: 1410–1411.

Roy, S. K. 1963. The variation of organs of individual plants. Journal of Genetics 58: 147–176.

Rubio de Casas, R., P. Vargas, E. Pérez-Corona, E. Manrique, J. R. Quintana, C. García-Verdugo, and L. Balaguer. 2007. Field patterns of leaf plasticity in adults of the long-lived evergreen *Quercus coccifera*. Annals of Botany 100: 325–334.

Ruel, J. J., and M. P. Ayres. 1999. Jensen's inequality predicts effects of environmental variation. Trends in Ecology and Evolution 14: 361–366.

Ruiz de Clavijo, E. 1995. The ecological significance of fruit heteromorphism in the amphicarpic species *Catananche lutea* (Asteraceae). International Journal of Plant Sciences 156: 824–833.

Ruiz de Clavijo, E. 2001. The role of dimorphic achenes in the biology of the annual weed *Leontodon longirrostris*. Weed Research 41: 275–286.

Ruiz de Clavijo, E., and M. J. Jiménez. 1998. The influence of achene type and plant density on growth and biomass allocation in the heterocarpic annual *Catananche lutea* (Asteraceae). International Journal of Plant Sciences 159: 637–647.

Sakai, K. I., and Y. Shimamoto. 1965. Developmental instability in leaves and flowers of *Nicotiana tabacum*. Genetics 51: 801–813.

Sakai, S., A. Sakai, and H. S. Ishii. 1997. Patterns of wing size variation in seeds of the lily *Cardiocrinum cordatum* (Liliaceae). American Journal of Botany 84: 1275–1278.

Salisbury, E. J. 1942. The reproductive capacity of plants. G. Bell and Sons, London.

Salisbury, F. B., and C. W. Ross. 1992. Plant physiology. 4th edition. Wadsworth Publishing, Belmont, CA.

Sallabanks, R. 1993. Hierarchical mechanisms of fruit selection by an avian frugivore. Ecology 74: 1326–1336.

Sallabanks, R., and S. P. Courtney. 1992. Frugivory, seed predation, and insect-vertebrate interactions. Annual Review of Entomology 37: 377–400.

Salopek-Sondi, B., M. Kovac, N. Ljubesic, and V. Magnus. 2000. Fruit initiation in *Helleborus niger* L. triggers chloroplast formation and photosynthesis in the perianth. Journal of Plant Physiology 157: 357–364.

Sanders, M. L., and T. Ericsson. 1998. Vertical distribution of plant nutrients and heavy metals in *Salix viminalis* stems and their implications for sampling. Biomass and Bioenergy 14: 9–19.

Sapir, N., Z. Abramsky, E. Shochat, and I. Izhaki. 2004. Scale-dependent habitat selection in migratory frugivorous passerines. Naturwissenschaften 91: 544–547.

Saracco, J. F., J. A. Collazo, and M. J. Groom. 2004. How do frugivores track resources? Insights from spatial analyses of bird foraging in a tropical forest. Oecologia 139: 235–245.

Sarkar, S. 2004. From the *Reaktionsnorm* to the evolution of adaptive plasticity: a historical sketch, 1909–1999. In T. J. DeWitt and S. M. Scheiner, eds., Phenotypic plasticity: functional and conceptual approaches, pp. 10–30. Oxford University Press, Oxford.

Sarkar, S., and T. Fuller. 2003. Generalized norms of reaction for ecological developmental biology. Evolution and Development 5: 106–115.

Satina, S., and A. F. Blakeslee. 1941. Periclinal chimeras in *Datura stramonium* in relation to development of leaf and flower. American Journal of Botany 28: 862–871.

Satina, S., and A. F. Blakeslee. 1943. Periclinal chimeras in *Datura* in relation to the development of the carpel. American Journal of Botany 30: 453–462.

Saunders, E. R. 1941. The significance of certain morphological variations of common occurrence in flowers of *Primula*. New Phytologist 40: 64–85.

Sawhney, V. K. 1983. The role of temperature and its relationship with gibberellic acid in the development of floral organs in tomato (*Lycopersicon esculentum*). Canadian Journal of Botany 61: 1258–1265.

Saxton, V. P., G. L. Creasy, A. M. Paterson, and M. C. T. Trought. 2004. Response of blackbirds (*Turdus merula*) and silvereyes (*Zosterops lateralis*) to geraniol and 2-methoxy-3-isobutylpyrazine. American Journal of Enology and Viticulture 55: 292–294.

Saxton, V. P., G. J. Hickling, M. C. T. Trought, and G. L. Creasy. 2004. Comparative behavior of free-ranging blackbirds (*Turdus merula*) and silvereyes (*Zosterops lateralis*) with hexose sugars in artificial grapes. Applied Animal Behaviour Science 85: 157–166.

Schaal, B. A. 1980. Reproductive capacity and seed size in *Lupinus texensis*. American Journal of Botany 67: 703–709.

Schaefer, H. M., V. Schmidt, and F. Bairlein. 2003. Discrimination abilities for nutrients: which difference matters for choosy birds and why? Animal Behaviour 65: 531–541.

Scheiner, S. M., and T. J. DeWitt. 2004. Future research directions. In T. J. DeWitt and S. M. Scheiner, eds., Phenotypic plasticity: functional and conceptual approaches, pp. 1–9. Oxford University Press, Oxford.

Scheiner, S. M., K. Donohue, L. A. Dorn, S. J. Mazer, and L. M. Wolfe. 2002. Reducing environmental bias when measuring natural selection. Evolution 56: 2156–2167.

Schemske, D. W., and C. C. Horvitz. 1984. Variation among floral visitors in pollination ability: a precondition for mutualism specialization. Science 225: 519–521.

Schemske, D. W., and C. C. Horvitz. 1989. Temporal variation in selection on a floral character. Evolution 43: 461–465.

Schittko, U., and I. T. Baldwin. 2003. Constraints to herbivore-induced systemic responses: bidirectional signaling along orthostichies in *Nicotiana attenuata*. Journal of Chemical Ecology 29: 763–770.

Schlichting, C. D. 1986. The evolution of phenotypic plasticity in plants. Annual Review of Ecology and Systematics 17: 667–693.

Schlichting, C. D., and M. Pigliucci. 1998. Phenotypic evolution: a reaction norm perspective. Sinauer Associates, Sunderland, MA.

Schmalhausen, I. I. 1949. Factors of evolution: the theory of stabilizing selection. Blakiston, Philadelphia.

Schmidt, B. L., and W. F. Millington. 1968. Regulation of leaf shape in *Proserpinaca palustris*. Bulletin of the Torrey Botanical Club 95: 264–286.

Schoen, D. J., and D. G. Lloyd. 1984. The selection of cleistogamy and heteromorphic diaspores. Biological Journal of the Linnean Society 23: 303–322.

Schondube, J. E., and C. Martínez del Río. 2003. Concentration-dependent sugar preferences in nectar-feeding birds: mechanisms and consequences. Functional Ecology 17: 445–453.

Schoonhoven, L. M., J. J. A. van Loon, and M. Dicke. 2005. Insect-plant biology. 2nd edition. Oxford University Press, Oxford.

Schultz, J. C. 1983. Habitat selection and foraging tactics of caterpillars in heterogeneous trees. In R. F. Denno and M. S. McClure, eds., Variable plants and herbivores in natural and managed systems, pp. 61–90. Academic Press, New York.

Schuster, D. J. 1998. Intraplant distribution of immature lifestages of *Bemisia argentifolii* (Homoptera: Aleyrodidae) on tomato. Environmental Entomology 27: 1–9.

Schuster, W. S. F., and J. B. Mitton. 1991. Relatedness within clusters of a bird-dispersed pine and the potential for kin interactions. Heredity 67: 41–48.

Scott, G. W., and J. H. MacGillivray. 1940. Variation in solids of the juice from different regions in melon fruits. Hilgardia 13: 69–79.

Scott, J. K., and R. Black. 1981. Selective predation by white-tailed black cockatoos on fruit of *Banksia attenuata* containing the seed-eating weevil *Alphitopis nivea*. Australian Wildlife Research 8: 421–430.

Scott, K. D., N. Lawrence, C. L. Lange, L. J. Scott, K. S. Wilkinson, M. A. Merritt, M. Miles, D. Murray, and G. C. Graham. 2005. Assessing moth migration and population structuring in *Helicoverpa annigera* (Lepidoptera: Noctuidae) at the regional scale: example from the Darling Downs, Australia. Journal of Economic Entomology 98: 2210–2219.

Scott, W. R., M. Appleyard, G. Fellowes, and E. J. M. Kirby. 1983. Effect of genotype and position in the ear on carpel and grain-growth and mature grain weight of spring barley. Journal of Agricultural Science 100: 383–391.

Sculthorpe, C. D. 1967. The biology of aquatic vascular plants. Edward Arnold, London.

Searle, K. R., N. T. Hobbs, and L. A. Shipley. 2005. Should I stay or should I go? Patch departure decisions by herbivores at multiple scales. Oikos 111: 417–424.

Searle, K. R., N. T. Hobbs, B. A. Wunder, and L. A. Shipley. 2006. Preference in patchy landscapes: the influence of scale-specific intake rates and variance in reward. Behavioral Ecology 17: 315–323.

Searle, S. R., G. Casella, and C. E. McCulloch. 1992. Variance components. Wiley, New York.

Seburn, C. N. L., T. A. Dickinson, and S. C. H. Barrett. 1990. Floral variation in

Eichhornia paniculata (Spreng.) Solms (Pontederiaceae): 1, Instability of stamen position in genotypes from northeast Brazil. Journal of Evolutionary Biology 3: 103–123.

Senalik, D., and P. W. Simon. 1987. Quantifying intra-plant variation of volatile terpenoids in carrot. Phytochemistry 26: 1975–1979.

Senar, J. C. 1983. On the siskin's ability to discriminate between edible and aborted pine seeds. Miscellània Zoològica (Barcelona) 7: 224–226.

Senft, R. L., M. B. Coughenour, D. W. Bailey, L. R. Rittenhouse, O. E. Sala, and D. M. Swift. 1987. Large herbivore foraging and ecological hierarchies. BioScience 37: 789–799.

Seyffert, W. 1983. Homeostasis in defined genotypes of *Matthiola incana*. Theoretical and Applied Genetics 64: 205–212.

Shafir, S. 2000. Risk-sensitive foraging: the effect of relative variability. Oikos 88: 663–669.

Shafir, S., A. Bechar, and E. U. Weber. 2003. Cognition-mediated coevolution: context-dependent evaluations and sensitivity of pollinators to variability in nectar rewards. Plant Systematics and Evolution 238: 195–209.

Shafir, S., D. D. Wiegmann, B. H. Smith, and L. A. Real. 1999. Risk-sensitive foraging: choice behaviour of honeybees in response to variability in volume of reward. Animal Behaviour 57: 1055–1061.

Shea, M. M., and M. A. Watson. 1989. Patterns of leaf and flower removal: their effect on fruit growth in *Chamaenerion angustifolium* (fireweed). American Journal of Botany 76: 884–890.

Shelton, A. L. 2000. Variable chemical defences in plants and their effects on herbivore behaviour. Evolutionary Ecology Research 2: 231–249.

Shelton, A. L. 2004. Variation in chemical defences of plants may improve the effectiveness of defence. Evolutionary Ecology Research 6: 709–726.

Shelton, A. L. 2005. Within-plant variation in glucosinolate concentrations of *Raphanus sativus* across multiple scales. Journal of Chemical Ecology 31: 1711–1732.

Sherry, R. A., and E. M. Lord. 1996a. Developmental stability in flowers of *Clarkia tembloriensis* (Onagraceae). Journal of Evolutionary Biology 9: 911–930.

Sherry, R. A., and E. M. Lord. 1996b. Developmental stability in leaves of *Clarkia tembloriensis* (Onagraceae) as related to population outcrossing rates and heterozygosity. Evolution 50: 80–91.

Shibata, S., T. A. Ishida, F. Soeya, N. Morino, and K. Yoshida. 2001. Within-tree variation in density and survival of leafminers on oak *Quercus dentata*. Ecological Research 16: 135–143.

Shimada, M., H. Kurota, and Y. Toquenaga. 2001. Regular distribution of larvae and resource monopolization in the seed beetle *Bruchidius dorsalis* infesting seeds of the Japanese honey locust *Gleditsia japonica*. Population Ecology 43: 245–252.

Shore, J. S., and C. M. Obrist. 1992. Variation in cyanogenesis within and among populations and species of *Turnera* series *Canaligerae* (Turneraceae). Biochemical Systematics and Ecology 20: 9–15.

Shulaev, V., J. León, and I. Raskin. 1995. Is salicylic acid a translocated signal of systemic acquired resistance in tobacco? Plant Cell 7: 1691–1701.

Shull, G. H. 1902. A quantitative study of variation in the bracts, rays, and disk florets of *Aster shortii* Hook., *A. novae-angliae* L., *A. puniceus* L., and *A. prenanthoides* Muhl., from Yellow Springs, Ohio. American Naturalist 36: 111–152.

Silvertown, J. W. 1984. Phenotypic variety in seed germination behavior: the ontogeny and evolution of somatic polymorphism in seeds. American Naturalist 124: 1–16.

Simmen, B., B. Josseaume, and M. Atramentowicz. 1999. Frugivory and taste responses to fructose and tannic acid in a prosimian primate and a didelphid marsupial. Journal of Chemical Ecology 25: 331–346.

Simon, P. W. 1982. Genetic variation for volatile terpenoids in roots of carrot, *Daucus carota*, backcrosses and F2 generations. Phytochemistry 21: 875–879.

Simons, A. M., and M. O. Johnston. 1997. Developmental instability as a bet-hedging strategy. Oikos 80: 401–406.

Simons, A. M., and M. O. Johnston. 2000. Variation in seed traits of *Lobelia inflata* (Campanulaceae): sources and fitness consequences. American Journal of Botany 87: 124–132.

Simpson, J. J. 1914. Contribution to a statistical study of the Cruciferae: variation in the flowers of *Lepidium draba* Linnaeus. Biometrika 10: 215–268.

Singaravelan, N., G. Nee'man, M. Inbar, and I. Izhaki. 2005. Feeding responses of free-flying honeybees to secondary compounds mimicking floral nectars. Journal of Chemical Ecology 31: 2791–2804.

Sinnott, E. W. 1921. The relation between body size and organ size in plants. American Naturalist 55: 385–403.

Sipe, T. W., and A. R. Linnerooth. 1995. Intraspecific variation in samara morphology and flight behavior in *Acer saccharinum* (Aceraceae). American Journal of Botany 82: 1412–1419.

Sites, J. W., and H. J. Reitz. 1950. The variation in individual Valencia oranges from different locations on the tree as a guide to sampling methods and spot-picking for quality: 3, Vitamin C and juice content of the fruit. Proceedings of the American Society for Horticultural Science 56: 103–110.

Slansky, F., and G. S. Wheeler. 1992. Caterpillars compensatory feeding response to diluted nutrients leads to toxic allelochemical dose. Entomologia Experimentalis et Applicata 65: 171–186.

Smallwood, P. D. 1996. An introduction to risk sensitivity: the use of Jensen's inequality to clarify evolutionary arguments of adaptation and constraint. American Zoologist 36: 392–401.

Smillie, R. M., S. E. Hetherington, and W. J. Davies. 1999. Photosynthetic activity of the calyx, green shoulder, pericarp, and locular parenchyma of tomato fruit. Journal of Experimental Botany 50: 707–718.

Smith, C. C., and S. D. Fretwell. 1974. The optimal balance between size and number of offspring. American Naturalist 108: 499–506.

Smith, G. S., J. P. Curtis, and C. M. Edwards. 1992. A method for analyzing plant architecture as it relates to fruit-quality using three-dimensional computer graphics. Annals of Botany 70: 265–269.

Smith, H. B. 1941. Variation and correlation of stomatal frequency and respiration rate in *Phaseolus vulgaris*. American Journal of Botany 28: 722–725.

Smith, R. F. 1967. The leaf dimorphism of *Liquidambar styraciflua*. American Midland Naturalist 77: 42–50.

Snow, A. A., T. P. Spira, R. Simpson, and R. A. Klips. 1996. The ecology of geitonogamous pollination. In D. G. Lloyd and S. C. H. Barrett, eds., Floral biology, pp. 191–206. Chapman and Hall, New York.

Snow, B., and D. Snow. 1988. Birds and berries: a study of an ecological interaction. T&AD Poyser, Calton, UK.

Sokal, R. R. 1976. The Kluge-Kerfoot phenomenon reexamined. American Naturalist 110: 1077–1091.

Sokal, R. R., and C. A. Braumann. 1980. Significance tests for coefficients of variation and variability profiles. Systematic Zoology 29: 50–66.

Soltis, D. E., P. S. Soltis, M. W. Chase, M. E. Mort, D. C. Albach, M. Zanis, V. Savolainen, et al. 2000. Angiosperm phylogeny inferred from 18S rDNA, rbcL, and atpB sequences. Botanical Journal of the Linnean Society 133: 381–461.

Sosa-Gómez, D. R. 2004. Intraspecific variation and population structure of the velvetbean caterpillar, *Anticarsia gemmatalis* Hubner, 1818 (Insecta: Lepidoptera: Noctuidae). Genetics and Molecular Biology 27: 378–384.

Soulé, M. E. 1982. Allomeric variation: 1, The theory and some consequences. American Naturalist 120: 751–764.

Southwick, E. E. 1982. "Lucky hit" nectar rewards and energetics of plant and pollinators. Comparative Physiological Ecology 7: 51–55.

Southwick, E. E. 1983. Nectar biology and nectar feeders of common milkweed, *Asclepias syriaca* L. Bulletin of the Torrey Botanical Club 110: 324–334.

Souty, M., M. Génard, M. Reich, and G. Albagnac. 1999. Influence de la fourniture en assimilats sur la maturation et la qualité de la pêche (*Prunus persica* L., cv. Suncrest). Canadian Journal of Plant Science 79: 259–268.

Sparks, P. D., and S. N. Postlethwait. 1967. Physiological control of the dimorphic leaves of *Cyamopsis tetragonoloba*. American Journal of Botany 54: 286–290.

Sperens, U. 1997. Long-term variation in, and effects of fertiliser addition on, flower, fruit and seed production in the tree *Sorbus aucuparia* (Rosaceae). Ecography 20: 521–534.

Spiegel, O., and R. Nathan. 2007. Incorporating dispersal distance into the disperser effectiveness framework: frugivorous birds provide complementary dispersal to plants in a patchy environment. Ecology Letters 10: 718–728.

Spiegelman, C. H. 1985. Jensen's inequality for general location parameter. American Statistician 39: 54.

Sprugel, D. G., T. M. Hinckley, and W. Schaap. 1991. The theory and practice of branch autonomy. Annual Review of Ecology and Systematics 22: 309–334.

Stanley, M. C., and A. Lill. 2001. Response of silvereyes (*Zosterops lateralis*) to dietary tannins: the paradox of secondary metabolites in ripe fruit. Australian Journal of Zoology 49: 633–640.

Stanley, M. C., and A. Lill. 2002. Importance of seed ingestion to an avian frugivore: an experimental approach to fruit choice based on seed load. Auk 119: 175–184.

Stanley, M. C., E. Smallwood, and A. Lill. 2002. The response of captive silvereyes

(*Zosterops lateralis*) to the colour and size of fruit. Australian Journal of Zoology 50: 205–213.

Stanton, M. L. 1984. Developmental and genetic sources of seed weight variation in *Raphanus raphanistrum* L. (Brassicaceae). American Journal of Botany 71: 1090–1098.

Stanton, M. L. 1987. Reproductive biology of petal color variants in wild populations of *Raphanus sativus*: 1, Pollinator response to color morphs. American Journal of Botany 74: 178–187.

Stanton, M. L., and R. E. Preston. 1988. Ecological consequences and phenotypic correlates of petal size variation in wild radish, *Raphanus sativus* (Brassicaceae). American Journal of Botany 75: 528–539.

Staudt, M., N. Mandl, R. Joffre, and S. Rambal. 2001. Intraspecific variability of monoterpene composition emitted by *Quercus ilex* leaves. Canadian Journal of Forest Research 31: 174–180.

Stebbins, G. L. 1950. Variation and evolution in plants. Columbia University Press, New York.

Stebbins, G. L. 1970. Adaptive radiation of reproductive characteristics in angiosperms: 1, Pollination mechanisms. Annual Review of Ecology and Systematics 1: 307–326.

Stein, O. L., and E. B. Fosket. 1969. Comparative developmental anatomy of shoots of juvenile and adult *Hedera helix*. American Journal of Botany 56: 546–551.

Stein, W. E., F. Mannolini, L. V. Hernick, E. Landing, and C. M. Berry. 2007. Giant cladoxylopsid trees resolve the enigma of the earth's earliest forest stumps at Gilboa. Nature 446: 904–907.

Steinbauer, M. J. 2002. Oviposition preference and neonate performance of *Mnesampela privata* in relation to heterophylly in *Eucalyptus dunnii* and *E. globulus*. Agricultural and Forest Entomology 4: 245–253.

Steinbauer, M. J., F. P. Schiestl, and N. W. Davies. 2004. Monoterpenes and epicuticular waxes help female autumn gum moth differentiate between waxy and glossy *Eucalyptus* and leaves of different ages. Journal of Chemical Ecology 30: 1117–1142.

Steiner, K. E. 1979. Passerine pollination of *Erythrina megistophylla* Diels (Fabaceae). Annals of the Missouri Botanical Garden 66: 490–502.

Steingraeber, D. A. 1982. Heterophylly and neoformation of leaves in sugar maple (*Acer saccharum*). American Journal of Botany 69: 1277–1282.

Stephenson, A. G. 1980. Fruit set, herbivory, fruit reduction, and the fruiting strategy of *Catalpa speciosa* (Bignoniaceae). Ecology 61: 57–64.

Stephenson, A. G. 1981. Flower and fruit abortion: proximate causes and ultimate functions. Annual Review of Ecology and Systematics 12: 253–279.

Stevens, P. T., C. A. Huether, and T. K. Wilson. 1972. Apical size in the determination of corolla lobe number in *Linanthus androsaecus* ssp. *androsaecus*. American Journal of Botany 59: 989–992.

Stewart, R. N., and L. G. Burk. 1970. Independence of tissues derived from apical layers in ontogeny of the tobacco leaf and ovary. American Journal of Botany 57: 1010–1016.

Stieber, J., and H. Beringer. 1984. Dynamic and structural relationships among leaves, roots, and storage tissue in the sugar beet. Botanical Gazette 145: 465–473.

Stinchcombe, J. R., M. T. Rutter, D. S. Burdick, P. Tiffin, M. D. Rausher, and R. Mauricio. 2002. Testing for environmentally induced bias in phenotypic estimates of natural selection: theory and practice. American Naturalist 160: 511–523.

Stockhoff, B. A. 1993. Diet heterogeneity: implications for growth of a generalist herbivore, the gypsy moth. Ecology 74: 1939–1949.

Stöcklin, J., and P. Favre. 1994. Effects of plant size and morphological constraints on variation in reproductive components in two related species of *Epilobium*. Journal of Ecology 82: 735–746.

Stout, J. C., and D. Goulson. 2001. The use of conspecific and interspecific scent marks by foraging bumblebees and honeybees. Animal Behaviour 62: 183–189.

Strauss, S. Y., J. K. Conner, and K. P. Lehtilä. 2001. Effects of foliar herbivory by insects on the fitness of *Raphanus raphanistrum*: damage can increase male fitness. American Naturalist 158: 496–504.

Strauss, S. Y., J. K. Conner, and S. L. Rush. 1996. Foliar herbivory affects floral characters and plant attractiveness to pollinators: implications for male and female plant fitness. American Naturalist 147: 1098–1107.

Stromberg, M. R., and P. B. Johnsen. 1990. Hummingbird sweetness preferences: taste or viscosity? Condor 92: 606–612.

Stryker, R. B., J. W. Gilliam, and W. A. Jackson. 1974. Nonuniform transport of phosphorus from single roots to the leaves of *Zea mays*. Physiologia Plantarum 30: 231–239.

Sugayama, R. L., E. S. Branco, A. Malavasi, A. Kovaleski, and I. Nora. 1997. Oviposition behavior of *Anastrepha fraterculus* in apple and diel pattern of activities in an apple orchard in Brazil. Entomologia Experimentalis et Applicata 83: 239–245.

Sultan, S. E. 1987. Evolutionary implications of phenotypic plasticity in plants. Evolutionary Biology 21: 127–178.

Sultan, S. E. 1995. Phenotypic plasticity and plant adaptation. Acta Botanica Neerlandica 44: 363–383.

Sultan, S. E. 1996. Phenotypic plasticity for offspring traits in *Polygonum persicaria*. Ecology 77: 1791–1807.

Sultan, S. E. 2000. Phenotypic plasticity for plant development, function and life history. Trends in Plant Science 5: 537–542.

Sultan, S. E. 2003. Phenotypic plasticity in plants: a case study in ecological development. Evolution and Development 5: 25–33.

Sumner, E. R., and S. V. Avery. 2002. Phenotypic heterogeneity: differential stress resistance among individual cells of the yeast *Saccharomyces cerevisiae*. Microbiology 148: 345–351.

Suomela, J. 1996. Within-tree variability of mountain birch leaves causes variation in performance for *Epirrita autumnata* larvae. Vegetatio 127: 77–83.

Suomela, J., and M. P. Ayres. 1994. Within-tree and among-tree variation in leaf characteristics of mountain birch and its implications for herbivory. Oikos 70: 212–222.

Suomela, J., P. Kaitaniemi, and A. Nilson. 1995. Systematic within-tree variation in mountain birch leaf quality for a geometrid, *Epirrita autumnata*. Ecological Entomology 20: 283–292.

Suomela, J., and A. Nilson. 1994. Within-tree and among-tree variation in growth of *Epirrita autumnata* on mountain birch leaves. Ecological Entomology 19: 45–56.

Suomela, J., V. Ossipov, and E. Haukioja. 1995. Variation among and within mountain birch trees in foliage phenols, carbohydrates, and amino-acids, and in growth of *Epirrita autumnata* larvae. Journal of Chemical Ecology 21: 1421–1446.

Susko, D. J., and L. Lovett-Doust. 1999. Effects of resource availability, and fruit and ovule position on components of fecundity in *Alliaria petiolata* (Brassicaceae). New Phytologist 144: 295–306.

Susko, D. J., and L. Lovett-Doust. 2000. Patterns of seed mass variation and their effects on seedling traits in *Alliaria petiolata* (Brassicaceae). American Journal of Botany 87: 56–66.

Svensson, L. 1992. Estimates of hierarchical variation in flower morphology in natural populations of *Scleranthus annuus* (Caryophyllaceae), an inbreeding annual. Plant Systematics and Evolution 180: 157–180.

Swaddle, J. P. 2003. Fluctuating asymmetry, animal behavior, and evolution. Advances in the Study of Behavior 32: 169–205.

Syafaruddin, Y. Yoshioka, A. Horisaki, S. Niikura, and R. Ohsawa. 2006. Intraspecific variation in floral organs and structure in *Brassica rapa* L. analyzed by principal component analysis. Breeding Science 56: 189–194.

Sydes, M. A., and R. Peakall. 1998. Extensive clonality in the endangered shrub *Haloragodendron lucasii* (Haloragaceae) revealed by allozymes and RAPDs. Molecular Ecology 7: 87–93.

Szentesi, A. 2003. Resource assessment and clutch size in the bean weevil, *Acanthoscelides obtectus*. Pest Management Science 59: 431–436.

Szymkowiak, E. J., and I. M. Sussex. 1992. The internal meristem layer (L3) determines floral meristem size and carpel number in tomato periclinal chimeras. Plant Cell 4: 1089–1100.

Szymkowiak, E. J., and I. M. Sussex. 1996. What chimeras can tell us about plant development. Annual Review of Plant Physiology and Plant Molecular Biology 47: 351–376.

Tackenberg, O., C. Romermann, K. Thompson, and P. Poschlod. 2006. What does diaspore morphology tell us about external animal dispersal? Evidence from standardized experiments measuring seed retention on animal-coats. Basic and Applied Ecology 7: 45–58.

Tamm, S., and C. L. Gass. 1986. Energy intake rates and nectar concentration preferences by hummingbirds. Oecologia 70: 20–23.

Tanowitz, B. D., P. F. Salopek, and B. E. Mahall. 1987. Differential germination of ray and disc achenes in *Hemizonia increscens* (Asteraceae). American Journal of Botany 74: 303–312.

Tansley, A. G. 1948. The nature and range of variation in the floral symmetry of *Potentilla erecta* (L.) Hampe. New Phytologist 47: 95–110.

Tashiro, T., Y. Fukuda, and T. Osawa. 1991. Oil contents of seeds and minor components in the oil of sesame, *Sesamum indicum* L., as affected by capsule position. Japanese Journal of Crop Science 60: 116–121.

Tébar, F. J., and L. Llorens. 1993. Heterocarpy in *Thymelaea velutina* (Poiret ex Camb.) Endl.: a case of phenotypic adaptation to Mediterranean selective pressures. Botanical Journal of the Linnean Society 111: 295–300.

Telenius, A., and P. Torstensson. 1989. The seed dimorphism of *Spergularia marina* in relation to dispersal by wind and water. Oecologia 80: 206–210.

Temeles, E. J. 1996. A new dimension to hummingbird-flower relationships. Oecologia 105: 517–523.

Temeles, E. J., and W. M. Roberts. 1993. Effect of sexual dimorphism in bill length on foraging behavior: an experimental analysis of hummingbirds. Oecologia 94: 87–94.

Temesgen, H. 2003. Evaluation of sampling alternatives to quantify tree leaf area. Canadian Journal of Forest Research 33: 82–95.

Tenhumberg, B., A. J. Tyre, and B. Roitberg. 2000. Stochastic variation in food availability influences weight and age at maturity. Journal of Theoretical Biology 202: 257–272.

Theophrastus. 1916. Enquiry into plants. Books 1–5. Translated by A. Hort. Loeb Classical Library. Harvard University Press, Cambridge, MA.

Thom, C., P. G. Guerenstein, W. L. Mechaber, and J. G. Hildebrand. 2004. Floral CO_2 reveals flower profitability to moths. Journal of Chemical Ecology 30: 1285–1288.

Thomas, D. W., D. Cloutier, M. Provencher, and C. Houle. 1988. The shape of bird- and bat-generated seed shadows around a tropical fruiting tree. Biotropica 20: 347–348.

Thompson, J. N. 1983a. Selection of plant parts by *Depressaria multifidae* (Lep., Oecophoridae) on its seasonally-restricted hostplant, *Lomatium grayi* (Umbelliferae). Ecological Entomology 8: 203–211.

Thompson, J. N. 1983b. The use of ephemeral plant parts on small host plants: how *Depressaria leptotaeniae* (Lepidoptera: Oecophoridae) feeds on *Lomatium dissectum* (Umbelliferae). Journal of Animal Ecology 52: 281–291.

Thompson, J. N. 1984. Variation among individual seed masses in *Lomatium grayi* (Umbelliferae) under controlled conditions: magnitude and partitioning of the variance. Ecology 65: 626–631.

Thompson, J. N. 1988. Evolutionary ecology of the relationship between oviposition preference and performance of offspring in phytophagous insects. Entomologia Experimentalis et Applicata 47: 3–14.

Thompson, J. N., and O. Pellmyr. 1989. Origins of variance in seed number and mass: interaction of sex expression and herbivory in *Lomatium salmoniflorum*. Oecologia 79: 395–402.

Thomson, J. D., S. Dent-Acosta, P. Escobar-Paramo, and J. D. Nason. 1997. Within-crown flowering synchrony in strangler figs, and its relationship to allofusion. Biotropica 29: 291–297.

Thomson, J. D., E. A. Herre, J. L. Hamrick, and J. L. Stone. 1991. Genetic mosaics in strangler fig trees: implications for tropical conservation. Science 254: 1214–1216.

Thorp, R. W., D. L. Briggs, J. R. Estes, and E. H. Erickson. 1975. Nectar fluorescence under ultraviolet irradiation. Science 189: 476–478.

Tibelius, A. C., and H. R. Klinck. 1987. Effects of artificial reduction in panicle size on weight of secondary seeds in oats (*Avena sativa* L.). Canadian Journal of Plant Science 67: 621–628.

Tilney-Bassett, R. A. E. 1963. The structure of periclinal chimeras. Heredity 18: 265–285.

Tilney-Bassett, R. A. E. 1969. Plant chimeras. 2nd edition. Methuen, London.

Ting, S. V. 1969. Distribution of soluble components and quality factors in the edible portion of citrus fruits. Journal of the American Society of Horticultural Science 94: 515–519.

Titman, P. W., and R. H. Wetmore. 1955. The growth of long and short shoots in *Cercidiphyllum*. American Journal of Botany 42: 364–372.

Titus, J. E., and P. G. Sullivan. 2001. Heterophylly in the yellow waterlily, *Nuphar variegata* (Nymphaeaceae): effects of $[CO_2]$, natural sediment type, and water depth. American Journal of Botany 88: 1469–1478.

Tomback, D. F. 1982. Dispersal of whitebark pine seeds by Clark's nutcracker: a mutualism hypothesis. Journal of Animal Ecology 51: 451–467.

Tomback, D. F., F. K. Holtmeier, H. Mattes, K. S. Carsey, and M. L. Powell. 1993. Tree clusters and growth form distribution in *Pinus cembra*, a bird-dispersed pine. Arctic and Alpine Research 25: 374–381.

Torick, L. L., D. F. Tomback, and R. Espinoza. 1996. Occurrence of multi-genet tree clusters in "wind-dispersed" pines. American Midland Naturalist 136: 262–266.

Torimaru, T., N. Tomaru, N. Nishimura, and S. Yamamoto. 2003. Clonal diversity and genetic differentiation in *Ilex leucoclada* M. patches in an old-growth beech forest. Molecular Ecology 12: 809–818.

Torres, C., and L. Galetto. 2002. Are nectar sugar composition and corolla tube length related to the diversity of insects that visit Asteraceae flowers? Plant Biology 4: 360–366.

Torres, J. B., C. A. Faria, W. S. Evangelista, and D. Pratissoli. 2001. Within-plant distribution of the leaf miner *Tuta absoluta* (Meyrick) immatures in processing tomatoes, with notes on plant phenology. International Journal of Pest Management 47: 173–178.

Torres Boeger, M. R., and M. E. Poulson. 2003. Morphological adaptations and photosynthetic rates of amphibious *Veronica anagallis-aquatica* L. (Scrophulariaceae) under different flow regimes. Aquatic Botany 75: 123–135.

Totland, Ø. 1999. Effects of temperature on performance and phenotypic selection on plant traits in alpine *Ranunculus acris*. Oecologia 120: 242–251.

Totland, Ø., H. L. Andersen, T. Bjelland, V. Dahl, and W. Eide. 1998. Variation in pollen limitation among plants and phenotypic selection on floral traits in an early-spring flowering herb. Oikos 82: 491–501.

Tower, W. L. 1902. Variation in the ray-flowers of *Chrysanthemum leucanthemum* L. at Yellow Springs, Greene Co., with remarks upon the determination of modes. Biometrika 1: 309–315.

Traveset, A. 1993. Deceptive fruits reduce seed predation by insects in *Pistacia terebinthus* L. (Anacardiaceae). Evolutionary Ecology 7: 357–361.

Traveset, A. 1994. Influence of type of avian frugivory on the fitness of *Pistacia terebinthus* L. Evolutionary Ecology 8: 618–627.

Traveset, A., N. Riera, and R. E. Mas. 2001. Passage through bird guts causes interspecific differences in seed germination characteristics. Functional Ecology 15: 669–675.

Traveset, A., M. F. Willson, and J. C. Gaither. 1995. Avoidance by birds of insect-infested fruits of *Vaccinium ovalifolium*. Oikos 73: 381–386.

Traw, M. B., and D. D. Ackerly. 1995. Leaf position, light levels, and nitrogen allocation in five species of rain forest pioneer trees. American Journal of Botany 82: 1137–1143.

Trewavas, A. 1986. Resource allocation under poor growth conditions: a major role for growth substances in developmental plasticity. In D. H. Jennings and A. Trewavas, eds., Symposia of the Society for Experimental Biology, symposium 40, pp. 31–76. Company of Biologists, Cambridge.

Trivers, R. L. 1974. Parent-offspring conflict. American Zoologist 14: 249–264.

Tsukaya, H., K. Shoda, G. T. Kim, and H. Uchimiya. 2000. Heteroblasty in *Arabidopsis thaliana* (L.) Heynh. Planta 210: 536–542.

Tucker, G. F., J. P. Lassoie, and T. J. Fahey. 1993. Crown architecture of stand-grown sugar maple (*Acer saccharum* Marsh.) in the Adirondacks Mountains. Tree Physiology 13: 297–310.

Tucker, S. C. 1988. Heteromorphic flower development in *Neptunia pubescens*, a mimosoid legume. American Journal of Botany 75: 205–224.

Tuomi, J., and T. Vuorisalo. 1989. Hierarchical selection in modular organisms. Trends in Ecology and Evolution 4: 209–213.

Tuskan, G. A., K. E. Francis, S. L. Russ, W. H. Romme, and M. G. Turner. 1996. RAPD markers reveal diversity within and among clonal and seedling stands of aspen in Yellowstone National Park, U.S.A. Canadian Journal of Forest Research 26: 2088–2098.

Tustin, D. S., P. M. Hirst, and I. J. Warrington. 1988. Influence of orientation and position of fruiting laterals on canopy light penetration, yield, and fruit quality of 'Granny Smith' apple. Journal of the American Society for Horticultural Science 113: 693–699.

Tyas, J. A., P. J. Hofman, S. J. R. Underhill, and K. L. Bell. 1998. Fruit canopy position and panicle bagging affects yield and quality of 'Tai So' lychee. Scientia Horticulturae 72: 203–213.

Uemura, A., H. Harayama, N. Koike, and A. Ishida. 2006. Coordination of crown structure, leaf plasticity and carbon gain within the crowns of three winter-deciduous mature trees. Tree Physiology 26: 633–641.

Uma Shaanker, R., K. N. Ganeshaiah, and K. S. Bawa. 1988. Parent-offspring conflict, sibling rivalry, and brood size patterns in plants. Annual Review of Ecology and Systematics 19: 177–205.

Umemoto, T., Y. Nakamura, and N. Ishikura. 1994. Effect of grain location on the panicle on activities involved in starch synthesis in rice endosperm. Phytochemistry 36: 843–847.

Umemoto, T., and K. Terashima. 2002. Activity of granule-bound starch synthase is an important determinant of amylose content in rice endosperm. Functional Plant Biology 29: 1121–1124.

Ushimaru, A., and F. Hyodo. 2005. Why do bilaterally symmetrical flowers orient vertically? Flower orientation influences pollinator landing behaviour. Evolutionary Ecology Research 7: 151–160.

Ushimaru, A., and A. Imamura. 2002. Large variation in flower size of the myco-heterotrophic plant, *Monotropastrum globosum*: effect of floral display on female reproductive success. Plant Species Biology 17: 147–153.

Ushimaru, A., T. Itagaki, and H. S. Ishii. 2003. Variation in floral organ size depends on function: a test with *Commelina communis*, an andromonoecious species. Evolutionary Ecology Research 5: 615–622.

Valburg, L. K. 1992a. Eating infested fruits: interactions in a plant-disperser-pest triad. Oikos 65: 25–28.

Valburg, L. K. 1992b. Feeding preferences of common bush-tanagers for insect-infested fruits: avoidance or attraction? Oikos 65: 29–33.

Valentine, H. T., and S. J. Hilton. 1977. Sampling oak foliage by the randomized-branch method. Canadian Journal of Forest Research 7: 295–298.

Valladares, F. 2003. Light heterogeneity and plants: from ecophysiology to species coexistence and biodiversity. Progress in Botany 64: 439–471.

Valladares, F., L. Balaguer, E. Martínez-Ferri, E. Pérez-Corona, and E. Manrique. 2002. Plasticity, instability and canalization: is the phenotypic variation in seedlings of sclerophyll oaks consistent with the environmental unpredictability of Mediterranean ecosystems? New Phytologist 156: 457–467.

Valladares, F., E. Martínez-Ferri, L. Balaguer, E. Pérez-Corona, and E. Manrique. 2000. Low leaf-level response to light and nutrients in Mediterranean ever-green oaks: a conservative resource-use strategy? New Phytologist 148: 79–91.

Valladares, F., and R. W. Pearcy. 1999. The geometry of light interception by shoots of *Heteromeles arbutifolia*: morphological and physiological consequences for individual leaves. Oecologia 121: 171–182.

Vallius, E. 2000. Position-dependent reproductive success of flowers in *Dactylorhiza maculata* (Orchidaceae). Functional Ecology 14: 573–579.

van Dam, N. M., T. J. de Jong, Y. Iwasa, and T. Kubo. 1996. Optimal distribution of defences: are plants smart investors? Functional Ecology 10: 128–136.

van Dam N. M., R. Verpoorte, and E. van der Meijden. 1994. Extreme differences in pyrrolizidine alkaloid levels between leaves of *Cynoglossum officinale*. Phytochemistry 37: 1013–1016.

Vander Kloet, S. P., and D. Tosh. 1984. Effects of pollen donors on seed production, seed weight, germination and seedling vigor in *Vaccinium corymbosum* L. American Midland Naturalist 112: 392–396.

van der Meij, M. A. A., and R. G. Bout. 2000. Seed selection in the Java sparrow (*Padda oryzivora*): preference and mechanical constraint. Canadian Journal of Zoology 78: 1668–1673.

van Dongen, S., G. Molenberghs, and E. Matthysen. 1999. The statistical analysis of fluctuating asymmetry: REML estimation of a mixed regression model. Journal of Evolutionary Biology 12: 94–102.

van Kleunen, M., and M. Fischer. 2005. Constraints on the evolution of adaptive phenotypic plasticity in plants. New Phytologist 166: 49–60.

van Kleunen, M., M. Fischer, and B. Schmid. 2000. Clonal integration in *Ranunculus reptans*: by-product or adaptation? Journal of Evolutionary Biology 13: 237–248.

van Schaik, C. P., J. W. Terborgh, and S. J. Wright. 1993. The phenology of tropical forests: adaptive significance and consequences for primary consumers. Annual Review of Ecology and Systematics 24: 353–377.

Van Valen, L. 1978. The statistics of variation. Evolutionary Theory 4: 33–43.

Van Zandt, P. A., and S. Mopper. 1998. A meta-analysis of adaptive deme formation in phytophagous insect populations. American Naturalist 152: 595–604.

Vaughton, G., and M. Ramsey. 1997. Seed mass variation in the shrub *Banksia spinulosa* (Proteaceae): resource constraints and pollen source effects. International Journal of Plant Sciences 158: 424–431.

Vaughton, G., and M. Ramsey. 1998. Sources and consequences of seed mass variation in *Banksia marginata* (Proteaceae). Journal of Ecology 86: 563–573.

Velasco, L., L. M. Martín, and A. de Haro. 1998. Within-plant variation for seed weight and seed quality traits in white lupin (*Lupinus albus* L.). Australian Journal of Agricultural Research 49: 59–62.

Velasco, L., and C. Möllers. 2002. Nondestructive assessment of protein content in single seeds of rapeseed (*Brassica napus* L.) by near-infrared reflectance spectroscopy. Euphytica 123: 89–93.

Vemmos, S. N., and G. K. Goldwin. 1994. The photosynthetic activity of Cox's Orange Pippin apple flowers in relation to fruit setting. Annals of Botany 73: 385–391.

Venable, D. L. 1985. The evolutionary ecology of seed heteromorphism. American Naturalist 126: 577–595.

Venable, D. L., A. Búrquez, G. Corral, E. Morales, and F. Espinosa. 1987. The ecology of seed heteromorphism in *Heterosperma pinnatum* in central Mexico. Ecology 68: 65–76.

Venable, D. L., E. Dyreson, and E. Morales. 1995. Population dynamic consequences and evolution of seed traits of *Heterosperma pinnatum* (Asteraceae). American Journal of Botany 82: 410–420.

Venable, D. L., and D. A. Levin. 1985. Ecology of achene dimorphism in *Heterotheca latifolia*: 1, Achene structure, germination and dispersal. Journal of Ecology 73: 133–145.

Verdú, M., and P. García-Fayos. 1998. Ecological causes, function, and evolution of abortion and parthenocarpy in *Pistacia lentiscus* (Anacardiaceae). Canadian Journal of Botany 76: 134–141.

Verdú, M., and P. García-Fayos. 2001. The effect of deceptive fruits on predispersal seed predation by birds in *Pistacia lentiscus*. Plant Ecology 156: 245–248.

Via, S., R. Gomulkiewicz, G. De Jong, S. M. Scheiner, C. D. Schlichting, and P. H. Van Tienderen. 1995. Adaptive phenotypic plasticity: consensus and controversy. Trends in Ecology and Evolution 10: 212–217.

Via, S., and R. Lande. 1985. Genotype-environment interaction and the evolution of phenotypic plasticity. Evolution 39: 505–522.

Vicari, M., P. E. Hatcher, and P. G. Ayres. 2002. Combined effect of foliar and mycorrhizal endophytes on an insect herbivore. Ecology 83: 2452–2464.

Villarreal, A. G., and C. E. Freeman. 1990. Effects of temperature and water stress on some floral nectar characteristics in *Ipomopsis longiflora* (Polemoniaceae) under controlled conditions. Botanical Gazette 151: 5–9.

Viswanathan, D. V., and J. S. Thaler. 2004. Plant vascular architecture and within-plant spatial patterns in resource quality following herbivory. Journal of Chemical Ecology 30: 531–543.

Voesenek, L. A. C. J., and C. W. P. M. Blom. 1996. Plants and hormones: an ecophysiological view on timing and plasticity. Journal of Ecology 84: 111–119.

Vogler, D. W., S. Peretz, and A. G. Stephenson. 1999. Floral plasticity in an iteroparous plant: the interactive effects of genotype, environment, and ontogeny in *Campanula rapunculoides* (Campanulaceae). American Journal of Botany 86: 482–494.

von Helversen, D., and O. von Helversen. 2003. Object recognition by echolocation: a nectar-feeding bat exploiting the flowers of a rain forest vine. Journal of Comparative Physiology 189A: 327–336.

Vourc'h, G., J. L. Martin, P. Duncan, J. Escarre, and T. P. Clausen. 2001. Defensive adaptations of *Thuja plicata* to ungulate browsing: a comparative study between mainland and island populations. Oecologia 126: 84–93.

Vuorisalo, T., and M. J. Hutchings. 1996. On plant sectoriality, or how to combine the benefits of autonomy and integration. Vegetatio 127: 3–8.

Vuorisalo, T., and P. Mutikainen. 1999. Modularity and plant life histories. In T. Vuorisalo and P. K. Mutikainen, eds., Life history evolution in plants, pp. 1–25. Kluwer, Dordrecht, Netherlands.

Waddington, C. H. 1941. Evolution of developmental systems. Nature 147: 108–110.

Waddington, C. H. 1942. Canalization of development and the inheritance of acquired characters. Nature 150: 563–565.

Waddington, C. H. 1959. Canalization of development and genetic assimilation of acquired characters. Nature 183: 1654–1655.

Waddington, K. D. 1980. Flight patterns of foraging bees relative to density of artificial flowers and distribution of nectar. Oecologia 44: 199–204.

Waddington, K. D., T. Allen, and B. Heinrich. 1981. Floral preferences of bumblebees (*Bombus edwardsii*) in relation to intermittent versus continuous rewards. Animal Behaviour 29: 779–784.

Wagner, G. P., and L. Altenberg. 1996. Complex adaptations and the evolution of evolvability. Evolution 50: 967–976.

Wagner, G. P., G. Booth, and H. Bagheri-Chaichian. 1997. A population genetic theory of canalization. Evolution 51: 329–347.

Walbot, V. 1996. Sources and consequences of phenotypic and genotypic plasticity in flowering plants. Trends in Plant Science 1: 27–32.

Waller, D. M. 1982. Factors influencing seed weight in *Impatiens capensis* (Balsaminaceae). American Journal of Botany 69: 1470–1475.

Wallin, K. F., and K. F. Raffa. 1998. Association of within-tree jack pine budworm feeding patterns with canopy level and within-needle variation of water, nutrient, and monoterpene concentrations. Canadian Journal of Forest Research 28: 228–233.

Wang, Z. J., J. H. Wang, C. J. Zhao, M. Zhao, W. J. Huang, and C. Z. Wang. 2005. Vertical distribution of nitrogen in different layers of leaf and stem and their relationship with grain quality of winter wheat. Journal of Plant Nutrition 28: 73–91.

Wang, Z. Y., F. Q. Zheng, G. Z. Shen, J. P. Gao, D. P. Snustad, M. G. Li, J. L. Zhang, and M. M. Hong. 1995. The amylose content in rice endosperm is related to the post-transcriptional regulation of the *waxy* gene. Plant Journal 7: 613–622.

Wardlaw, I. F. 1968. The control and pattern of movement of carbohydrates in plants. Botanical Review 34: 79–105.

Waring, R. H., and W. B. Silvester. 1994. Variation in foliar $\delta^{13}C$ values within the crowns of *Pinus radiata* trees. Tree Physiology 14: 1203–1213.

Warringa, J. W., R. de Visser, and A. D. H. Kreuzer. 1998. Seed weight in *Lolium perenne* as affected by interactions among seeds within the inflorescence. Annals of Botany 82: 835–841.

Warringa, J. W., P. C. Struik, R. de Visser, and A. D. H. Kreuzer. 1998. The pattern of flowering, seed set, seed growth and ripening along the ear of *Lolium perenne*. Australian Journal of Plant Physiology 25: 213–223.

Waser, N. M. 1998. Task-matching and short-term size shifts in foragers of the harvester ant, *Messor pergandei* (Hymenoptera: Formicidae). Journal of Insect Behavior 11: 451–462.

Waser, N. M., L. Chittka, M. V. Price, N. M. Williams, and J. Ollerton. 1996. Generalization in pollination systems, and why it matters. Ecology 77: 1043–1060.

Waser, N. M., and J. A. McRobert. 1998. Hummingbird foraging at experimental patches of flowers: evidence for weak risk-aversion. Journal of Avian Biology 29: 305–313.

Waser, N. M., and M. V. Price. 1984. Experimental studies of pollen carryover: effects of floral variability in *Ipomopsis aggregata*. Oecologia 62: 262–268.

Watkinson, A. R., and J. White. 1985. Some life-history consequences of modular construction in plants. Philosophical Transactions of the Royal Society of London B 313: 31–51.

Watson, M. A. 1986. Integrated physiological units in plants. Trends in Ecology and Evolution 1: 119–123.

Watson, M. A., and B. B. Casper. 1984. Morphogenetic constraints on patterns of carbon distribution in plants. Annual Review of Ecology and Systematics 15: 233–258.

Watson, M. A., M. A. Geber, and C. S. Jones. 1995. Ontogenetic contingency and the expression of phenotypic plasticity. Trends in Ecology and Evolution 10: 474–475.

Webb, C. J. 1984. Heterophylly in *Eryngium vesiculosum* (Umbelliferae). New Zealand Journal of Botany 22: 29–33.

Weber, E. U., S. Shafir, and A. R. Blais. 2004. Predicting risk sensitivity in humans and lower animals: risk as variance or coefficient of variation. Psychological Review 111: 430–445.

Weberling, F. 1989. Morphology of flowers and inflorescences. Translated by R. J. Pankhurst. Cambridge University Press, Cambridge.

Webster, R. 1985. Quantitative spatial analysis of soil in the field. Advances in Soil Science 3: 1–70.

Weisberg, P. J., and H. Bugmann. 2003. Forest dynamics and ungulate herbivory: from leaf to landscape. Forest Ecology and Management 181: 1–12.

Weldon, W. F. R. 1901. Change in organic correlation of *Ficaria ranunculoides* during the flowering season. Biometrika 1: 125–128.

Wells, C. L., and M. Pigliucci. 2000. Adaptive phenotypic plasticity: the case of heterophylly in aquatic plants. Perspectives in Plant Ecology, Evolution and Systematics 3: 1–18.

Wenny, D. G., and D. J. Levey. 1998. Directed seed dispersal by bellbirds in a tropical cloud forest. Proceedings of the National Academy of Sciences of the USA 95: 6204–6207.

West-Eberhard, M. J. 1989. Phenotypic plasticity and the origins of diversity. Annual Review of Ecology and Systematics 20: 249–278.

West-Eberhard, M. J. 2003. Developmental plasticity and evolution. Oxford University Press, Oxford.

West-Eberhard, M. J. 2005. Phenotypic accommodation: adaptive innovation due to developmental plasticity. Journal of Experimental Zoology (Molecular and Developmental Evolution) 304B: 610–618.

Westman, W. E. 1981. Seasonal dimorphism of foliage in Californian coastal sage scrub. Oecologia 51: 385–388.

Wetherwax, P. B. 1986. Why do honeybees reject certain flowers? Oecologia 69: 567–570.

Whaley, W. G. 1939. Developmental changes in apical meristems. Proceedings of the National Academy of Sciences of the USA 25: 445–448.

Whaley, W. G., and C. Y. Whaley. 1942. A developmental analysis of inherited leaf patterns in *Tropaeolum*. American Journal of Botany 29: 195–200.

Wheeler, B. R., and M. J. Hutchings. 2002. Biological flora of the British Isles: 223, *Phyteuma spicatum* L. Journal of Ecology 90: 581–591.

Wheeler, J. K., J. S. Sperry, U. G. Hacke, and N. Hoang. 2005. Inter-vessel pitting and cavitation in woody Rosaceae and other vesselled plants: a basis for a safety versus efficiency trade-off in xylem transport. Plant Cell and Environment 28: 800–812.

Wheelwright, N. T. 1985. Fruit size, gape width, and the diets of fruit-eating birds. Ecology 66: 808–818.

Wheelwright, N. T. 1993. Fruit size in a tropical tree species: variation, preference by birds, and heritability. Vegetatio 107/108: 163–174.

White, D. W., and E. W. Stiles. 1985. The use of refractometry to estimate nutrient rewards in vertebrate-dispersed fruits. Ecology 66: 303–307.

White, J. 1979. The plant as a metapopulation. Annual Review of Ecology and Systematics 10: 109–145.

White, J. 1984. Plant metamerism. In R. Dirzo and J. Sarukhán, eds., Perspectives on plant population ecology, pp. 15–47. Sinauer, Sunderland, MA.

White, P. S. 1983. Corner's rules in eastern deciduous trees: allometry and its implications for the adaptive architecture of trees. Bulletin of the Torrey Botanical
Club 110: 203–212.

Whitehead, H. 1902. Variation in the moscatel (*Adoxa moschatellina*, L.).
Biometrika 2: 108–113.

Whitham, T. G. 1978. Habitat selection by *Pemphigus* aphids in response to resource
limitation and competition. Ecology 59: 1164–1176.

Whitham, T. G. 1980. The theory of habitat selection: examined and extended using
Pemphigus aphids. American Naturalist 115: 449–466.

Whitham, T. G. 1981. Individual trees as heterogeneous environments: adaptation
to herbivory or epigenetic noise? In R. F. Denno and H. Dingle, eds., Insect life
history patterns: habitat and geographic variation, pp. 9–27. Springer-Verlag,
New York.

Whitham, T. G. 1983. Host manipulation of parasites: within-plant variation as a
defense against rapidly evolving pests. In R. F. Denno and M. S. McClure, eds.,
Variable plants and herbivores in natural and managed systems, pp. 15–41. Academic Press, New York.

Whitham, T. G., and C. N. Slobodchikoff. 1981. Evolution by individuals, plant-
herbivore interactions, and mosaics of genetic variability: the adaptive significance of somatic mutations in plants. Oecologia 49: 287–292.

Whitham, T. G., A. G. Williams, and A. M. Robinson. 1984. The variation principle:
individual plants as temporal and spatial mosaics of resistance to rapidly evolving pests. In P. W. Price, C. N. Slobodchikoff, and W. S. Gaud, eds., A new ecology: novel approaches to interactive systems, pp. 15–51. Wiley, New York.

Wilkens, R. T., D. W. Vanderklein, and R. W. Lemke. 2005. Plant architecture and
leaf damage in bear oak: 2, Insect usage patterns. Northeastern Naturalist 12:
153–168.

Williams, J. H. 1962. Influence of plant spacing and flower position on oil content of
safflower, *Carthamus tinctorius*. Crop Science 2: 475–477.

Williams, J. H. H., B. E. Collis, C. J. Pollock, M. L. Williams, and J. F. Farrar. 1993.
Variability in the distribution of photoassimilates along leaves of temperate
Gramineae. New Phytologist 123: 699–703.

Williams, J. L, and J. K. Conner. 2001. Sources of phenotypic variation in floral
traits in wild radish, *Raphanus raphanistrum* (Brassicaceae). American Journal
of Botany 88: 1577–1581.

Williams, K., G. W. Koch, and H. A. Mooney. 1985. The carbon balance of flowers
of *Diplacus aurantiacus* (Scrophulariaceae). Oecologia 66: 530–535.

Williams, M. L., B. J. Thomas, J. F. Farrar, and C. J. Pollock. 1993. Visualizing the distribution of elements within barley leaves by energy dispersive X-ray image
maps (EDX maps). New Phytologist 125: 367–372.

Willson, M. F. 1983. Plant reproductive ecology. Wiley, New York.

Willson, M. F., and T. A. Comet. 1993. Food choices by northwestern crows: experiments with captive, free-ranging and hand-raised birds. Condor 95: 596–615.

Wilson, D., and S. H. Faeth. 2001. Do fungal endophytes result in selection for leaf-
miner ovipositional preference? Ecology 82: 1097–1111.

Wilson, E. O. 1971. The insect societies. Harvard University Press, Cambridge, MA.

Wilson, P., and J. D. Thomson. 1991. Heterogeneity among floral visitors leads to discordance between removal and deposition of pollen. Ecology 72: 1503–1507.

Wilson, S. L., and G. I. H. Kerley. 2003. Bite diameter selection by thicket browsers: the effect of body size and plant morphology on forage intake and quality. Forest Ecology and Management 181: 51–65.

Winn, A. A. 1991. Proximate and ultimate sources of within-individual variation in seed mass in *Prunella vulgaris* (Lamiaceae). American Journal of Botany 78: 838–844.

Winn, A. A. 1996a. Adaptation to fine-grained environmental variation: an analysis of within-individual leaf variation in an annual plant. Evolution 50: 1111–1118.

Winn, A. A. 1996b. The contributions of programmed developmental change and phenotypic plasticity to within-individual variation in leaf traits in *Dicerandra linearifolia*. Journal of Evolutionary Biology 9: 737–752.

Winn, A. A. 1999a. The functional significance and fitness consequences of heterophylly. International Journal of Plant Sciences 160 (suppl.): S113–S121.

Winn, A. A. 1999b. Is seasonal variation in leaf traits adaptive for the annual plant *Dicerandra linearifolia*? Journal of Evolutionary Biology 12: 306–313.

Winn, A. A. 2004. Natural selection, evolvability and bias due to environmental covariance in the field in an annual plant. Journal of Evolutionary Biology 17: 1073–1083.

Witmer, M. C. 1998. Ecological and evolutionary implications of energy and protein requirements of avian frugivores eating sugary diets. Physiological Zoology 71: 599–610.

Wolf, L. L., F. R. Hainsworth, T. Mercier, and R. Benjamin. 1986. Seed size variation and pollinator uncertainty in *Ipomopsis aggregata* (Polemoniaceae). Journal of Ecology 74: 361–371.

Wolfe, L. M. 1992. Why does the size of reproductive structures decline through time in *Hydrophyllum appendiculatum* (Hydrophyllaceae)? Developmental constraints vs. resource limitation. American Journal of Botany 79: 1286–1290.

Wolfe, L. M. 1995. The genetics and ecology of seed size variation in a biennial plant, *Hydrophyllum appendiculatum* (Hydrophyllaceae). Oecologia 101: 343–352.

Wolfe, L. M., and W. Denton. 2001. Morphological constraints on fruit size in *Linaria canadensis*. International Journal of Plant Sciences 162: 1313–1316.

Wolfe, L. M., and J. L. Krstolic. 1999. Floral symmetry and its influence on variance in flower size. American Naturalist 154: 484–488.

Wolfe, L. M., and S. E. Sellers. 1997. Polymorphic floral traits in *Linaria canadensis* (Scrophulariaceae). American Midland Naturalist 138: 134–139.

Wood, P. J. 1972. Sampling systems to assess variability in the needles of twelve Mexican pines. New Phytologist 71: 925–936.

Worley, A. C., A. M. Baker, J. D. Thompson, and S. C. H. Barrett. 2000. Floral display in *Narcissus*: variation in flower size and number at the species, population, and individual levels. International Journal of Plant Sciences 161: 69–79.

Worman, C. O., and C. A. Chapman. 2005. Seasonal variation in the quality of a tropical ripe fruit and the response of three frugivores. Journal of Tropical Ecology 21: 689–697.

Wunderle, J. M., and T. G. O'Brien. 1985. Risk aversion in hand-reared bananaquits. Behavioral Ecology and Sociobiology 17: 371–380.

Wyatt, R. 1982. Inflorescence architecture: how flower number, arrangement, and phenology affect pollination and fruit-set. American Journal of Botany 69: 585–594.

Wykes, G. R. 1952. The influence of variations in the supply of carbohydrate on the process of nectar secretion. New Phytologist 51: 294–300.

Wylie, R. B. 1951. Principles of foliar organization shown by sun-shade leaves from ten species of deciduous dicotyledonous trees. American Journal of Botany 38: 355–361.

Yamada, M., H. Yamane, Y. Takano, and Y. Ukai. 1997. Estimating the proportion of offspring having soluble solids content in fruit exceeding a given critical value in Japanese persimmon. Euphytica 93: 119–126.

Yamasaki, M., and K. Kikuzawa. 2003. Temporal and spatial variations in leaf herbivory within a canopy of Fagus crenata. Oecologia 137: 226–232.

Yano, S., and I. Terashima. 2001. Separate localization of light signal perception for sun or shade type chloroplast and palisade tissue differentiation in Chenopodium album. Plant and Cell Physiology 42: 1303–1310.

Yonemori, K., K. Hirano, and A. Sugiura. 1995. Growth inhibition of persimmon fruit caused by calyx lobe removal and possible involvement of endogenous hormones. Scientia Horticulturae 61: 37–45.

Young, D. R., and J. B. Yavitt. 1987. Differences in leaf structure, chlorophyll, and nutrients for the understory tree Asimina triloba. American Journal of Botany 74: 1487–1491.

Young, H. J., and M. L. Stanton. 1990. Temporal patterns of gamete production within individuals of Raphanus sativus (Brassicaceae). Canadian Journal of Botany 68: 480–486.

Yule, G. U. 1902. Variation of the number of sepals in Anemone nemorosa. Biometrika 1: 307–309.

Zamora, R. 1995. The trapping success of a carnivorous plant, Pinguicula vallisneriifolia: the cumulative effects of availability, attraction, retention and robbery of prey. Oikos 73: 309–322.

Zangerl, A. R., M. R. Berenbaum, and J. K. Nitao. 1991. Parthenocarpic fruits in wild parsnip: decoy defence against a specialist herbivore. Evolutionary Ecology 5: 136–145.

Zanne, A. E., K. Sweeney, M. Sharma, and C. M. Orians. 2006. Patterns and consequences of differential vascular sectoriality in 18 temperate tree and shrub species. Functional Ecology 20: 200–206.

Zar, J. H. 1999. Biostatistical analysis. 4th edition. Prentice Hall, London.

Zhang, X. M., C. H. Shi, J. G. Wu, H. Hisamitus, T. Katsura, S. Y. Feng, G. L. Bao, and S. H. Ye. 2003. Analysis of variations in the amylose content of grains located at different positions in the rice panicle and the effect of milling. Starch 55: 265–270.

Zhang, X. S. 2005. Evolution and maintenance of the environmental component of the phenotypic variance: benefit of plastic traits under changing environments. American Naturalist 166: 569–580.

Zhang, X. S., and W. G. Hill. 2005. Evolution of the environmental component of the phenotypic variance: stabilizing selection in changing environments and the cost of homogeneity. Evolution 59: 1237–1244.

Zimmerman, D. C., and G. N. Fick. 1973. Fatty acid composition of sunflower (*Helianthus annuus* L.) oil as influenced by seed position. Journal of the American Oil Chemists Society 50: 273–275.

Zimmerman, M. 1988. Pollination biology of montane plants: relationship between rate of nectar production and standing crop. American Midland Naturalist 120: 50–57.

Zimmerman, M., and G. H. Pyke. 1986. Reproduction in *Polemonium*: patterns and implications of floral nectar production and standing crops. American Journal of Botany 73: 1405–1415.

Ziv, Y., and J. L. Bronstein. 1996. Infertile seeds of *Yucca schottii*: a beneficial role for the plant in the yucca-yucca moth mutualism? Evolutionary Ecology 10: 63–76.

Zohary, M. 1952. A monographical study of the genus *Pistacia*. Palestine Journal of Botany, Jerusalem Series 5: 187–228.

Zucker, W. V. 1982. How aphids choose leaves: the role of phenolics in host selection by a galling aphid. Ecology 63: 972–981.

Zwieniecki, M. A., C. M. Orians, P. J. Melcher, and N. M. Holbrook. 2003. Ionic control of the lateral exchange of water between vascular bundles in tomato. Journal of Experimental Botany 54: 1399–1405.

Index

abscisic acid, 128–29
Acacia greggii, 242
Acer, 144, 146, 147
Acer pseudoplatanus, 152
Acer saccharum, 45, 96, 157, 237, 238
achenes, 27, 35
Ackerly, D. D., 82–83, 85
Actaea rubra, 21, 173
Actinidia deliciosa, 32, 80, 84, 104
adaptive deme formation hypothesis,
    279–80
Adoxa moschatellina, 10
Aegilops, 32
Aesculus hippocastanum, 142
alkaloid content, 196
Allium sativum, 104
Alnus rubra, 34, 56, 58
Alonso, C., 72, 93–94
Alstad, D. N., 279
Amaranthus retroflexus, 200
Amelanchier arborea, 30
American Journal of Botany, 10
American Naturalist, 10
amino acids, 17, 45
amplified fragment length polymorphism
    (AFLP) markers, 103, 105, 106
Anchusa strigosa, 223, 244, 282
Andersen, M. C., 58
androecium, 21
Anemone nemorosa, 10
Anemonopsis macrophylla, 68
aneusomaty, 100–1
angiosperms, 21, 33
Annona cherimola, 78, 180, 197

annual plants. *See specific plants*
antagonistic flower visitors, 27
Anten, N. P. R., 275
anther: length of, 50, 79, 153; number of, 68;
    size of, 153
anthocyanin content, 204
Anthoxanthum odoratum, 115
Antirrhinum majus, 282
aphids, 237, 238, 239, 241, 242
apples and apple trees, 32, 56, 77, 80, 84, 88
aquatic plants, 76, 122, 128, 268; hetero-
    phylly, 13, 15
Aquilegia, 55, 94, 221–22
Aquilegia canadensis, 68, 148, 156, 164
Aquilegia cazorlensis, 115
Aquilegia coerulea, 115
Arabidopsis thaliana, 99, 109, 129, 148, 153,
    160, 167, 202
architectural effects: direct and indirect,
    152–55, 182, 184, 188, 201, 328; and
    extrinsic-intrinsic gradients, 83–85, 88;
    gradients in organ features, 82; and
    leaf size and shape, 78–79; organ and
    organ-group levels, 147–50; and pattern
    of variation, 83; proximate mechanisms
    of direct effects, 155–61; and seed size,
    80; space constraints, 150, 151–52; spa-
    tial organization of variation, 94–95,
    134–35; suppression of growth, 150–51;
    whole-plant level, 138–47
Armbruster, W. S., 48
Armesto, J. J., 34, 193
Arnold, S. J., 317, 318
Arum italicum, 80